Natural Nanoparticle for Cancer Diagnosis and Treatment

Natural Nanoparticle for Cancer Diagnosis and Treatment

Editor

Serena Mazzucchelli

MDPI • Basel • Beijing • Wuhan • Barcelona • Belgrade • Manchester • Tokyo • Cluj • Tianjin

Editor
Serena Mazzucchelli
Università di Milano
Italy

Editorial Office
MDPI
St. Alban-Anlage 66
4052 Basel, Switzerland

This is a reprint of articles from the Special Issue published online in the open access journal *Pharmaceutics* (ISSN 1999-4923) (available at: https://www.mdpi.com/journal/pharmaceutics/special_issues/Natural_Nanoparticle).

For citation purposes, cite each article independently as indicated on the article page online and as indicated below:

LastName, A.A.; LastName, B.B.; LastName, C.C. Article Title. *Journal Name* **Year**, *Volume Number*, Page Range.

ISBN 978-3-0365-7432-5 (Hbk)
ISBN 978-3-0365-7433-2 (PDF)

Cover image courtesy of Raffaele Allevi

Contents

About the Editor

Serena Mazzucchelli

Serena Mazzucchelli, PhD, is a researcher in Biochemistry at the University of Milan. She is a member of the board of the PhD School in Translational Medicine, at the University of Milan, and a tutor of several PhD students. Mazzucchelli is also the scientific manager of the Pediatric Clinical Research Center in Invernizzi's imaging facility. SM's research is focused on ferritin nanocages' development, as a platform to specifically deliver drugs and theranostics in cancer and to drive tumor-targeted fluorescence-image-guided surgery. She has received some recognition and awards, such as the first prize for Best Paper Presentation and the third prize at the first "Gianni Bonadonna Prize" for young researchers in oncology. SM is a speaker at international conferences and a reviewer for the main journals in the nanomedicine field. Scopus ID: 35223720300.

pharmaceutics

MDPI

Editorial

Natural Nanoparticles: A Safe Bullet for Treatment and Detection of Solid Tumors

Serena Mazzucchelli

Dipartimento di Scienze Biomediche e Cliniche, Università di Milano, 20157 Milan, Italy; serena.mazzucchelli@unimi.it

Citation: Mazzucchelli, S. Natural Nanoparticles: A Safe Bullet for Treatment and Detection of Solid Tumors. *Pharmaceutics* **2022**, *14*, 1126. https://doi.org/10.3390/pharmaceutics14061126

Received: 11 May 2022
Accepted: 23 May 2022
Published: 25 May 2022

In the last couple of decades, nanoparticles have been extensively studied as carriers for cancer imaging agents and as drug delivery platforms, due to their ability to positively affect the distribution and tumor-targeting properties of delivered compounds [1,2]. However, nanoparticle off-target accumulation, immunogenicity, and sequestration by macrophages are issues that strongly limit their clinical translation [3,4]. These issues could be overcome by the use of natural nanoparticles, since they are mainly constituted by proteins or by cell membrane portions, resulting in high biocompatibility, stability in biological fluids and bioinvisibility [5]. This Special Issue depicts the current landscape of natural nanoparticles for cancer diagnosis and treatment, collecting articles able to better highlight any feature relating to this topic. Not only natural nanoparticles, such as exosomes and protein nanocages, but also organic nanoformulations of natural compounds, such as the flavonoid luteolin and the xanthone derivate α-mangostin [6–9], represent attractive options to improve knowledge in this area and to accelerate the translation of nanotechnological products to the clinic, as summarized by Ion. D. et al. [10]. Among natural nanoparticles, ferritins (HFn) are the most fascinating natural drug delivery system, and have been extensively investigated in relation to cancer due to their intrinsic tumor-targeting capability, as reported by Mainini F. et al., in this Special Issue [11]. HFn display the targeted recognition of cancer cells and tissues thanks to their ability to specifically recognize the transferrin receptor 1 (TfR1), which is overexpressed in tumors [11]. HFn-based nanoparticles have been engineered by DNA recombinant technology or by chemical surface modification, obtaining products exploitable for the treatment, diagnosis and theranostics of cancer, already tested both in vitro and in vivo with very promising results [11]. In addition to ferritin-based nanoparticles, an important position is occupied by exosomes, a class of natural nanosized vesicles derived by cells for intracellular communication and recently developed and assessed for drug delivery purposes [12]. They display a typical tropism determined by the cell subtype from which they are generated. Here, we report a study about exosomes from M1 macrophages loaded with the combination of gemcitabine and the oral iron chelator deferasirox, developed and tested for the treatment of gemcitabine-resistant pancreatic cancer [13]. This work demonstrated how drug entrapment into M1-macrophage-derived exosomes is an effective strategy to bypass drug resistance in this pancreatic cancer model, suggesting that it could be successfully exploited for resistant pancreatic cancer and investigated in other cancer subtypes [13]. To date, the clinical exploitation of exosomes derived from human or mammalian sources is challenging, since the set-up of large-scale human or mammalian cultures to produce exosomes seems to limit the feasibility [14]. Recently, exosomes derived from the budding of bacterial membranes have been explored as a valuable alternative for large-scale productions [14]. The review of S. Fazal and R. Lee describes bacterial membrane vesicles (BMV), focusing attention on the sources used for their production and on techniques for separation, purification, characterization and drug loading. Moreover, also discussed was their exploitation as drug delivery platforms in cancer therapy, with some in vivo applications reported [14]. Finally, this Special Issue is completed by a literature review of nanotechnological solutions adopted for the delivery of

RNA-i therapeutics in pancreatic cancers [15] and by natural carbon nanodots [16], a field of study that creates a bridge between synthetic and natural nanoparticles, since it employs classical chemical methods of synthesis starting from natural carbon sources [16].

Overall, the articles and reviews collected in this Special Issue describe a prolific and dynamic field of study, which promises to bring natural nanocarriers to cancer clinical trials, accelerating clinical translation.

Conflicts of Interest: The author declares no conflict of interest.

References

1. Anselmo, A.C.; Mitragotri, S. An overview of clinical and commercial impact of drug delivery systems. *J. Control. Release* **2014**, *190*, 15–28. [CrossRef] [PubMed]
2. Anselmo, A.C.; Mitragotri, S. Nanoparticles in the clinic: An update. *Bioeng. Transl. Med.* **2019**, *4*, e10143. [CrossRef] [PubMed]
3. Wong, J.K.L.; Mohseni, R.; Hamidieh, A.A.; MacLaren, R.E.; Habib, N.; Seifalian, A.M. Limitations in Clinical Translation of Nanoparticle-Based Gene Therapy. *Trends Biotechnol.* **2017**, *35*, 1124. [CrossRef] [PubMed]
4. Đorđević, S.; Gonzalez, M.M.; Conejos-Sánchez, I.; Carreira, B.; Pozzi, S.; Acúrcio, R.C.; Satchi-Fainaro, R.; Florindo, H.F.; Vicent, M.J. Current hurdles to the translation of nanomedicines from bench to the clinic. *Drug Deliv. Transl. Res.* **2022**, *12*, 500–525. [CrossRef] [PubMed]
5. Hong, S.; Choi, D.W.; Kim, H.N.; Park, C.G.; Lee, W.; Park, H.H. Protein-Based Nanoparticles as Drug Delivery Systems. *Pharmaceutics* **2020**, *12*, 604. [CrossRef] [PubMed]
6. Jafar, M.; Salahuddin, M.; Khan, M.S.A.; Alshehry, Y.; Alrwaili, N.R.; Alzahrani, Y.A.; Imam, S.S.; Alshehri, S. Preparation and In Vitro-In Vivo Evaluation of luteolin Loaded Gastroretentive microsponge for the Eradication of *Helicobacter pylori* Infections. *Pharmaceutics* **2021**, *13*, 2094. [CrossRef] [PubMed]
7. Kazmi, I.; Alshehri, S.; Al-Abbasi, F.A.; Nadeem, M.S.; Altayb, H.N.; Imam, S.S. Formulation, Optimization and Evaluation of Luteolin-Loaded Topical Nanoparticulate Delivery System for the Skin Cancer. *Pharmaceutics* **2021**, *13*, 1749. [CrossRef] [PubMed]
8. Meylina, L.; Muchtaridi, M.; Joni, I.M.; Mohammed, A.F.A.; Wathoni, N. Nanoformulations of α-Mangostin for Cancer Drug Delivery System. *Pharmaceutics* **2021**, *13*, 1993. [CrossRef] [PubMed]
9. Chiang, C.-S.; Huang, B.-J.; Chen, J.-Y.; Chieng, W.W.; Lim, S.H.; Lee, W.; Shyu, W.-C.; Jeng, L.-B. Fucoidan-Based Nanoparticles with Inherently Therapeutic Efficacy for Cancer Treatment. *Pharmaceutics* **2021**, *13*, 1986. [CrossRef] [PubMed]
10. Ion, D.; Niculescu, A.-G.; Păduraru, D.N.; Andronic, O.; Mușat, F.; Grumezescu, A.M.; Bolocan, A. An Up-to-Date Review of Natural Nanoparticles for Cancer Management. *Pharmaceutics* **2022**, *14*, 18. [CrossRef] [PubMed]
11. Mainini, F.; Bonizzi, A.; Sevieri, M.; Sitia, L.; Truffi, M.; Corsi, F.; Mazzucchelli, S. Protein-Based Nanoparticles for the Imaging and Treatment of Solid Tumors: The Case of Ferritin Nanocages, a Narrative Review. *Pharmaceutics* **2021**, *13*, 2000. [CrossRef] [PubMed]
12. Nam, G.H.; Choi, Y.; Kim, G.B.; Kim, S.; Kim, S.A.; Kim, I.S. Emerging Prospects of Exosomes for Cancer Treatment: From Conventional Therapy to Immunotherapy. *Adv. Mater.* **2020**, *32*, 2002440. [CrossRef] [PubMed]
13. Zhao, Y.; Zheng, Y.; Zhu, Y.; Zhang, Y.; Zhu, H.; Liu, T. M1 Macrophage-Derived Exosomes Loaded with Gemcitabine and Deferasirox against Chemoresistant Pancreatic Cancer. *Pharmaceutics* **2021**, *13*, 1493. [CrossRef] [PubMed]
14. Fazal, S.; Lee, R. Biomimetic Bacterial Membrane Vesicles for Drug Delivery Applications. *Pharmaceutics* **2021**, *13*, 1430. [CrossRef] [PubMed]
15. Kim, M.J.; Chang, H.; Nam, G.; Ko, Y.; Kim, S.H.; Roberts, T.M.; Ryu, J.H. RNAi-Based Approaches for Pancreatic Cancer Therapy. *Pharmaceutics* **2021**, *13*, 1638. [CrossRef] [PubMed]
16. Chan, M.-H.; Chen, B.-G.; Ngo, L.T.; Huang, W.-T.; Li, C.-H.; Liu, R.-S.; Hsiao, M. Natural Carbon Nanodots: Toxicity Assessment and Theranostic Biological Application. *Pharmaceutics* **2021**, *13*, 1874. [CrossRef] [PubMed]

pharmaceutics

MDPI

Article

Fucoidan-Based Nanoparticles with Inherently Therapeutic Efficacy for Cancer Treatment

Chih-Sheng Chiang [1,2,3,*], Bo-Jie Huang [4,†], Jui-Yu Chen [4,†], Wee Wei Chieng [4,†], Seh Hong Lim [5], Wei Lee [4], Weoi-Cherng Shyu [1,3,4,6,7,*] and Long-Bin Jeng [2,8,*]

1 Graduate Institute of Biomedical Sciences, China Medical University, Taichung 404, Taiwan
2 Cell Therapy Center, China Medical University, Taichung 404, Taiwan
3 Neuroscience and Brain Disease Center, China Medical University, Taichung 404, Taiwan
4 Translational Medicine Research Center, China Medical University, Taichung 404, Taiwan; bojie@nanoriginal.onmicrosoft.com (B.-J.H.); juiyu0314@nanoriginal.onmicrosoft.com (J.-Y.C.); mcsherry@nanoriginal.onmicrosoft.com (W.W.C.); T24266@mail.cmuh.org.tw (W.L.)
5 Department of Physician Assistant Studies, MGH Institute of Health Professions, Boston, MA 02129-4557, USA; chrislim@mghihp.edu
6 Department of Neurology, China Medical University, Taichung 404, Taiwan
7 Department of Occupational Therapy, Asia University, Taichung 404, Taiwan
8 Organ Transplantation Center, China Medical University, Taichung 404, Taiwan
* Correspondence: brian.chiang@mail.cmu.edu.tw (C.-S.C.); shyu9423@gmail.com (W.-C.S.); longbin@mail.cmuh.org.tw (L.-B.J.)
† Huang B.-J., Chen J.-Y. and Chieng, W.W. contributed equally.

Citation: Chiang, C.-S.; Huang, B.-J.; Chen, J.-Y.; Chieng, W.W.; Lim, S.H.; Lee, W.; Shyu, W.-C.; Jeng, L.-B. Fucoidan-Based Nanoparticles with Inherently Therapeutic Efficacy for Cancer Treatment. *Pharmaceutics* **2021**, *13*, 1986. https://doi.org/10.3390/pharmaceutics13121986

Academic Editor: Mazzucchelli Serena

Received: 19 October 2021
Accepted: 18 November 2021
Published: 23 November 2021

Abstract: The anticancer properties of fucoidan have been widely studied in cancer research. However, the lack of safety information on the parenteral administration of fucoidan and its rapid clearance from the system have limited its application. Herein, we assessed the therapeutic efficacy and safety of fucoidan and developed fucoidan nanoparticles (FuNPs) to enhance their therapeutic effect in the mouse model of breast cancer. FuNPs were synthesized through the emulsion method, and the stable colloid has an average size of 216.3 nm. FuNPs were efficiently internalized into breast cancer cells in vitro, demonstrating an enhanced antitumor activity in comparison with free form fucoidan. After the treatment of FuNPs, the tumor progression was significantly inhibited in viv. The tumor volume was reduced by 2.49-fold compared with the control group. Moreover, the inhibition of the invasion of tumor cells into the lungs revealed the antimetastatic properties of the FuNPs. FuNPs, as naturally marine polysaccharide-based nanoparticles, have shown their broader therapeutic window and enhanced antimetastatic ability, opening an avenue to the development of the inherently therapeutic nanomedicines.

Keywords: fucoidan nanoparticle; marine polysaccharide; safety; antitumor effect; anti-metastasis; parenteral administration

1. Introduction

Fucoidan is a fucose-containing sulphated polysaccharide extracted from brown seaweeds. The biological functions of this marine polysaccharide include anticancer, antivirus, anti-coagulant, and modulation on diabetic and metabolic syndrome, have been extensively studied [1]. With advantages such as being high solubility in aqueous solution and a favorable safety profile, fucoidan has been applied to cancer treatments in preclinical and clinical studies [2–4].

The therapeutic effects of fucoidan have been demonstrated in abundant cancer types including acute leukemia [5], lymphoma [6], head and neck [7], lung [8], breast [9], hepatoblastoma [10], prostate [11], and ovarian cancer [10] in animal studies. The mechanisms or reactions discovered in these cancers were mostly associated with the induction of apoptosis and the inhibition of cell cycle. Moreover, fucoidan possesses antimetastatic ability [12], inhibiting the formation of tumor nodules in the lungs of metastatic 4T1 tumor-bearing animal model [9]. Fucoidan has also been used as a complementary therapy for patients in clinical studies, and the results demonstrated that fucoidan may augment the therapeutic index [2].

In current applications, fucoidan is mostly administered via oral delivery. However, obstacles such as the relatively low bioavailability [13] and the fast clearance rate [14] have restricted the effective accumulation of fucoidan within the tumor, and thus limited its therapeutic efficacy. Engineered nanoparticles with the tailorable properties such as size, zeta potential, surface coating, and shape can optimize the pharmacokinetic behavior and the ability of one material to accumulate in a tumor [15,16]. As reported by Abdollah et al., the coating of fucoidan on the surface of nanoparticles can significantly extend its circulation time in the bloodstream [17], thus offering more opportunities for the nanomedicines to accumulate within the tumor.

In this study, we developed a fucoidan-based nanoparticle (FuNP), explored its safety profile and its potential to inhibit tumor progression via parenteral administration route. Herein, we synthesized the colloidally stable FuNPs using the emulsion method. The compositions we used to stabilize the structure are listed as approved excipients by USFDA. The safety profiles of fucoidan and FuNPs injected via parenteral route were evaluated in mice, and the therapeutic efficacy of FuNPs was assessed in MDA-MB-231-tumor bearing mice. We demonstrated that FuNPs possessed a favorable safety profile and the potency to inhibit tumor progression as well as metastasis. This evidence paves the way for further translational applications of FuNPs.

2. Materials and Methods

2.1. Materials and Reagents

Poly(D,L-lactide-co-glycolide) (PLGA), soybean oil, and Poly(lactide-co-glycolide)-Rhodamine B were purchased from Merck (Germany). Anti-F-actin antibody and 2-(4-Amidinophenyl)-6-indolecarbamidine dihydrochloride, 4′,6-Diamidino-2-phenylindole dihydrochloride (DAPI) were purchased from BD Biosciences (USA). Fucoidan from fucus vesiculosus was purchased from Marinova (Tasmania, Australia).

2.2. Preparation of Fucoidan Nanoparticles (FuNPs)

FuNPs were synthesized by emulsification process. The aqueous phase with fucoidan and organic phase (i.e., dichloromethane, DCM) with PLGA plus soybean oil were mixed and sonicated using the ultrasonic homogenizer (UP200S with S2 probe, Hielscher, Germany) under ice bath. After forming a one-phase emulsion, the sample was subjected to the rotary evaporator to remove the DCM. The aqueous solution after evaporation was purified using a centrifuge.

2.3. Size Selection of FuNPs Using Centrifuge

The as-synthesized FuNPs were subjected to size selection to separate the large and small FuNPs. The group without the exclusion of large FuNPs was centrifuged at 14,000× g for 15 min. We then removed the suspension and redispersed them in water. To collect the large FuNPs, the as-synthesized FuNPs were centrifuged at 800× g for 5 min. The pellets represented the larger FuNPs were collected and redispersed in water, while the suspension was further centrifuged at 14,000× g for 15 min. The pellet as the smaller FuNPs was redispersed in water and stored in 4 °C for further use.

2.4. Characterization of FuNPs

The morphology and structure of FuNPs were characterized using a scanning electron microscopy (SEM, JSM-6700F, JOEL, Akishima, Japan) and a transmission electron microscopy (TEM, JEOL JEM-2100F, JEOL, Akishima, Japan). Dynamic light scattering was used (DLS, Litesizer 500, Anton Paar, Graz, Austria) to measure the size and zeta potential of FuNPs.

2.5. Stability Test

FuNPs were transferred from water to PBS for the stability test. In brief, the FuNPs were stored in 4 °C and 37 °C, respectively, and the size and zeta potential of the FuNPs were analyzed at least 3 times for 7 days post PBS incubation.

2.6. Cell Culture

MDA-MB-231 breast cancer cell line was purchased from Bioresource Collection and Research Center, Food Industry Research and Development Institute, Taiwan. The cells were cultured using Dulbecco's Modified Eagle Medium (DMEM, Thermo Fisher Scientific, MA, USA) containing 10% fetal bovine serum (FBS, from Thermo Fisher Scientific, MA, USA) and 1% penicillin/streptomycin (Thermo Fisher Scientific, MA, USA). The 4T1 was purchased from ATCC (Manassas, Virginia, USA), and the culture method was identical to MDA-MB-231 cells.

2.7. In Vitro Antitumor Effects

MDA-MB-231 cells were treated with fucoidan or FuNPs at different concentrations and the cell viabilities at 24 h and 48 h were measured using CCK-8 assay. In brief, 5×10^3 cells were seeded in 96 well plates and cultured with culture medium (100 µL) at 37 °C overnight. Fucoidan and FuNPs with different concentrations were then added to the cells. After incubation for 24 h or 48 h, cells were washed with PBS, and CCK-8 solution was added to the medium and allowed reaction for 30 min. The plate was then subjected to Elisa reader (SpectraMax iD3, Molecular Devices, San Jose, CA, USA) and the absorbance was measured at 450 nm.

2.8. Monitoring Cell Internalization

We analyzed the cell internalization using flow cytometry (Attune Nxt flow cytometry, ThermoFisher, USA) and fluorescence microscopy (AxioImager. A1, Zeiss, Germany). To facilitate the observation, PLGA-Rhodamine was mixed with PLGA at 1:20 w/w ratio during the synthesis process to obtain the fluorescence-labeled FuNPs. After the incubation of FuNPs with 4T1 and MDA-MB-231 cells for 24 h and 48 h in a 6-well plate, the cells (1×10^5) were collected and measured using flow cytometry.

For fluorescence microscopy, 4T1 cells were seeded and allowed to grow on Millicell EZ SLID (Merck, Germany) overnight. After that, we incubated the cells with FuNPs for 24 h and washed the cells twice with PBS. To generate cellular permeability, triton X-100 (0.3% in PBS buffer) was added for 30 min. The cells were then blocked using FBS buffer (30 min), immobilized using anti-F-actin at 4 °C overnight, and further stained using a secondary antibody (1:250) for 1 h. After an 1 h incubation, the cells were washed twice with PBS and stained with DAPI (1 µg mL^{-1}) for 5 min. The cells were then observed using a fluorescent microscope.

2.9. Animals

ICR mice were used in the safety study, in which the animal protocol has been reviewed and approved by the Institute of Animal Care and Use Committee (IACUC) of Agriculture Technology Research Institute, Taiwan, approval number of 109113. For therapeutic and safety studies using balc/c nude mice, the animal experiments were performed in compliance with the guidelines of the Animal Care and Use Committee (IACUC) of the China Medical University under the approval number of CMUIACUC-2021-299.

2.10. Safety of Fucoidan via IP Injection

A 14-day acute toxicity study of fucoidan via intraperitoneal (IP) injection was carried out to determine the safety and the adverse effect of fucoidan on 12 male and 12 female adult ICR mice (7–9 weeks old). Mice were randomly assigned to either Control, Low-dose, Mid-dose or High-dose groups in which they intraperitoneally (IP) received a 0, 50, 275 or 500 mg kg^{-1} of single dose fucoidan, respectively. During the 14-day observation period, clinical observation including body weight, mortality, clinical symptoms including exterior, behavior, breath, mouth and nose, eyes, skin, digestion, and metabolism were monitored and recorded. Following the monitoring for 14 days, the mice were sacrificed and were subjected to assessments including hematology, serum chemistry and histopathology. For hematology, the whole blood was collected from abdominal aorta through syringe aspiration and preserved in clean K^2-EDTA centrifuge tubes at room temperature before the analysis of red blood cells (RBC), hemoglobin (HGB), white blood cells (WBC), hematocrit (HCT), blood platelet (PLT), lymphocyte composition, eosinophil composition, neutrophil composition, monocyte composition, and basophil composition.

For serum chemistry, the whole blood collected from the heart through syringe aspiration was preserved in clean centrifuge tubes at room temperature for 30 min and centrifuged to harvest the serum. Aspartate aminotransferase (AST), alanine aminotransferase (ALT), creatinine, blood urea nitrogen (BUN), albumin, and glucose of the mice were measured using clinical chemistry analyzer.

For histopathology, the harvested organs include adrenal glands, heart, lungs with trachea, kidneys, spleen, liver with gallbladder, salivary glands, and mandibular lymph were preserved in 10% neutral buffered formalin (NBF) at room temperature for 72~96 h. After fixation, the tissues were trimmed, dehydrated with ethanol and infiltrated by paraffin. Thin sections (4–6 μm in thickness) were cut using a microtome and were subjected to hematoxylin and eosin (H&E) stain. Histopathological alterations were documented by a pathologist.

2.11. Therapeutic Effect of FuNPs

To evaluate the therapeutic effect, the MDA-MB-231 tumor-bearing mice were treated with saline (control) or FuNPs via intravenous (i.v, 100 mg kg^{-1}) injection (q3dx6) at 19 days after tumor inoculation ($n = 4$). The tumor volumes were monitored using a digital caliper (Mitutoyo) three times a week using the following Equation (1):

$$\text{Tumor volume (V)} = \frac{W \times L^2}{2} \tag{1}$$

in which W is the width of the tumor, and the L is the length of the tumor (W < L).

2.12. Safety of FuNPs

Following the monitoring of the tumor progression, the mice were later sacrificed on day 49 and were subjected to histopathological assessments. In brief, the harvested organs include adrenal glands, heart, lungs with trachea, kidneys, spleen, liver with gallbladder, salivary glands, mandibular lymph node, and inoculated neoplasm (human breast cancer model, MDA-MB-231 cell line) were preserved in 10% neutral buffered formalin (NBF) at room temperature for 72~96 h. After fixation, the tissues were trimmed, dehydrated with ethanol, and infiltrated by paraffin. The microtome was used to cut the sample with multiple sections with 4–6 μm in thickness These sections were subjected to hematoxylin and eosin (H&E) stains to observe histological alteration by the study pathologist.

2.13. Statistical Analysis

Results are expressed as mean ± s.d. unless otherwise noted. Non-parametric Kruskal–Wallis test was used to evaluate the significance of mean differences between the control and the fucoidan-treated group in body weights and clinical chemistry values. One-way ANOVA was used to evaluate the statistical differenceusing GraphPad Prism version 9.2.0, USA. p value < 0.05 was considered statistically significant. If a significant result was determined, post hoc tests were used to seek the differences with the control group.

3. Results

3.1. Safety of Fucoidan

The high biocompatibility of fucoidan has been recognized by the Generally Recognized as Safe (GRAS) of the Food and Drug Administration (FDA). While most studies focus on the safety of fucoidan using the oral administration route, scarce information on the dose-related toxicity via parenteral administration has been revealed.

Herein, a 14-day acute toxicity study of fucoidan via intraperitoneal (IP) injection was carried out to determine the safety and the adverse effect of fucoidan on 12 male and 12 female adult ICR mice (7–9 weeks old). Mice were randomly assigned to either Control, Low-dose, Mid-dose or High-dose groups in which their IP received a 0, 50, 275 or 500 mg kg^{-1} of single dose fucoidan, respectively. All mice survived to the end of the study without any abnormal clinical manifestation on their skin, eyes, nose, and mouth. They exhibited normal breathing, behavioral, digestion and metabolism patterns. There was no statistical difference in body weight across all groups before the injection, 7 days (Kruskal–Wallis; $p = 0.514$ for male, $p = 0.182$ for female) and 14 days (Kruskal–Wallis; $p = 0.304$ for male, $p = 0.227$ for female) after the IP injection of fucoidan (Figure 1a). Both the clinical chemistry (Supplementary Table S1a) and hematology values (Supplementary Table S1b) of all groups were within normal reference range [18]. Histopathological examination revealed that cerebrum, cerebellum, heart, lung, spleen, and kidney remained healthy and normal across groups (Figure 1b) except for liver and lymphatic organs.

Their abnormalities were only to be observed in the High-dose group. The abnormal histopathological findings were observed in multiple lymphatic organs. Interestingly, when comparing with the Control, Low-dose and Mid-dose groups, the thymus of the High-dose group was significantly heavier (Supplementary Figure S1) and demonstrated a mild increase of corticomedullary ratio (Supplementary Figure S2a). Mild to moderate hyperplasia were found in gut-associated lymphoid tissue (GALT, Supplementary Figure S2b) and mesenteric lymph nodes, mostly in their follicular (Supplementary Figure S2c), indicating the immune enhancement effect at this area was dominated by T cells. Besides the lymphatic organs, the males' livers were weighted heavier in the High-dose group than those in the Control, Low-dose and Mid-dose groups (Supplementary Figure S1). In addition, mild to moderate hepatic lesions along with microvascular foamy microphages aggregation were found only in some of the female mice in the High-dose group (Supplementary Figure S2d). Based on the result of this acute toxicity test, the decision was made to the dose below Mid-dose (i.e., 275 mg/kg) in the subsequent studies to minimize the interference of adverse effects.

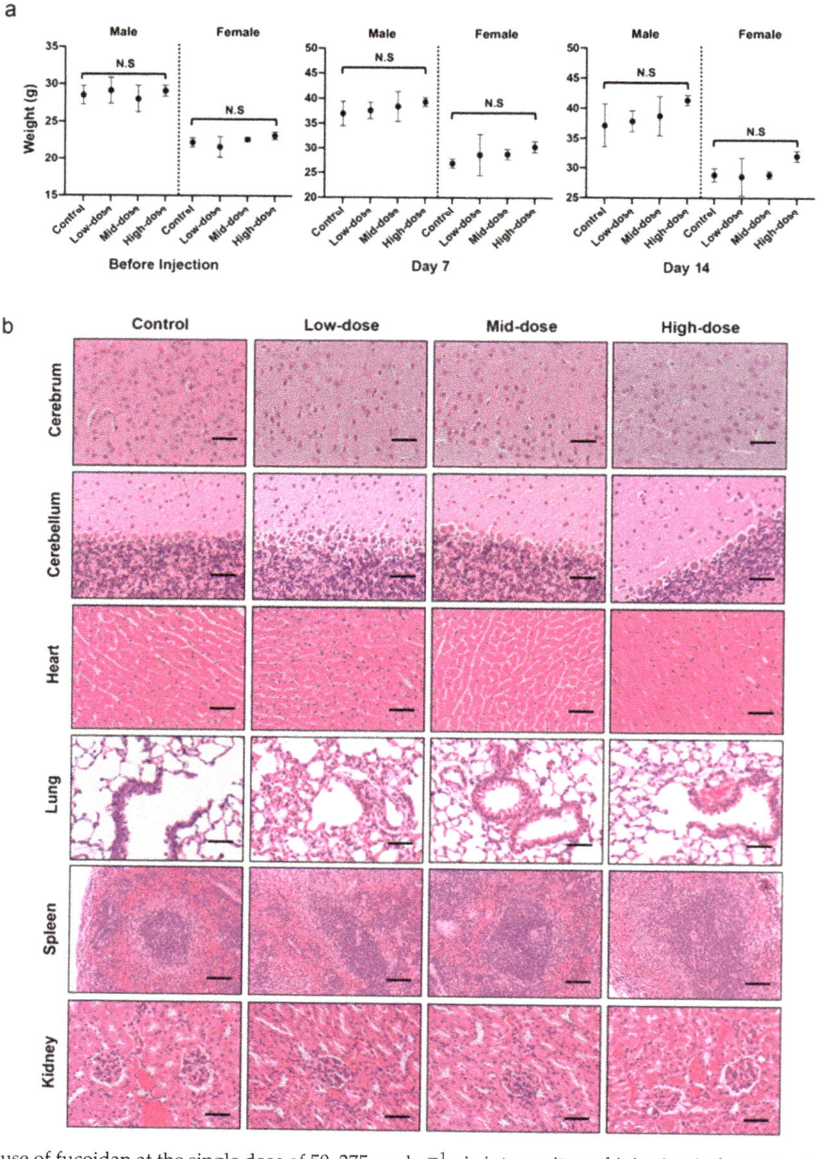

Figure 1. The use of fucoidan at the single dose of 50–275 mg kg^{-1} via intraperitoneal injection is determined safe in mice. (**a**) Left panel: there was no significant difference in body weight of the males and females when they were randomly assigned to each group before the injection. Middle panel: The body weight of the mice was not affected 14 days after the exposure to a single dose of Fucoidan. Right panel: Low-, mid- and high-dose Fucoidan did not affect the body weight of the mice at the end point of the toxicity test. (**b**) Representative microscopic images depict no observable impact of fucoidan to the brain, heart, lung, spleen, and kidney. All data are shown in means ± s.d. Exact p-values are reported in text. Scale bar = 50 μm for cerebrum, cerebellum, heart, lung, and kidney = 100 μm for spleen.

3.2. Characterization of Fucoidan Nanoparticles (FuNPs)

We synthesized FuNPs using the emulsion method. The size and the structural stability of FuNPs can be controlled by adjusting the composition, solvent, and sonication parameters including the intensity and the process time. Although emulsion is a well-established method to synthesize nanoparticles, the size distribution of the produced nanoparticles is broad. The size of a nanoparticle affects its interaction, internalization, and degradation with a cell, which would consequently impact its heterogenous performance on efficacy and safety [15]. Therefore, after emulsification, we performed size selection using the centrifugation and the selection strategy is illustrated in Figure 2a. The FuNPs without exclusion of large particles were abbreviated as L+S, while the FuNPs obtained via size selection process were categorized into either "Large" or "Small" groups.

The sizes of the FuNPs were then verified using scanning electron microscopy (SEM), transmission electron microscopy (TEM), and dynamic light scattering (DLS). Heterogeneous size distribution of the L+S group was observed (Figure 2b). After size selection, the homogeneous FuNPs of Large (Figure 2c) and Small groups (Figure 2d) were observed using SEM. They were further analyzed and verified using DLS (Figure 2f). The size of the FuNP in the L+S, Large, and Small groups were 223.6, 410.1, and 216.3 nm, respectively, indicating the vast majority of the produced FuNPs were in the smaller size. The homogenous FuNPs in the Small group were then observed using TEM (Figure 2e). These FuNPs were strongly negatively charged due to the sulphate groups present in the fucoidan structure (Figure 2g). FuNPs with larger size demonstrated a relatively stronger charge, possibly due to the fact that more fucoidan molecules were coated on the surface.

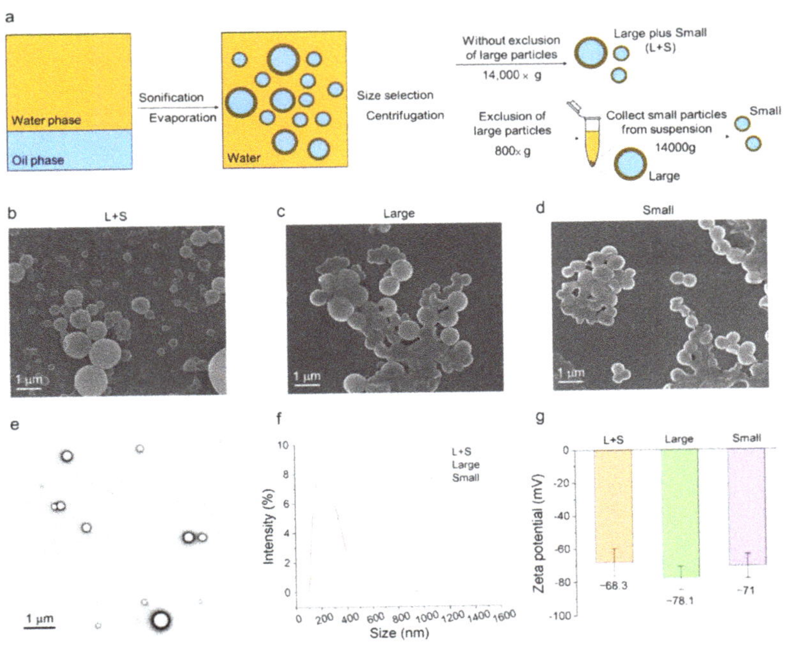

Figure 2. Illustration of the size selection strategy and the characterization of FuNPs. (**a**) Schematic illustration of size selection for FuNP using centrifugation. SEM images of L+S (**b**), Large (**c**), and Small (**d**) FuNPs groups; (**e**) TEM image of the Small FuNPs group; The size (**f**) and zeta potential (**g**) of the different groups were measured using DLS.

3.3. The Effect of the Size of Fucoidan Nanoparticles on Cell Uptake

The size of a nanoparticle can affect the efficiency of cellular interaction and internalization. After breast cancer cell line 4T1 and MDA-MB-231 were incubated with the FuNPs from L+S, Large, and Small groups for 24 h or 48 h, the cell internalization efficiency was assessed using flow cytometry. Rhodamine-modified PLGA were introduced in the emulsion process to form the dye-labeled FuNPs to facilitate the observation.

The percentage of the cells that internalized FuNPs increased following the time-dependent manner in both cells. At 24 h incubations, 31.4%, 26.4%, and 42.4% of the 4T1 cells, and 53.5%, 32.8%, and 56.5% of the MDA-MB-231 cells treated with L+S, Large, and Small FuNPs, respectively, were found to associate with the FuNPs (Figure 3). FuNPs from L+S and Small groups demonstrated a higher fluorescent intensity compared with the Large group, indicating that the smaller FuNPs possessed a higher cell internalization efficiency compared with the larger ones. FuNPs from the Small groups also exhibited a higher cell internalization efficiency compared with those of L+S and Large groups at 48 h post incubation in both cells (Figure 3). These results were confirmed using a fluorescent microscope. The higher red fluorescence intensity (rhodamine-labeled FuNPs) was found in the cancer cells for the Small group (Supplementary Figure S3). We also found that FuNPs from the Large group were less stable and tended to precipitate in PBS and culture media after 24 h of incubation. In contrast, the Small group was colloidally stable during the assessments, which might provide a higher opportunity to interact with the cells to facilitate cellular internalization. Thus, FuNPs from the Small group were chosen to perform subsequent in vitro and in vivo studies.

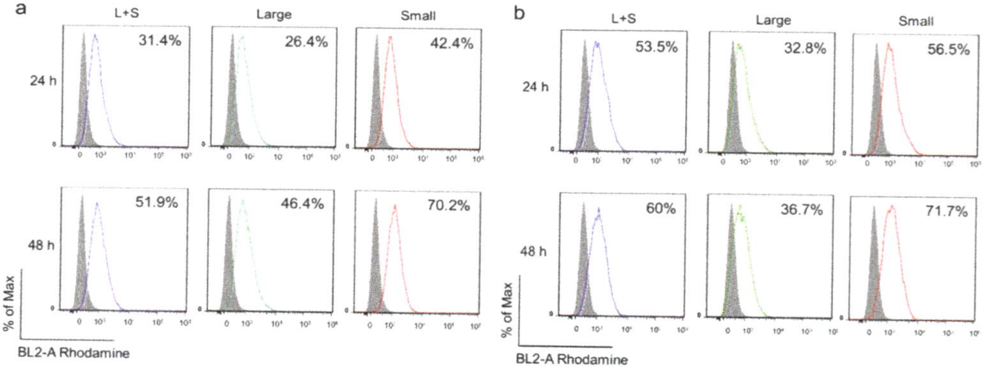

Figure 3. Internalization of FuNPs with different sizes into cancer cell lines. The percentage of 4T1 (**a**) and MDA-MB-231 cells (**b**) internalizing FuNPs from L+S, Large, and Small groups at 24 h and 48 h incubation was evaluated using flow cytometry.

3.4. Stability of the FuNPs

The stability of FuNPs were assessed by incubating them in PBS at 4 °C and 37 °C, respectively. FuNPs at 4 °C were found to be stable in PBS, demonstrating identical size (Figure 4a) and zeta potential (Figure 4b) from day 0 to day 7. The polydispersity index (PDI) at 0.05, 0.02, and 0.012 for 2-, 4-, and 7-day incubation demonstrated that FuNPs sustained homogenous size distribution. Of note, since the ion environment in PBS is more complex, the zeta potential of FuNPs measured in this experiment was higher than the value in water (Figure 2g). On the other hand, at 37 °C, the size of FuNPs increased significantly on day 7 and the zeta potential of FuNPs increased significantly at day 2 and day 7. These changes happened probably due to the degradation of PLGA.

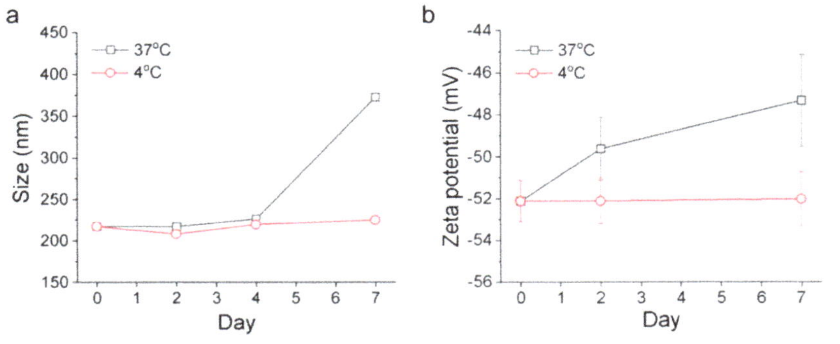

Figure 4. Monitoring the stability of FuNPs in PBS by size (**a**) and zeta potential (**b**) using DLS from day 0 (as synthesis) to day 7.

3.5. Cytotoxicity of Fucoidan and FuNPs on Cancer Cells In Vitro

To evaluate the cytotoxicity of fucoidan and FuNPs, different concentrations were used to treat MDA-MB-231 cells. Interestingly, we found that FuNPs showed stronger antitumor effect while compared with the fucoidan (Figure 5). MDA-MB-231 cells sustained a viability of more than 80% in the exposure to fucoidan (0.5 to 2 mg mL^{-1}). In contrast, the FuNPs significantly decreased the viability of MDA-MB-231 cells at a concentration as low as 0.3125 mg mL^{-1}. At 0.5 mg mL^{-1}, cell viability was approximately 7 times lower when treated with FuNPs than with fucoidan, suggesting the anticancer potency of FuNPs was significantly stronger than free form fucoidan. Based on its superior efficacy in cytotoxicity studies, FuNPs were assessed in the MDA-MB-231 animal model to further evaluate its anticancer effects.

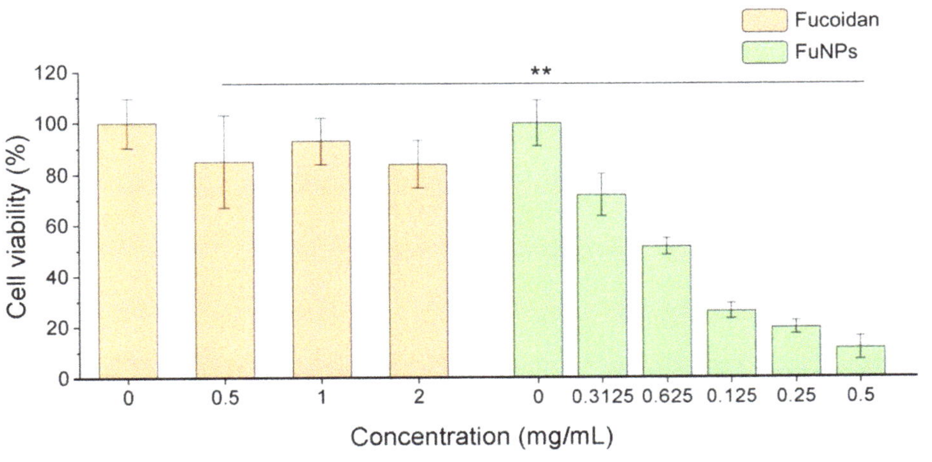

Figure 5. The cell viability of MDA-MB-231 after treating fucoidan or FuNPs with various concentrations. The experiments were performed with at least 6 biological independent samples. One-way ANOVA was performed to evaluate the statistical difference of cell viability between the cells treated with fucoidan and FuNPs at the concentration of 0.5 mg mL^{-1}; ** $p < 0.01$.

3.6. Tumor Inhibition Effect

FuNPs demonstrated therapeutic effect in MDA-MB-231 tumor-bearing mice in a q3dx6 treatment regimen, showing a superior tumor inhibition effect compared with the control group ($p = 0.003$, day 49, Figure 6a). The body weight of the mice treated with

FuNPs slowly increased over time, showing no significant difference in comparison with the control group (Figure 6b). The results indicated that the administration of FuNPs inhibited the growth of the MDA-MB-231 tumor, while no abnormal clinical manifestation was found throughout the treatment.

On day 49 post tumor inoculation, the mice were sacrificed, and the major organs were subjected to pathological analysis. The lungs of the mice treated with saline (control groups) were found to be invaded by the metastatic tumor cells (Figure 6c). Importantly, the sections of the arterial lumen of the FuNPs-treated groups were dominated with embolus, which comprised spindle cells, fibrin, immune cells, and tumor cells. This evidence was a heritage that the tumor cells failed to extravasate into lung parenchyma, thus preventing the tumor metastasis (Figure 6c).

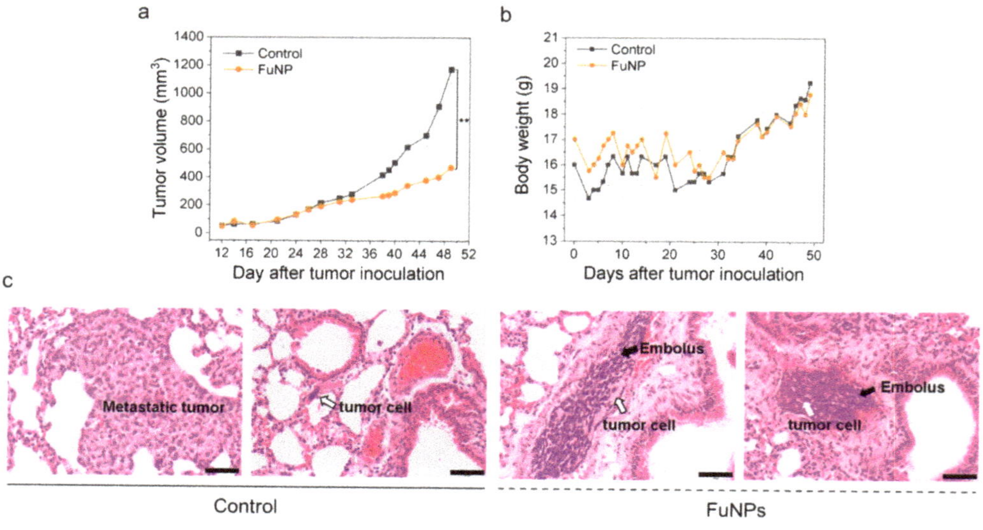

Figure 6. Therapeutic efficacy of FuNPs in MDA-MB-231-tumor bearing mice. Tumor volume (**a**) and body weight change (**b**) of the MDA-MB-231-tumor-bearing mice treated with saline (control) and FuNPs following the treatment course of q3dx6; the section of lung in control and FuNPs-treated groups. (**c**) H&E stains of the mice's lungs showed metastatic tumors in the control group, while the invasion was inhibited in the FuNPs-treated group. Scale bar = 50 μm. The experiments were performed with 4 biological independent animals. One-way ANOVA was performed in Figure 6a to evaluate the statistical difference between the tumor volume at day 49.

3.7. Safety of Fucoidan Nanoparticles

After sacrificing the mice, the organs including hearts, livers with gallbladders, spleens, lungs, kidneys, and salivary glands as well as tumors were collected, weighted, sectioned, stained, and subjected to pathological assessments. There were no significant differences in the organs from the mice treated with saline and FuNPs (Figure 7a). As the tumor growth was effectively inhibited as shown in Figure 6a, the average weight of the tumor from the mice treated with FuNPs was significantly lighter compared with the control group (Figure 7a). We did not observe any abnormality from the H&E stains of heart, liver, spleen, kidney, and salivary gland of the mice treated with FuNPs (Figure 7b). With all the evidence, we concluded that FuNPs are safe under current treatment regimen and dose.

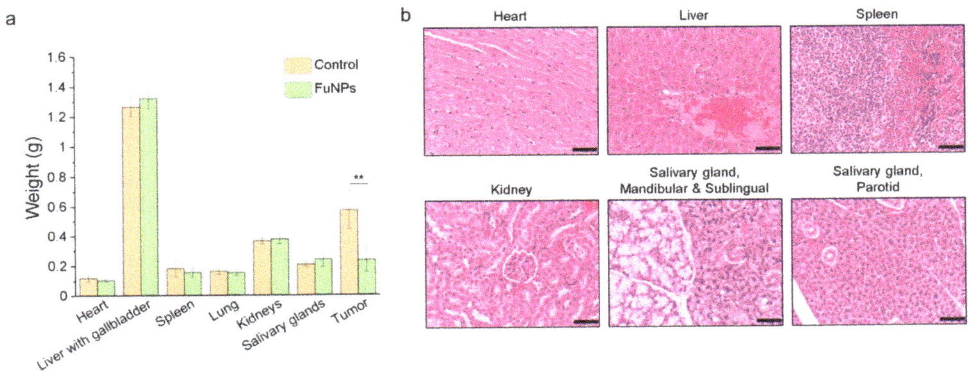

Figure 7. Safety assessments of FuNPs. (**a**) The organs and tumors weight of the mice treated with saline (control) and FuNPs; (**b**) H&E stain for the pathological analysis of the mice treated with FuNPs. Scale bar = 50 µm. The experiments were performed with 4 biological independent animals. One-way ANOVA was performed in Figure 7a to evaluate the statistical difference between the tumor weight between control and FuNP-treated mice; ** $p < 0.01$.

4. Discussion

Fucoidan from fucus vesiculosus was demonstrated to be safe at the doses ranging from 50 to 500 mg kg^{-1} and 50 to 275 mg kg^{-1} via the parenteral IP administration route in male and female mice, respectively (Figure 1 and Supplementary Figure S1). While not every female mouse that received 500 mg kg^{-1} fucoidan showed liver lesion, further studies involving more female mice are required to draw conclusions. Generally, our study provides solid evidence that fucoidan is a substance with a favorable safety profile for the applications associated with parenteral administration.

With the simple emulsion process, we synthesized the FuNPs with an inherently therapeutic effect on tumor inhibition. In Figure 2a–d, we showed that larger FuNPs could be collected via centrifugation at a lower speed (i.e., 800× *g*). By excluding the FuNPs with the particle size larger than 400 nm, the smaller ones presented a more efficient cell internalization efficacy in MDA-MB-231 and 4T1 breast cancer cell lines (Figure 3). The storage of FuNPs in PBS at 4 °C was demonstrated to be stable for at least one week. In contrast, zeta potential and size of FuNPs significantly increased at 2- and 7-day post incubation in PBS at 37 °C (Figure 4), indicating the structural instability took place within 48 h when subjecting the environment. The instability might be attributed to the hydrolytic degradation of PLGA in aqueous environment [19], and lead to the detachment of fucoidan from the particle surface. The degradation of FuNPs at 37 °C (i.e., the biological temperature) addresses the potential of FuNP being a vesicle to encapsulate, deliver, and release active pharmaceutical ingredients to a tumor tissue, and such controlled drug release application is worth exploring in the future studies.

The anti-cancer activity and biological effects of fucoidan have been studied extensively. The regulation of cell cycle, inhibition for the receptors such as epidermal growth factor receptor (EGFR), transforming growth factor-beta receptors (TGFF), and Toll-like receptor 4 (TLR4) for triggering downstream apoptotic pathways [2,20] are among the common mechanisms that fucoidan activates to suppress the proliferation of cancer cells. Noteworthily, FuNPs we synthesized had a significantly stronger anti-cancer activity compared with the free form fucoidan (Figure 5). The stronger anti-cancer activity could be attributed to the nanoparticle-related factors such as the enhanced cell uptake efficiency and the induction of higher-level reactive oxygen species (ROS) in cancer cells [21].

FuNPs significantly inhibited the tumor progress in MDA-MB-231-bearing mice, indicating the nanoscale formulation can successfully accumulate in the tumors after being

IV injected into the bloodstream (Figure 6a). Importantly, fucoidan has been demonstrated to inhibit metastasis in multiple cancer types including breast, lung, and liver cancer [12,20]. Encouragingly, FuNPs also demonstrated antimetastatic activities, where the tumor cells were found to lose the ability to invade into the lung after receiving the FuNPs treatments (Figure 6c). In our study, multiple doses were administered in a 3-day treatment plan, and no adverse effect was observed in the histological assessments (Figure 7). This indicates that FuNPs are a safe formulation with the dual effects of inhibiting tumor growth and metastasis under the current treatment regimen and dose. A further exploration of FuNPs dose escalation study is valuable since FuNPs have such a favorable therapeutic index and broad safety margin.

The aim of this study is to explore the possibility of forming fucoidan-based nanoparticles and evaluate their anti-cancer activity. Our results demonstrated that the stable FuNPs with appropriate size have a far superior tumor progress inhibition ability. The anti-metastasis function of FuNPs was also observed in the mouse cancer model. Noteworthily, under the intensive treatment regimen, FuNPs did not induce adverse effects as confirmed by histological analysis. However, future studies elucidating the mechanisms on tumor and metastasis inhibition are needed. While FuNPs with biodegradable ability can also act as drug carriers to achieve controlled drug release, exploring the mechanisms of FuNPs and discovering the ideal drugs to facilitate the synergistic combination for cancer therapy can open an avenue for a new class of therapeutic nanomedicines.

Supplementary Materials: The following are available online at https://www.mdpi.com/article/10.3390/pharmaceutics13121986/s1. Table S1a: Clinical chemistry values (mean ± SD) for male and female mice, Table S1b: Hematology values (mean ± SD) for male and female mice, Figure S1: High-dose fucoidan significantly impacts the liver and lymphatic organs. Graphs show that high-dose Fucoidan significantly increases the weight of the female spleen, the male liver and the thymus of both the male and female. All data are shown in means ± S.D. * $p < 0.05$; ** $p < 0.01$ (two-tailed one-way ANOVA), Figure S2: The impact of a High-dose fucoidan to the thymus (**a**), GALT (**b**), mesenteric lymph node (**c**), and liver (**d**) is depicted in these representative microscopic images. (**a**) High-dose Fucoidan causes the corticomedullary ratio (black arrows) increase in the thymus. (**b**) High-dose Fucoidan also induces the follicular hyperplasia (grey circle) in the GALT. (**c**) Foamy microphages are shown (white arrows) in the mesenteric lymph nodes of the High-dose group. (**d**) The aggregation of foamy microphages (white arrow) and the apoptosis of the liver cells (black arrows) are presented 14 days after the high-dose Fucoidan injection. Scale bar = 100 μm for (**b**,**d**); 200 μm for (**a**) and 20 μm for (**c**) and Figure S3: Monitoring the internalization of L+S, Large, and Small FuNPs groups into 4T1 cells at 48 h incubation using fluorescent microscope. Scale bar = 50 μm.

Author Contributions: Conceptualization, C.-S.C.; methodology, L.-B.J., W.-C.S., B.-J.H., J.-Y.C., W.W.C., W.L., S.H.L. and C.-S.C.; writing, review and editing, S.H.L. and C.-S.C.; supervision, L.-B.J., W.-C.S. and C.-S.C. All authors have read and agreed to the published version of the manuscript.

Funding: This research was funded by China Medical University and Hospital (DMR-110-182), and Minister of Science and Technology, Taiwan (MOST-110-2314-B-039-043-MY3).

Institutional Review Board Statement: The animal study was conducted with the approval by the Institute of Animal Care and Use Committee (IACUC) of Agriculture Technology Research Institute, Taiwan, under the approval number of 109113 and Animal Care and Use Committee (IACUC) of the China Medical University under the approval number of CMUIACUC-2021-299.

Informed Consent Statement: Not applicable.

Data Availability Statement: The data presented in this study are available on request from the corresponding author.

Acknowledgments: The authors thank for the supported research grants from China Medical University and Hospital (DMR-110-182) and Minister of Science and Technology, Taiwan (MOST-110-2314-B-039-043-MY3). We truly appreciate the professional technical supports, suggestions, and insights from Frank Wu and Lepto Biotech Pte Ltd., (Taipei, Taiwan) on in vivo studies including histology and safety assessments. The authors would like to thank Wan-Zhen (Uni) Hung for the administrative support and thus the teams can focus on exploring the interesting science and studies.

Conflicts of Interest: The authors declare no conflict of interest.

References

1. Luthuli, S.; Wu, S.; Cheng, Y.; Zheng, X.; Wu, M.; Tong, H. Therapeutic Effects of Fucoidan: A Review on Recent Studies. *Mar. Drugs* **2019**, *17*, 487. [CrossRef] [PubMed]
2. Hsu, H.Y.; Hwang, P.A. Clinical applications of fucoidan in translational medicine for adjuvant cancer therapy. *Clin. Transl. Med.* **2019**, *8*, 15. [CrossRef]
3. Oliveira, C.; Neves, N.M.; Reis, R.L.; Martins, A.; Silva, T.H. A review on fucoidan antitumor strategies: From a biological active agent to a structural component of fucoidan-based systems. *Carbohydr. Polym.* **2020**, *239*, 116131. [CrossRef]
4. Kwak, J.-Y. Fucoidan as a Marine Anticancer Agent in Preclinical Development. *Mar. Drugs* **2014**, *12*, 851–870. [CrossRef]
5. Jin, J.O.; Song, M.G.; Kim, Y.N.; Park, J.I.; Kwak, J.Y. The mechanism of fucoidan-induced apoptosis in leukemic cells: Involvement of ERK1/2, JNK, glutathione, and nitric oxide. *Mol. Carcinog.* **2010**, *49*, 771–782. [CrossRef] [PubMed]
6. Yang, G.; Zhang, Q.; Kong, Y.; Xie, B.; Gao, M.; Tao, Y.; Xu, H.; Zhan, F.; Dai, B.; Shi, J.; et al. Antitumor activity of fucoidan against diffuse large B cell lymphoma in vitro and in vivo. *Acta Biochim. Biophys. Sin.* **2015**, *47*, 925–931. [CrossRef] [PubMed]
7. Mizrachi, A.; Shamay, Y.; Shah, J.; Brook, S.; Soong, J.; Rajasekhar, V.K.; Humm, J.L.; Healey, J.H.; Powell, S.N.; Baselga, J.; et al. Tumour-specific PI3K inhibition via nanoparticle-targeted delivery in head and neck squamous cell carcinoma. *Nat. Commun.* **2017**, *8*, 14292. [CrossRef]
8. Lee, H.; Kim, J.-S.; Kim, E. Fucoidan from Seaweed Fucus vesiculosus Inhibits Migration and Invasion of Human Lung Cancer Cell via PI3K-Akt-mTOR Pathways. *PLoS ONE* **2012**, *7*, e50624. [CrossRef]
9. Hsu, H.-Y.; Lin, T.-Y.; Hwang, P.-A.; Tseng, L.-M.; Chen, R.-H.; Tsao, S.-M.; Hsu, J. Fucoidan induces changes in the epithelial to mesenchymal transition and decreases metastasis by enhancing ubiquitin-dependent TGFβ receptor degradation in breast cancer. *Carcinogenesis* **2013**, *34*, 874–884. [CrossRef]
10. Fukahori, S.; Yano, H.; Akiba, J.; Ogasawara, S.; Momosaki, S.; Sanada, S.; Kuratomi, K.; Ishizaki, Y.; Moriya, F.; Yagi, M.; et al. Fucoidan, a major component of brown seaweed, prohibits the growth of human cancer cell lines in vitro. *Mol. Med. Rep.* **2008**, *1*, 537–542. [CrossRef]
11. Choo, G.S.; Lee, H.N.; Shin, S.A.; Kim, H.J.; Jung, J.Y. Anticancer Effect of Fucoidan on DU-145 Prostate Cancer Cells through Inhibition of PI3K/Akt and MAPK Pathway Expression. *Mar. Drugs* **2016**, *14*, 126. [CrossRef]
12. Alekseyenko, T.V.; Zhanayeva, S.Y.; Venediktova, A.A.; Zvyagintseva, T.N.; Kuznetsova, T.A.; Besednova, N.N.; Korolenko, T.A. Antitumor and antimetastatic activity of fucoidan, a sulfated polysaccharide isolated from the Okhotsk sea Fucus evanescens brown alga. *Bull. Exp. Biol. Med.* **2007**, *143*, 730–732. [CrossRef]
13. Pozharitskaya, O.N.; Shikov, A.N.; Faustova, N.M.; Obluchinskaya, E.D.; Kosman, V.M.; Vuorela, H.; Makarov, V.G. Pharmacokinetic and Tissue Distribution of Fucoidan from Fucus vesiculosus after Oral Administration to Rats. *Mar. Drugs* **2018**, *16*, 132. [CrossRef]
14. Tokita, Y.; Nakajima, K.; Mochida, H.; Iha, M.; Nagamine, T. Development of a Fucoidan-Specific Antibody and Measurement of Fucoidan in Serum and Urine by Sandwich ELISA. *Biosci. Biotechnol. Biochem.* **2010**, *74*, 350–357. [CrossRef]
15. Mitchell, M.J.; Billingsley, M.M.; Haley, R.M.; Wechsler, M.E.; Peppas, N.A.; Langer, R. Engineering precision nanoparticles for drug delivery. *Nat. Rev. Drug Discov.* **2021**, *20*, 101–124. [CrossRef]
16. Shi, J.; Kantoff, P.W.; Wooster, R.; Farokhzad, O.C. Cancer nanomedicine: Progress, challenges and opportunities. *Nat. Rev. Cancer* **2017**, *17*, 20–37. [CrossRef]
17. Abdollah, M.R.A.; Carter, T.J.; Jones, C.; Kalber, T.L.; Rajkumar, V.; Tolner, B.; Gruettner, C.; Zaw-Thin, M.; Baguña Torres, J.; Ellis, M.; et al. Fucoidan Prolongs the Circulation Time of Dextran-Coated Iron Oxide Nanoparticles. *ACS Nano* **2018**, *12*, 1156–1169. [CrossRef]
18. Serfilippi, L.M.; Pallman, D.R.; Russell, B. Serum clinical chemistry and hematology reference values in outbred stocks of albino mice from three commonly used vendors and two inbred strains of albino mice. *J. Am. Assoc. Lab. Anim. Sci.* **2003**, *42*, 46–52.
19. Keles, H.; Naylor, A.; Clegg, F.; Sammon, C. Investigation of factors influencing the hydrolytic degradation of single PLGA microparticles. *Polym. Degrad. Stab.* **2015**, *119*, 228–241. [CrossRef]
20. Lin, Y.; Qi, X.; Liu, H.; Xue, K.; Xu, S.; Tian, Z. The anti-cancer effects of fucoidan: A review of both in vivo and in vitro investigations. *Cancer Cell Int.* **2020**, *20*, 154. [CrossRef] [PubMed]
21. Yu, Z.; Li, Q.; Wang, J.; Yu, Y.; Wang, Y.; Zhou, Q.; Li, P. Reactive Oxygen Species-Related Nanoparticle Toxicity in the Biomedical Field. *Nanoscale Res. Lett.* **2020**, *15*, 115. [CrossRef] [PubMed]

pharmaceutics

MDPI

Article

Formulation, Optimization and Evaluation of Luteolin-Loaded Topical Nanoparticulate Delivery System for the Skin Cancer

Imran Kazmi [1,*], Fahad A. Al-Abbasi [1], Muhammad Shahid Nadeem [1], Hisham N. Altayb [1], Sultan Alshehri [2] and Syed Sarim Imam [2,*]

1 Department of Biochemistry, Faculty of Science, King Abdulaziz University, Jeddah 23443, Saudi Arabia; fabbasi@kau.edu.sa (F.A.A.-A.); mhalim@kau.edu.sa (M.S.N.); hdemmahom@kau.edu.sa (H.N.A.)

2 Department of Pharmaceutics, College of Pharmacy, King Saud University, Riyadh 11451, Saudi Arabia; salshehri1@ksu.edu.sa

* Correspondence: ikazmi@kau.edu.sa (I.K.); simam@ksu.edu.sa (S.S.I.); Tel.: +966-543970731 (I.K.); +966-554543058 (S.S.I.)

Abstract: In the present study, luteolin (LT)-loaded nanosized vesicles (LT-NVs) were prepared by a solvent evaporation–hydration method using phospholipid and edge activator. The formulation was optimized using three factors at a three-level Box–Behnken design. The formulated LT-NVs were prepared using the three independent variables phospholipid (A), edge activator (B) and sonication time (C). The effect of used variables was assessed on the vesicle size (Y_1) and encapsulation efficiency (Y_2). The selection of optimum composition (LT-NVopt) was based on the point prediction method of the software. The prepared LT-NVopt showed the particle size of 189.92 ± 3.25 nm with an encapsulation efficiency of 92.43 ± 4.12% with PDI and zeta potential value of 0.32 and −21 mV, respectively. The formulation LT-NVopt was further converted into Carbopol 934 gel (1% w/v) to enhance skin retention. LT-NVoptG was further characterized for viscosity, spreadability, drug content, drug release, drug permeation and antioxidant, antimicrobial and cytotoxicity assessment. The evaluation result revealed optimum pH, viscosity, spreadability and good drug content. There was enhanced LT release (60.81 ± 2.87%), as well as LT permeation (128.21 ± 3.56 $\mu g/cm^2/h$), which was found in comparison to the pure LT. The antioxidant and antimicrobial study results revealed significantly ($p < 0.05$) better antioxidant potential and antimicrobial activity against the tested organisms. Finally, the samples were evaluated for cytotoxicity assessment using skin cancer cell line and results revealed a significant difference in the viability % at the tested concentration. LT-NVoptG showed a significantly lower IC_{50} value than the pure LT. From the study, it can be concluded that the prepared LT-NVoptG was found to be an alternative to the synthetic drug as well as conventional delivery systems.

Keywords: luteolin; vesicles; irritation study; optimization; topical gel

Citation: Kazmi, I.; Al-Abbasi, F.A.; Nadeem, M.S.; Altayb, H.N.; Alshehri, S.; Imam, S.S. Formulation, Optimization and Evaluation of Luteolin-Loaded Topical Nanoparticulate Delivery System for the Skin Cancer. *Pharmaceutics* **2021**, *13*, 1749. https://doi.org/10.3390/pharmaceutics13111749

Academic Editor: Mazzucchelli Serena

Received: 23 August 2021
Accepted: 19 October 2021
Published: 20 October 2021

1. Introduction

Cancer is a major health issue all over the globe caused by abnormal cell growth with invasive potentials [1]. There are multiple influencing factors such as genetic factors, environmental factors, alcohol consumption, smoking, exposure to radiation and heredity. Melanoma is a type of skin cancer with the highest metastatic effect rate. It can spread to the other sites of the body by entering into the lymphatic system and bloodstream [2]. It can originate from the malignant transformation of melanocytes and is the most aggressive skin cancer. It has a low survival rate, high multidrug resistance and common relapse. Nowadays, nanoformulations are the most widely explored delivery systems for skin-related disease. They can bypass the effect of the first pass through the liver, with high stability and low dose, and can target the affected area [3,4].

Nowadays much attention has been given to the bioactive compounds with antioxidant properties in the treatment of cancer. The flavonoid luteolin (LT) is an important natural antioxidant that has potent anticancer effects. It is a natural flavonoid, present in different plant species. It has been reported to have a wide range of pharmacological actions such as anti-inflammatory, anti-allergic, antioxidant and anticancer properties. The anti-inflammatory activity of LT is related to its anticancer properties [5]. There are numerous research studies that reported cell line activity against different cancers [1]. It acts by exhibiting cell cycle arrest during the G1 phase linked to suppression of CDK2 activity [6]. It helps to reduce the epidermal growth factor-induced markers as well as restoration of cell–cell junctions [1,7]. LT-loaded nanoformulations such as folacin-modified nanoparticle [8], nanoparticles [9,10], NLCs [11], folic acid-modified ROS-responsive nanoparticles [12], and nanospheres [13] have been prepared and enhanced bioavailability and efficacy have been reported. However, the application of LT vesicles in the skin cancer has not been reported.

The application of nanoformulations has been found effective in the enhancement of solubility of poorly soluble drugs [14]. Over the last decades, lipid-based nanovesicles for topical delivery have been used to improve therapeutic efficacy. There are different types of nanovesicles such as transferosomes, ethosomes, niosomes and cubosomes which are used as topical delivery [15]. These vesicles are composed of cholesterol, phospholipids, surfactants and water. These carrier systems can encapsulate both hydrophilic and hydrophobic drugs. They can deliver drugs to both topical as well as systemic circulation [16]. The use of an edge activator in the formulation of lipid vesicles gives flexibility to the lipid bilayer and can permeate into a very low skin pore size [17]. It is also termed an ultra-deformable vesicle with an aqueous core surrounded by the lipid bilayer. Due to the ultra-flexibility, it can penetrate into the intact human skin and act as non-invasive targeting. The vesicles have the ability to protect a drug from unfavorable absorption into the cutaneous blood vessels. This helps to retain the drug at the skin site [18,19]. The edge activator promotes skin permeation through an intercellular lipid matrix by mixing with stratum corneum as well as by altering the lamellae [18,20,21]. It can permeate with low-, medium- and high-molecular-weight drugs [22].

The object of the present study was to prepare luteolin-loaded nanovesicles (LT-NVs) and characterize them for different parameters. The present delivery systems were optimized by using three factors, phospholipid 90 G (A), edge activator (B) and sonication time (C), at three levels (−, 0, +). The different formulation compositions were assessed on the particle size (PS) and encapsulation efficiency (EE) to select the optimized formulation. From the formulation design approach, optimized luteolin nanovesicles (LT-NVopt) were selected and characterized for permeation, drug release, antioxidant activity and cytotoxicity activity.

2. Material and Methods

2.1. Materials

Luteolin was purchased from Beijing Mesochem Technology Co. Pvt. Ltd. (Beijing, China). Phospholipid 90G was received as a gift sample from Lipoid GmbH, Ludwigshafen, Germany. Sodium cholate, methyl paraben, triethanolamine and tween 80 were purchased from Sigma Aldrich, St Louis, MO, USA and Loba Chemie, Mumbai, India. Optimization of the formulation was performed using Design Expert software (Stat-Ease, Minneapolis, MN, USA). The cells were procured from the National Centre for Cell Science, Pune, India. Carbopol 934 and disodium hydrogen phosphate were procured from Loba Chemie Pvt. Ltd. Mumbai, India and Planet Science, Vadodra, India. The solvents methanol, chloroform, ethanol and water were used at analytical grade.

2.2. Optimization

The prepared LT-NVs were statistically optimized by a three factors, three level Box–Behnken design. LT-NVs were prepared using the variables phospholipid (A), edge activator (B) and sonication time (C) at three-levels, i.e., low (−), medium (0) and high (+) (Table 1). The design showed fifteen formulation runs with three center points (the same composition to check the error) and their effects were assessed on PS (Y_1) and EE (Y_2).

The polynomial equation and 3D and contour plots were generated for all the variables to evaluate the individual as well as combined effects. The actual and predicted values were also generated from the software to confirm the results. The selection of formulation variables was done based on the preliminary study. The dose of LT was fixed in all compositions.

Table 1. Independent variables used to optimize luteolin nanovesicles (LT-NVopt) using a Box–Behnken design.

Independent Variables	Code	Low (−1)	Medium (0)	High (+1)
Phospholipid 90 G (% *w*/*v*)	A	70	80	90
Edge activator (% *w*/*v*)	B	10	20	30
Sonication time (min)	C	3	6	9
Dependent Variables				
Particle size (nm)	Y_1			
Encapsulation efficiency (%)	Y_2			

2.3. Formulation of Luteolin Nanovesicles (LT-NVs)

LT-NVs were prepared using a solvent evaporation hydration method as per a reported procedure with slight modifications [23]. The ingredients phospholipid 90 G (A) and edge activator (B, sodium cholate—tween 80 blends, 50:50) were taken in specified amount as shown in Table 2. Each ingredient including LT (25 mg) was accurately weighed and transferred to a round bottom flask containing chloroform:methanol (10 mL, 1:1). The flask was attached to a rotary evaporator and the organic solvent was removed at low temperature (40 °C) to form a thin lipid film. The flask was removed from the evaporator and kept overnight to remove the traces of organic solvent. LT-loaded thin lipid film was hydrated with phosphate buffer saline (10 mL, pH 6.8) at 100 rpm for 30 min. The prepared luteolin lipid vesicles (LT-NVs) were kept overnight at room temperature for stabilization of vesicles. Finally, the LT-NVs were probe sonicated in ice condition (4 °C) for different time points with 5 min interval to reduce the size. The prepared samples were transferred to a vial and then stored for further characterization.

Table 2. Low, medium and high levels of experimental independent variables phospholipid 90 G (A), edge activator (B), sonication time (C) with their effects on size (Y_1, nm) and encapsulation efficiency (Y_2, %).

Code	A (%, *w*/*v*)	B (%, *w*/*v*)	C (min)	Y_1 (nm)	Y_2 (%)
1	80.00	20.00	6.00	213.8 ± 1.1	81.11 ± 3.3
2	80.00	20.00	6.00	215.1 ± 1.9	80.24 ± 3.9
3	90.00	10.00	6.00	222.4 ± 4.3	73.56 ± 2.6
4	90.00	30.00	6.00	214.2 ± 1.5	89.87 ± 4.1
5	80.00	20.00	6.00	212.6 ± 2.1	81.21 ± 3.5
6	70.00	10.00	6.00	148.5 ± 3.7	81.11 ± 3.2
7	80.00	30.00	3.00	155.2 ± 2.9	76.87 ± 4.1
8	90.00	20.00	3.00	187.2 ± 1.8	77.28 ± 4.7
9	90.00	20.00	9.00	254.6 ± 2.3	69.11 ± 3.2
10	80.00	10.00	9.00	166.6 ± 4.1	63.56 ± 1.9
11	70.00	30.00	6.00	202.4 ± 3.2	73.44 ± 2.1
12	80.00	30.00	9.00	231.1 ± 1.5	80.12 ± 1.7
13	80.00	10.00	3.00	164.7 ± 1.7	77.32 ± 2.3
14	70.00	20.00	9.00	182.4 ± 2.3	73.32 ± 3.2
15	70.00	20.00	3.00	173.3 ± 3.5	75.65 ± 2.7

2.4. Vesicle Characterization

The prepared LT-NVs were evaluated for particle size (PS), size distribution (PDI) and surface charge (ZP). The samples were analyzed by a particle size analyzer (Malvern zeta sizer, Malvern, UK). The samples (0.1 mL) were taken, diluted with double distilled water and scanned for PS and PDI. ZP was also evaluated using a similar method with an electrode-containing cuvette to measure the surface charge. The ideal ZP and PDI value must be ±30 mV and less than 0.7 to get uniform vesicle size distribution [24,25].

2.5. Encapsulation Efficiency (EE)

LT encapsulation from the prepared LT-NVs was evaluated by an ultracentrifugation method [26]. The formulations (5 mL) were taken in centrifuge tube and centrifuged at 10,000 rpm for 1 h. The supernatant containing LT was collected and diluted further to evaluate LT content in each sample by using UV spectrophotometer (Shimadzu 1800, Kyoto, Japan). The concentration of LT was calculated by the equation:

$$\text{EE }(\%) = \left(\frac{Wa - Wb}{Wa}\right) \times 100 \quad (1)$$

Wa: Initial LT content; *Wb*: Free LT content.

2.6. Formulation of Luteolin Nanovesicles Based Gel (LT-NVoptG)

The optimized formulation (LT-NVopt) was converted into the semisolid gel formulation using carbopol 934 as gelling agent. The previously optimized gelling agent Carbopol (1%, *w/v*) was dispersed into the distilled water and kept aside for 24 h for complete swelling. The prepared LT-NVopt was added to polymer dispersion with continuous stirring to get a uniform homogenous gel. Triethanolamine and methyl paraben was added to maintain the pH and preservation to the gel system [27].

2.7. LT-NVoptG Characterization

The prepared LTNVoptG was characterized for different parameters to evaluate the characteristics of the gel. The different parameters such as drug content, pH, viscosity and spreadability were evaluated. The drug content was evaluated to calculate the LT amount in the gel formulation. The difference in the amount of LT added and the amount of LT present was calculated. The prepared LT-NVoptG (50 mg) was taken and dissolved in methanol. The sample was centrifuged at 10,000 rpm for 10 min and the supernatant was collected. The supernatant was further diluted, filtered and the drug content was estimated by UV spectrophotometer. The pH of prepared gel was evaluated using a digital pH meter [28]. The gel sample was taken in a small beaker and the pH meter was dipped into it until it showed a stable value. The viscosity was evaluated to check the flow property of the prepared LT-NVoptG by a viscometer at room temperature [29]. The gel was further evaluated for extrudability and spreadability. The gel sample was taken and kept on the glass slide with a pre-marked area. Then, another slide was placed over the sample and weight was applied. The spread of gel after application of weight was noted and the difference between the initial area and final area was calculated [30]. The extrudability of the gel was evaluated by filling the gel sample into the tube and weight was applied. The tube was pressed from the crimp side end and the extruded gel was collected to calculate the extrudability.

2.8. Drug Release

The release study from the prepared LT-NVoptG, LT-NVopt and pure LT were evaluated using a dialysis bag [31]. A 2 mL (5 mg LT) sample was filled in the dialysis bag and both the ends were tied. The bags were dipped into the dissolution medium (500 mL with 1% tween 80, pH 6.8) and temperature was set at 37 ± 0.5 °C with stirring speed of 50 rpm. At specific time intervals, the released LT content (5 mL) was collected and replenished with the same volume to maintain the uniform condition throughout the study. The released content from the samples was evaluated by UV spectrophotometer.

2.9. Permeation Study

The comparative permeation study was performed using egg membrane following a reported procedure with slight modifications [32]. Egg membrane has similar properties to stratum corneum of human skin [33,34]. The different samples of pure LT, LT-NVopt and LT-NVoptG were taken and filled to the diffusion cell with effective surface area of 3 cm^2 and receptor volume of 20 mL. The egg membrane was carefully removed and checked for any damage [35]. The samples (~5 mg LT) were filled in the donor compartment and the receptor compartment was filled with phosphate buffer saline. The study was performed at 37 °C with continuous stirring. After specific time points, the permeated content (1 mL) was collected and replenished with fresh release media. The permeated contents were filtered and diluted further with appropriate solvent. The drug content at each time point was measured using spectrophotometer in triplicate (n = 3).

2.10. Antioxidant Assessment

The prepared LT-NVopt, LT-NVoptG and pure LT were evaluated for DPPH-based antioxidant activity. The samples were reacted with ά,ά-diphenyl-β-picrylhydrazyl (DPPH) standard to change the color from violet to colorless [36]. The antioxidant has the property to donate the hydrogen ion and decrease the absorbance of the test compounds. The pure LT and LT-NVoptG were prepared in different concentration ranges and sample volume 100 μL was transferred to small glass vials. The samples were incubated with standard DPPH solution and kept aside for 30 min to complete the reaction mixture. Finally, the sample plate was assessed at 517 nm. The study was performed in triplicate and the effects were calculated using the equation:

$$AA\ \% = \frac{(\text{Absorbance of control} - \text{Absorbance of test})}{\text{Absorbance of control}} \times 100 \qquad (2)$$

2.11. Antimicrobial Activity

The prepared LT-NVoptG was evaluated for antibacterial activity and results were compared with pure LT. The study was performed using the microdilution test with a slightly modified reported method [37]. The samples were tested against the microorganisms *S. Aureus*, *E. coli* and *B. subtilis*. The organism's broth culture was added to the growth medium and transferred to a clean sterilized Petriplate at 121 °C (15 PSI). The plates were kept aside for the solidification of the media. The wells were prepared with a sterilized stainless-steel borer. Each sample was transferred to the well and plates were kept aside at room temperature to diffuse the sample into the medium. The plates were kept in an incubator and the zone of inhibition was measured to check the effect of sample.

2.12. Cytotoxicity Study

The cytotoxicity assessment was evaluated to check the effect of the prepared LT-NVoptG, LT-NVopt and pure LT on the skin cancer cell line (B16F1). The cells were collected and stored in CO_2 (5%) and oxygen (95%) at 37 °C by using Dulbecco Modified Eagle media with the support of serum of fetal calf (5%). The experiments were performed with asynchronous populations in the phase of exponential and rapid growth, 24 h after the plating of a sample [38]. B16F1 cells (3×10^3) were added to DMEM (200 μL) and placed in the microplate (96 plate). The fresh medium was replaced after 24 h incubation time with serum-free DMEM. The cell line was incubated with pure LT, LT-NVopt and LT-NVoptG in the media corresponding to a concentration between 10–1000 μM for 24 h. Then, MTT was added into the well of the microplate and further incubated for 4 h at 37 °C. The formazan crystals were formed after the lysis of cells and then dissolved using DMSO (100 μL). The absorbance of the pure LT and LT-NVoptG were evaluated at 570 nm using the microplate reader. The IC_{50} values of the samples were calculated to compare the difference between them. IC_{50} was expressed as the concentration of drug needed to kill 50% of the cells. The study was performed in triplicate.

2.13. Irritation Study

The chorioallantoic membrane (HET-CAM) method was used to study the irritation [39]. This method is commonly used because no animal is required to perform the study. It is a sensitive alternative to the Draize test [40]. The study was performed with negative control, positive control and prepared LT-NVoptG to compare the results. Hen eggs were taken and kept in an incubator for 10 days at 37 °C with 55 ± 2% RH. The eggs were regularly rotated after 24 h for 10 days. On the 10th day, the eggs were taken out from the chamber and then the outer eggshell was removed from the air chamber side. For clear visibility of CAM from the air chamber side, sterilized normal saline solution was added. The samples negative control, positive control and prepared LT-NVoptG were added to the CAM and the cumulative scoring was noted at different time points. The cumulative irritation score from each treated egg was compared with the standard irritant. The scoring of irritation was done as per the scale of hemorrhage, lysis and coagulation. The score was calculated between 0 (no reaction) to 3 (strong reaction). The irritation score was classified as slight irritation (≤); moderate irritation (>0.8–<1.2); irritation (≥1.2–≤2); severe irritation (≥2) [39].

2.14. Statistical Analysis

The data are presented in triplicate and shown as mean ± SD. Graph pad Instat was used to analyze the data (GraphPad Software Inc., La Jolla, CA, USA). Data were subjected to one-way ANOVA followed by Bonferroni multiple tests to analyze statistically significant differences between samples.

3. Results and Discussion

3.1. Optimization

The prepared luteolin nanovesicles (LT-NVs) were prepared by a solvent evaporation–film hydration method. BBD optimization techniques give the maximum variables at different levels along with a lower number of experimental runs [41]. The formulations were optimized using phospholipid (A), edge activator (B) and sonication time (C) as independent variables. The lower and upper levels of the independent variables were taken as phospholipid 90 G (70–90% *w*/*v*), edge activator (10–30% *w*/*v*) and sonication time (3–9 min). The design had fifteen different formulation compositions with three common compositions to check the error in the results. The used formulation variables showed a significant effect on the size (Y_1) and encapsulation efficiency (Y_2). The effect of formulation variables was observed by the application of the polynomial equation and response surface plot (Figures 1 and 2). The independent variables showed individual as well as a combined effect on the size (Y_1) and encapsulation efficiency (Y_2). The different statistical parameters such as linear, cubic, quadratic and 2F models were evaluated and the best fit model was found to be quadratic. The maximum R^2 values were found to be for the quadratic model. From the results, a closer value of predicted as well as practical value confirms that the used method composition is ideal for the prepared delivery systems. The closeness of the actual and predicted value is also shown graphically in Figure 3. Regression analysis was used to analyze the different models for dependent variables and was found to be quadratic as shown in Table 3.

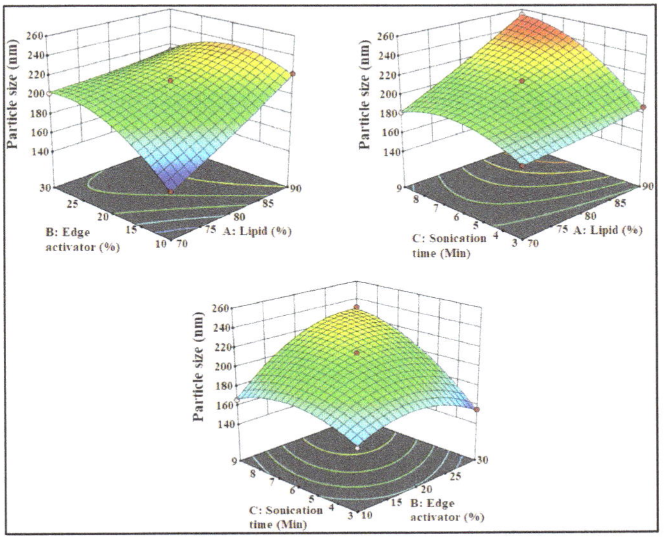

Figure 1. Effect of lipid (A), edge activator (B) and sonication time (C) on size.

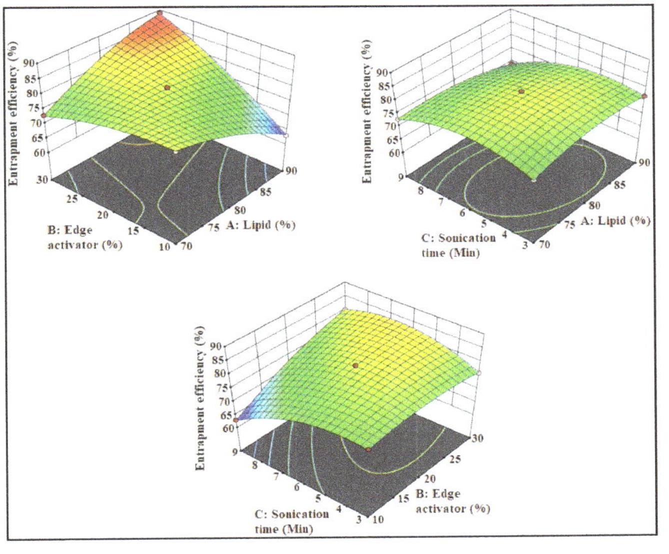

Figure 2. Effect of phospholipid (A), edge activator (B) and sonication time (C) on encapsulation efficiency.

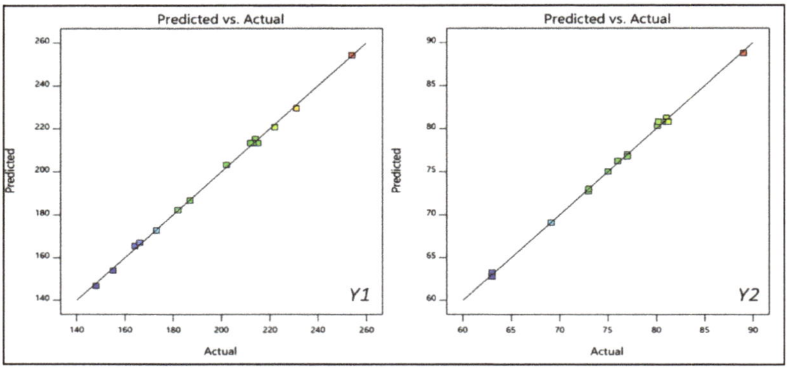

Figure 3. Effect of actual and predicted value of independent variables on PS (Y_1) and EE (Y_2).

Table 3. Regression analysis summary for responses Y_1 (PS) and Y_2 (EE).

Model	R^2	Adjusted R^2	Predicted R^2	SD
(Y_1)				
Linear	0.6021	0.4924	0.2149	21.91
2F1	0.8406	0.7210	0.4411	16.25
Quadratic	0.9987	0.9963	0.9835	1.88
(Y_2)				
Linear	0.2857	0.0909	−0.4251	6.68
2F1	0.8383	0.7170	0.5789	3.73
Quadratic	0.9985	0.9959	0.9876	0.45

R^2 = Coefficient of correlation; SD = Standard deviation.

3.2. *Effect of Formulation Factors on PS*

The prepared LT-NVs showed vesicle size in the range of 155.2 ± 2.9 nm (F7) to 254.6 ± 2.3 nm (F9). There was a significant ($p < 0.05$) difference in the size observed between the prepared LT-NVs. The formulation F7 with the composition of phospholipid 80%, edge activator 30% and sonication time 3 min showed the lowest size. The formulation F9 showed the highest size with the composition of phospholipid 90% *w*/*v*, edge activator 20% *w*/*v* and sonication time 9 min. The used composition depicted a significant effect on the size. The polynomial equation, 3D response surface plot and contour plot (Figure 1) showed the effect of phospholipid (A), edge activator (B) and sonication time (C) on the vesicle size:

Particle size: $+231.33 + 21.5\ A - 12.75\ B + 19.25\ C - 15.5\ AB + 14.5\ AC + 18.5\ BC + 1.58\ A^2 - 18.42\ B^2 - 15.92\ C^2$.

The used factors lipid (A) and sonication time (C) showed a positive effect on the size, whereas edge activator (B) showed a negative effect. As the lipid concentration (A) increases, the vesicle size (Y_1) increases. The enhancement in the size was thus due to the increase in the lipid concentration. At high concentration of lipid, the greater availability of lipid can entrap higher amounts of drug and the size increases. Therefore, the optimum concentration of lipid is important to get the optimum vesicle size. The lack of sufficient concentration of edge activator leads to low drug solubility and it also reduces the surface tension. With an increase in sonication time, the size of vesicles decreases. The phospholipid of the vesicles might rearrange to form smaller size vesicles. However, with a longer sonication time, the vesicles break into smaller size. These smaller size vesicles fold up into thermodynamically stable vesicles and the formed unstable vesicles during probe sonication may fuse together to form larger vesicles [42,43].

3.3. Effect of Formulation Factors on EE

The prepared LT-NVs showed the encapsulation efficiency in the range of 63.56 ± 1.9% (F10) to 89.87 ± 4.1% (F4). There was a significant ($p < 0.05$) difference in the encapsulation efficiency observed due to the variation in the composition. The formulation F10 with the composition of phospholipid 80% w/v, edge activator 10% w/v and sonication time 9 min showed the minimum encapsulation and the formulation F4 showed the maximum size with composition of lipid 90%, edge activator 30% and sonication time 6 min. The effects of used composition lipid (A), edge activator (B) and sonication time (C) are depicted by the polynomial equation, 3D response plot and contour plot (Figure 2). The below given polynomial equation showed the effect on the encapsulation efficiency:

Encapsulation efficiency: $+80.8 + 0.4875\ A - 4.26\ B - 2.48\ C + 8.5\ AB - 1.48\ AC + 4.52\ BC - 2.4\ A^2 - 1.9\ B^2 - 4.88\ C^2$.

The used factor phospholipid (A) showed a positive effect on the encapsulation efficiency (Y_2). With the increase in lipid concentration (A), the encapsulation efficiency of LT increases. The presence of a high concentration of lipid (A) accommodates a high concentration of lipophilic drugs, LT. The optimum concentration of lipid is important to get the optimum encapsulation efficiency because the linear increase in lipid concentration also affects the EE. The edge activator (B) showed a negative effect on the encapsulation efficiency. With the gradual increase in the edge activator concentration, the encapsulation of LT decreased. At high concentration, a greater amount of LT leaches out from the vesicles. In the case of the third factor, with the increase in sonication time (C), the encapsulation efficiency decreases. At longer sonication time, the disruption of vesicle structures as well as degradation of phospholipids takes place, resulting in higher amounts of drug leakage which leads to low EE [42].

3.4. Point Prediction

The selection of optimized formulation (LT-NVopt) was performed using the point prediction optimization method by further changing the independent variables. The slight change in composition also depicted changes in vesicle size and encapsulation efficiency. The optimized composition (LT-NVopt) was found to be lipid 85% w/v, edge activator 15% w/v and sonication time 4 min. This optimized composition revealed a vesicle size of 189.92 ± 3.25 nm with an encapsulation efficiency of 92.43 ± 4.12%. The software also gives predicted vesicle size of 185.11 nm with an encapsulation efficiency of 91.12%.

The used composition was also evaluated by the desirability value of the individual as well as combined independent variables. The desirability value of the prepared LT-NVs was found to be closer to 1 (0.991). The value closer to unity confirms that the used method is robust. Therefore, the LT-NVopt formulation was further converted into Carbopol gel and was characterized for different parameters.

3.5. Particle Size and Surface Charge

The prepared LT-NVs showed vesicle sizes in the range of 155.2 ± 2.9 nm (F7) to 254.6 ± 2.3 nm (F9). The optimized composition was a vesicle size of 189.92 ± 3.25 nm (Figure 4). PS less than 500 nm (100–500 nm) is ideal for cellular uptake via the endocytic pathway. In our study, prepared LT-NVopt size was found to be in the desired range of internalization by cancer cells [44,45]. The PDI and surface charge of the prepared LT-NVopt was 0.32 and −21 mV (Figure 5). Thus, less than 0.7 is considered as suitable for the delivery systems [24].

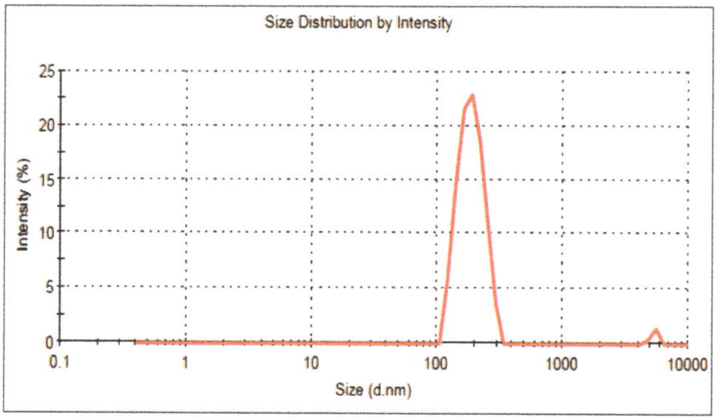

Figure 4. Vesicle size of optimized luteolin loaded nanovesicles (LT-NVopt).

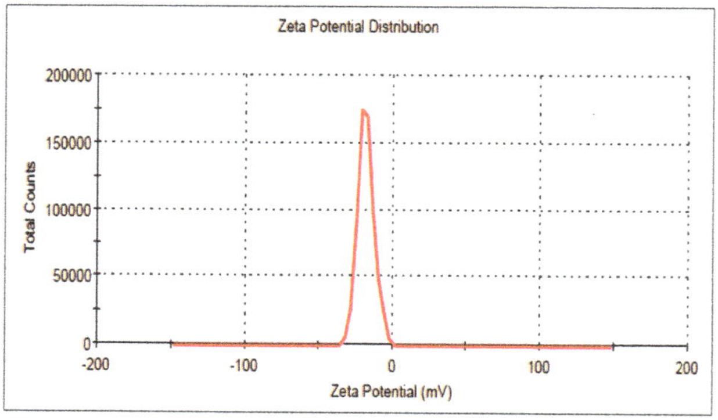

Figure 5. Vesicle size of optimized luteolin-loaded nanovesicles (LT-NVopt).

3.6. Formulation of Gel

The prepared LT-NVs had low viscosity and rheological properties so it was difficult to apply the skin layer. For better application, its viscosity was enhanced by the addition of Carbopol as a gelling agent. The semisolid gel system will better adhere to the skin layer. Carbopol is the most common and widely used gelling agent due to its compatibility with skin layers. The optimized concentration of Carbopol was found to be 1% w/v. At this concentration, LTNVoptG showed good rheological properties so was selected as the final gelling agent. Triethanolamine was used as a neutralizing agent to enhance the stability of the formulation [46].

3.7. Characterization of Gel

The prepared LT-NVopG was evaluated for different parameters such as pH, drug content, viscosity, spreadability and release study. The drug content results indicated the presence of LT in the prepared gel formulation. The prepared LT-NVoptG showed a high drug content (98.8 ± 2.12%). The high drug content is good for the adopted method as well as a delivery system. pH of the prepared LT-NVoptG was evaluated and the result was found to be closer to skin pH. The variation in pH value may lead to skin irritation. The prepared LT-NVoptG had a pH value of 6.6 ± 0.44, which is within the limit of skin formulation and does not produce any toxicity [47]. The viscosity is also one of the important parameters for the semisolid formulation. The viscosity was found to be 534 ± 1.22 cps. The particle size and PDI of a nanoformulation significantly affect the viscosity of the gel formulation. A higher particle size and PDI gives greater viscosity [48]. The prepared LT-NVoptG showed extrudability and spreadability values of 13.11 ± 1.23 g/cm^2 and 6.2 ± 0.41 cm, respectively. The optimum range gives better application to the affected skin membrane. The high viscosity of the gel formulation gives low spreadability and extrudability values. The optimum viscosity, spreadability and extrudability results will give better adherence to the skin [49].

3.8. Drug Release

The prepared LT-NVopt, LT-NVoptG and pure LT were evaluated for drug release and the results are presented in Figure 6. The pure LT showed poor drug release (23.98 ± 1.12%) in 12 h of the study. The poor release of LT is due to the poor solubility of LT. The prepared LT-NVopt and LT-NVoptG showed a significantly ($p < 0.001$) enhanced drug release profile with a maximum drug release of 79.81 ± 3.15% and 60.81 ± 2.12%, respectively. The enhanced drug release was achieved from both the prepared formulations, which may be due to the nano size and enhanced solubility of LT in the presence of the surfactant. The nanosized vesicle has a greater surface area available to solubilize in the presence of a surfactant. There was also a significant ($p < 0.01$) difference in the release observed between the LT-NVopt and LT-NVoptG. In the case of gel formulation, the drug release was found to be slower due to the presence of one extra layer of Carbopol which slowly diffuses the drug into the release media. Therefore, the LT release was found to be slower with LT-NVoptG than LT-NVopt. The slower release pattern is ideal for topical delivery because the initial fast release helps to achieve the drug concentration at the target site and later slow release helps to maintain the therapeutic concentration [50].

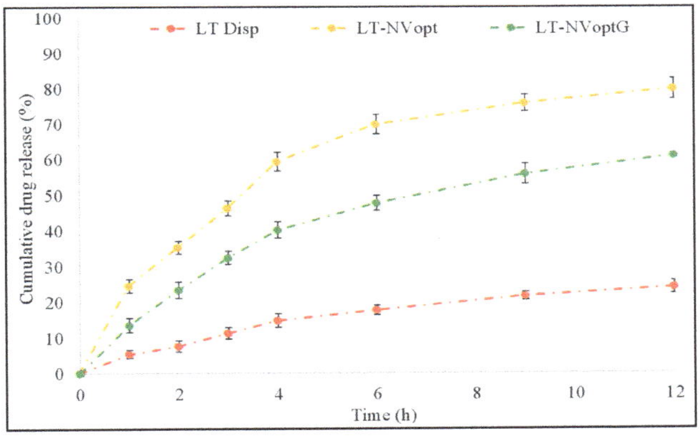

Figure 6. Drug release profile of pure luteolin, luteolin nanovesicles (LT-NVopt) and luteolin nanovesicles gel (LT-NVoptG). The study was performed in triplicate and data are shown as mean ± SD.

3.9. Permeation Study

The comparative permeation study of LT-NVoptG, LT-NVopt and pure LT was performed to check the amount of LT which permeated across the membrane (Figure 7). The result revealed significant variation in the drug permeation profile. LT-NVopt and LT-NVoptG showed the amount of drug permeated was 231.92 ± 3.23 µg/cm^2/h and 128.21 ± 3.56 µg/cm^2/h in the tested 6 h study, whereas the pure LT dispersion showed the flux value of 64.59 ± 2.11 µg/cm^2/h. There was about a 2–3.6-fold enhancement in the permeation flux achieved from LT-NVoptG and LT-NVopt. The poor permeability of pure LT is due to its poor solubility. LT-NVopt also had about 1.7-fold enhancement in the permeation compared to LT-NVoptG. LT-NVopt has enhanced permeation across the membrane due to the nanovesicle size which can easily penetrate the small pore size. LT-NVopt is prepared with sodium cholate (edge activator) which has the property to deform the membrane and penetrate across the membrane. It helps to maintain the structure of the vesicles during permeating through the tight junction of the membrane and carry the drug into the systemic circulation. [51,52]. NVoptG showed less permeation due to the slow release of LT from the gel which prolongs the drug permeation. This behavior suggests a long-lasting delivery due to the formation of a drug reservoir into the skin, which can reduce the frequency of application, thus improving patient compliance [53].

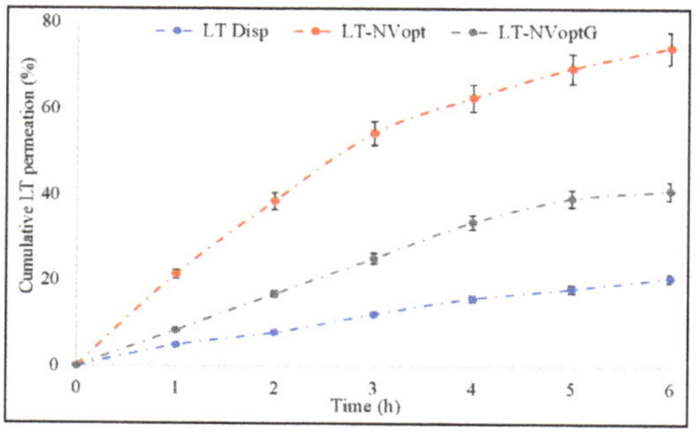

Figure 7. Drug permeation profile of pure luteolin, luteolin nanovesicles (LT-NVopt) and luteolin nanovesicles gel (LT-NVoptG). The study was performed in triplicate and data are shown as mean ± SD.

3.10. *Antioxidant Activity*

The antioxidant potential of the prepared LT-NVopt and LT-NVoptG was evaluated using the DPPH method and the results were compared with the pure LT (Figure 8). The antioxidant potential plays an important role in the biological activity of the bioactive compound. The comparison was performed to check the effect of excipients. A significant effect was observed in the tested groups. The result was found to be concentration-dependent, so as the concentration of LT increases antioxidant potential also increases. The formulations LT-NVopt and LT-NVoptG showed significantly higher activity than the pure LT. LT-NVopt, LT-NVoptG and pure LT showed the maximum antioxidant activity of 89.18 ± 3.95%, 84.43 ± 3.11% and 70.23 ± 2.98% at 500 μg/mL, respectively. There was a significant difference in the activity observed ($p < 0.001$) at the highest concentration (500 μg/mL) in comparison to pure LT. The result was also compared between LT-NVopt and LT-NVoptG and the difference was found to be non-significant. LT-NVopt showed slightly higher activity than LT-NVoptG. The presence of gelling agents in LT-NVoptG slows the release of LT and leads to lower activity. From the results, it was observed that the DPPH-scavenging activity of LT was increased after encapsulation into lipid vesicles [36].

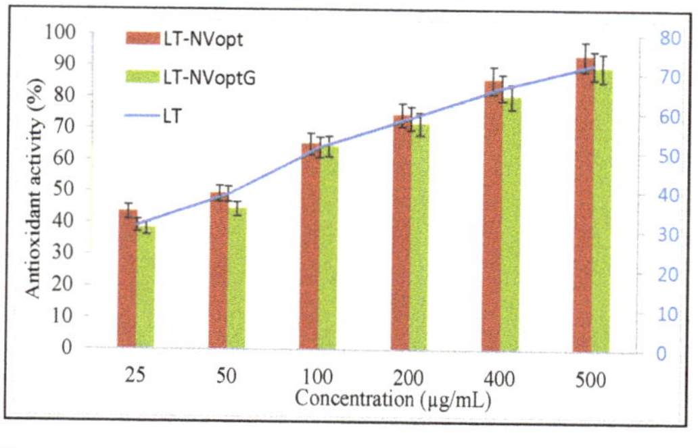

Figure 8. Antioxidant effects of pure luteolin (LT), luteolin nanovesicles (LT-NVopt) and luteolin nanovesicles gel (LT-NVoptG). The study was performed in triplicate and data are shown as mean ± SD.

3.11. Antibacterial Activity

The antibacterial activity of prepared LT-NVoptG was evaluated and results were compared with pure LT dispersion. The pure LT-treated well showed a zone of inhibition of 13.56 ± 1.8 mm, 14.23 ± 2.1 mm and 11.76 ± 1.1 mm against *S. aureus*, *E. coli* and *B. subtilis*, respectively. The LT-NVoptG-treated well showed higher ZOI of 16.11 ± 2.2 mm, 15.32 ± 1.4 mm and 15.22 ± 1.9 mm, respectively. There was marked enhancement in the ZOI observed from LT-NVoptG-treated organisms. The enhancement in the activity may be due to the higher solubility of LT in the presence of the used edge activator as well as the nano size of vesicles. Due to the higher solubility of LT, it showed activity by destroying the cell wall and cell membrane and inhibiting nucleic acid synthesis [54].

3.12. Cell Viability

The comparative cell viability study results showed greater activity in LT-NVoptG and LT-NVopt compared to pure LT (Figure 9). LT-NVoptG and LT-NVopt revealed significantly ($p < 0.01$) greater activity than pure LT. The comparison performed between LT-NVoptG and LT-NVopt also showed significant ($p < 0.05$) differences among them. LT-NVoptG showed slower activity than LT-NVopt due to slower LT release from gel. LT-NVoptG- and LT-NVopt-treated cells showed higher IC_{50} values of 428.11 µM and 380 µM compared to pure LT (780.55 µM) after 24 h of treatment. The effect on the cell viability is concentration dependent. With the increase in concentration of LT, the viability % changes. The pure LT-treated cells showed the cell viability % at different concentrations of 250 µM (92.11 ± 3.2), 500 µM (69.44 ± 4.4) and 1000 µM (38.11 ± 3.8). LT-NVopt showed the cell viability % at concentration of 100 µM (85.54 ± 2.9), 250 µM (71.74 ± 4.7), 500 µM (38.54 ± 5.9) and 1000 µM (21.87 ± 2.4). LT-NVoptG-treated cells showed significant ($p < 0.001$) effects at each concentration. LT-NVoptG showed the cell viability % at 100 µM (82.54 ± 2.9), 250 µM (73.65 ± 2.4), 500 µM (41.21 ± 3.6) and 1000 µM (31.11 ± 1.2). During the comparison, the difference was found to be significantly ($p < 0.001$) higher among both the groups. LT-NVoptG showed enhanced growth inhibitory effects in the treated cells at lower concentration in comparison to pure LT.

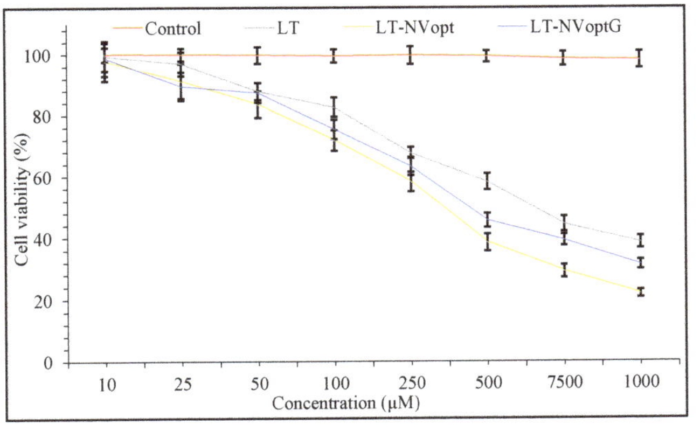

Figure 9. Cell viability study of pure luteolin, luteolin nanovesicles (LT-NVopt) and luteolin nanovesicles gel (LT-NVoptG). The study was performed in triplicate and data are shown as mean ± SD.

3.13. Irritation Study

The HET CAM method is a well-accepted method to check the irritation potential of a sample. The negative and positive control-treated sample showed a response that comes under the category of non-irritating, irritating and severely irritating. This method was applied to check the irritation potential of the prepared LT-NVoptG. The cumulative irritation score was calculated for LT-NVoptG, negative control (NaCl 0.9% *w*/*v*) and positive control (SLS, 1% *w*/*v*) to compare the results (Table 4). Visual observation was done for the lysis, hemorrhage and coagulation of blood vessels. A score between 0–0.8 is considered as non-irritant with no sign of abnormality after treatment. The negative control and prepared LT-NVoptG showed no sign of lysis, hemorrhage or coagulation. The negative control and LT-NVoptG-treated samples showed a cumulative score of 0 and 0.15 after the treatment with CAM. The positive control sample showed a cumulative score of 2.8 after treatment with CAM. The high score was shown by the positive control due to the hemorrhage and coagulation observed at different time points.

Table 4. HET-CAM irritation score treated with different groups.

Test Sample	Egg	Time (min)				Overall Score
		0	0.5	2	5	
LT-NVopG	Egg 1	0	0	0	0	0.15
	Egg 2	0	0	0	0	
	Egg 3	0	0	0.2	0.4	
	Egg 4	0	0	0	0	
	Mean score	0	0	0.05	0.1	
SLS, 1% *w*/*v* (Positive control)	Egg 1	0	0.8	0.8	2	2.8
	Egg 2	0	0.8	1.2	2	
	Egg 3	0	0.8	1.2	1	
	Egg 4	0	0	0.8	0	
	Mean score	0	0.6	1	1.2	
NaCl 0.9% *w*/*v* (Negative control)	Egg 1	0	0	0	0	0
	Egg 2	0	0	0	0	
	Egg 3	0	0	0	0	
	Egg 4	0	0	0	0	
	Mean score	0	0	0	0	

4. Conclusions

LT-NVs were prepared by solvent evaporation–hydration method using phospholipid, edge activator and sonication time as independent variables. The optimization of LT-NVs was performed using a Box–Behnken design with three factors at three levels. The prepared LT-NVs showed the nanometric vesicle size with high encapsulation efficiency. The optimized formulation LT-NVopt was converted to gel (LT-NVoptG) using Carbopol as a gelling agent. LT-NVoptG showed optimum viscosity, pH, spreadability, and prolonged-release profile with enhanced permeation compared to pure LT. The irritation, antibacterial and cytotoxicity results showed that the prepared LT-NVoptG was found to be non-irritating, with good antibacterial properties and lower cytotoxicity. The overall formulation design showed that the prepared LT-NVs-based gel delivery system acts as a potential delivery system in the treatment of skin diseases.

Author Contributions: Conceptualization, methodology, writing—original draft preparation, I.K. and S.S.I.; software, S.S.I.; formal analysis and validation, H.N.A.; resources, F.A.A.-A. and M.S.N.; writing—review and editing, S.A. and F.A.A.-A.; funding acquisition, I.K. All authors have read and agreed to the published version of the manuscript.

Funding: This research work was funded by Institutional Fund Projects under grant no. (IFPHI-006-130-2020). Therefore, authors gratefully acknowledge technical and financial support from the Ministry of Education and King Abdulaziz University, DSR, Jeddah, Saudi Arabia.

Institutional Review Board Statement: Not applicable.

Informed Consent Statement: Not applicable.

Acknowledgments: This research work was funded by Institutional Fund Projects under grant no. (IFPHI-006-130-2020). Therefore, authors gratefully acknowledge technical and financial support from the Ministry of Education and King Abdulaziz University, DSR, Jeddah, Saudi Arabia.

Conflicts of Interest: The authors declare no conflict of interest.

References

1. Imran, M.; Rauf, A.; Abu-Izneid, T.; Nadeem, M.; Ali, M.; Khan, I.A.; Imran, A.; Orhan, I.E.; Rizwan, M.; Atif, M.; et al. Luteolin, a flavonoid, as an anticancer agent: A review. *Biomed. Pharmacother.* **2019**, *112*, 108612. [CrossRef]
2. Naves, L.B.; Dhand, C.; Venugopal, J.R. Nanotechnology for the treatment of melanoma skin cancer. *Prog. Biomater.* **2017**, *6*, 13–26. [CrossRef]
3. Calienni, M.N.; Temprana, C.F.; Prieto, M.J. Nano-formulation for topical treatment of precancerous lesions: Skin penetration, in vitro, and in vivo toxicological evaluation. *Drug Deliv. Transl. Res.* **2018**, *8*, 496–514. [CrossRef]
4. Kryczyk-Poprawa, A.; Kwiecień, A.; Opoka, W. Photostability of Topical Agents Applied to the Skin: A Review. *Pharmaceutics* **2020**, *12*, 10. [CrossRef]
5. Lin, Y.; Shi, R.; Wang, X.H.; Shen, H. Luteolin, a flavonoid with potentials for cancer prevention and therapy. *Curr. Cancer Drug Targets* **2008**, *8*, 634–646. [CrossRef]
6. Martin, K.R. Targeting apoptosis with dietary bioactive agents. *Exp. Biol. Med.* **2006**, *231*, 117–129. [CrossRef]
7. Tai, Z.; Lin, Y.; He, Y.; Huang, J.; Guo, J.; Yang, L.; Zhang, G.; Wang, F. Luteolin sensitizes the antiproliferative effect of interferon α/β by activation of Janus kinase/ signal transducer and activator of transcription pathway signaling through protein kinase A-mediated inhibition of protein tyrosine phosphatase SHP-2 in cancer cells. *Cell. Signal.* **2014**, *26*, 619–628.
8. Wu, C.; Xu, Q.; Chen, X.; Liu, J. Delivery luteolin with folacin-modified nanoparticle for glioma therapy. *Int. J. Nanomed.* **2019**, *14*, 7515–7531. [CrossRef]
9. Wang, L.; Zhong, C.; Zu, Y.; Zhao, X.; Deng, Y.; Wu, W.; Sun, X.; Wang, L.; Wu, M. Preparation and characterization of luteolin nanoparticles for enhance bioavailability and inhibit liver microsomal peroxidation in rats. *J. Funct. Foods* **2019**, *55*, 57–64. [CrossRef]
10. Shinde, P.; Agraval, H.; Singh, A.; Yadav, U.C.S.; Kumar, U. Synthesis of luteolin loaded zein nanoparticles for targeted cancer therapy improving bioavailability and efficacy. *J. Drug Deliv. Sci. Technol.* **2019**, *52*, 369–378. [CrossRef]
11. Gilani, S.J.; Bin-Jumah, M.; Rizwanullah, M.; Imam, S.S.; Imtiyaz, K.; Alshehri, S.; Alam, M.R.M. Chitosan Coated Luteolin Nanostructured Lipid Carriers: Optimization, In Vitro-Ex Vivo Assessments and Cytotoxicity Study in Breast Cancer Cells. *Coatings* **2021**, *11*, 158. [CrossRef]
12. Wang, Y.; Wang, Q.; Feng, W.; Yuan, Q.; Qi, X.; Chen, S.; Yao, P.; Dai, Q.; Xia, P.; Zhang, D.; et al. Folic acid-modified ROS-responsive nanoparticles encapsulating luteolin for targeted breast cancer treatment. *Drug Deliv.* **2021**, *28*, 1695–1708. [CrossRef]
13. Taworntchat, P.; Pattarakankul, T.; Palaga, T.; Intasanta, V.; Wanichwecharungruang, S. Polymerized Luteolin Nanoparticles: Synthesis, Structure Elucidation, and Anti-Inflammatory Activity. *ACS Omega* **2021**, *6*, 2846–2855. [CrossRef]
14. Jangdey, M.S.; Gupta, A.; Saraf, S.; Saraf, S. Development and optimization of apigenin-loaded transfersomal system for skin cancer delivery: In vitro evaluation. *Artif. Cells Nanomed. Biotechnol.* **2017**, *45*, 1452–1462. [CrossRef]
15. Cristiano, M.C.; Froiio, F.; Spaccapelo, R.; Mancuso, A.; Nisticò, S.P.; Udongo, B.P.; Fresta, M.; Paolino, D. Sulforaphane-Loaded Ultradeformable Vesicles as A Potential Natural Nanomedicine for the Treatment of Skin Cancer Diseases. *Pharmaceutics* **2020**, *12*, 6. [CrossRef]
16. Ramezani, V.; Honarvar, M.; Seyedabadi, M.; Karimollah, A.; Ranjbar, A.M.; Hashemi, M. Formulation and optimization of transfersome containing minoxidil and caffeine. *J. Drug Deliv. Sci. Technol.* **2018**, *44*, 129–135. [CrossRef]
17. Omar, M.M.; Hasan, O.A.; El Sisi, A.M. Preparation and optimization of lidocaine transferosomal gel containing permeation enhancers: A promising approach for enhancement of skin permeation. *Int. J. Nanomed.* **2019**, *14*, 1551–1562. [CrossRef]
18. Chen, M.; Shamim, M.A.; Shahid, A.; Yeung, S.; Andresen, B.T.; Wang, J.; Nekkanti, V.; Meyskens, F.L., Jr.; Kelly, K.M.; Huang, Y. Topical Delivery of Carvedilol Loaded Nano-Transfersomes for Skin Cancer Chemoprevention. *Pharmaceutics* **2020**, *12*, 1151. [CrossRef]
19. Elsayed, M.M.; Abdallah, O.Y.; Naggar, V.F.; Khalafallah, N.M. Deformable liposomes and ethosomes as carriers for skin delivery of ketotifen. *Die Pharm.* **2007**, *62*, 133–137.
20. Lalotra, A.S.; Singh, V.; Khurana, B.; Agrawal, S.; Shrestha, S.; Arora, D.A. Comprehensive Review on Nanotechnology-Based Innovations in Topical Drug Delivery for Treatment of Skin Cancer. *Curr. Pharm. Des.* **2020**, *26*, 5720–5731. [CrossRef]
21. El Zaafarany, G.M.; Awad, G.A.; Holayel, S.M.; Mortada, N.D. Role of edge activators and surface charge in developing ultradeformable vesicles with enhanced skin delivery. *Int. J. Pharm.* **2010**, *397*, 164–172. [CrossRef] [PubMed]
22. Zhang, X.; Liu, J.; Qiao, H.; Liu, H.; Ni, J.; Zhang, W.; Shi, Y. Formulation optimization of dihydroartemisinin nanostructured lipid carrier using response surface methodology. *Powder Technol.* **2010**, *197*, 120–128. [CrossRef]
23. Imam, S.S.; Ahad, A.; Aqil, M.; Akhtar, M.; Sultana, Y.; Ali, A. Formulation by design based risperidone nano soft lipid vesicle as a new strategy for enhanced transdermal drug delivery: In-vitro characterization, and in-vivo appraisal. *Mater. Sci. Eng. C* **2017**, *75*, 1198–1205. [CrossRef] [PubMed]
24. Danaei, M.; Dehghankhold, M.; Ataei, S.; Hasanzadeh Davarani, F.; Javanmard, R.; Dokhani, A.; Khorasani, S.; Mozafari, M.R. Impact of Particle Size and Polydispersity Index on the Clinical Applications of Lipidic Nanocarrier Systems. *Pharmaceutics* **2018**, *10*, 57. [CrossRef] [PubMed]
25. Joseph, E.; Singhvi, G. Chapter 4—Multifunctional nanocrystals for cancer therapy: A potential nanocarrier. In *Nanomaterials for Drug Delivery and Therapy*; William Andrew Publishing: Norwich, NY, USA, 2019; pp. 91–116.

26. Abou Samra, M.M.; Salama, A.H.; Awad, G.E.A.; Mansy, S.S. Formulation and Evaluation of Novel Hybridized Nanovesicles for Enhancing Buccal Delivery of Ciclopirox Olamine. *AAPS PharmSciTech* **2020**, *21*, 283. [CrossRef]
27. Yang, X.; Trinh, H.M.; Agrahari, V.; Sheng, Y.; Pal, D.; Mitra, A.K. Nanoparticle-Based Topical Ophthalmic Gel Formulation for Sustained Release of Hydrocortisone Butyrate. *AAPS PharmSciTech* **2016**, *17*, 294–306. [CrossRef]
28. Nautiyal, U.; Mohammed, J.; Kazmi, I. Preparation and Evaluation of Antifungal Micro-Emulsion/Gel using reduce Dose of Silver, Supported by Ciprofloxacin. *Int. Pharm. Sci.* **2012**, *2*, 72–87.
29. Barot, B.S.; Parejiya, P.B.; Patel, H.K.; Gohel, H.C.; Shelat, P.K. Microemulsion-based gel of terbinafine for the treatment of onychomycosis: Optimization of formulation using D-optimal design. *AAPS PharmSciTech* **2012**, *13*, 184–192. [CrossRef]
30. Gaba, B.; Fazil, M.; Khan, S.; Ali, A.; Baboota, S.; Ali, J. Nanostructured lipid carrier system for topical delivery of terbinafine hydrochloride. *Bull. Fac. Pharm. Cairo Univ.* **2015**, *53*, 147–159. [CrossRef]
31. Byeon, J.C.; Lee, S.E.; Kim, T.H.; Ahn, J.B.; Kim, D.H.; Choi, J.S.; Park, J.S. Design of novel proliposome formulation for antioxidant peptide, glutathione with enhanced oral bioavailability and stability. *Drug Deliv.* **2019**, *26*, 216–225. [CrossRef] [PubMed]
32. Saleem, M.N.; Idris, M. Formulation Design and Development of a Unani Transdermal Patch for Antiemetic Therapy and Its Pharmaceutical Evaluation. *Scientifica* **2016**, *2016*, 5. [CrossRef] [PubMed]
33. Kunzi-Rapp, K.; Ruck, A.; Kaufmann, R. Characterization of the chick chorioallantoic membrane model as a short-term in vivo system for human skin. *Arch. Dermatol. Res.* **1999**, *291*, 290–295. [CrossRef] [PubMed]
34. Tay, S.L.M.; Heng, P.W.S.; Chan, L.W. An investigation of the chick chorioallantoic membrane as an alternative model to various biological tissues for permeation studies. *J. Pharm. Pharmacol.* **2011**, *63*, 1283–1289. [CrossRef] [PubMed]
35. Haigh, J.M.; Smith, E.W. The selection and use of natural and synthetic membranes for in vitro diffusion experiments. *Eur. J. Pharm. Sci.* **1994**, *2*, 311–330. [CrossRef]
36. Caddeo, C.; Pucci, L.; Gabriele, M.; Carbone, C.; Fernàndez-Busquetsd, X.; Valenti, D.; Pons, R.; Vassallo, A.; Fadda, A.M.; Manconi, M. Stability, biocompatibility and antioxidant activity of PEG-modified liposomes containing resveratrol. *Int. J. Pharm.* **2018**, *538*, 40–47. [CrossRef] [PubMed]
37. Alshehri, S.; Imam, S.S.; Hussain, A.; Altamimi, M.A. Formulation of Piperine Ternary Inclusion Complex Using β CD and HPMC: Physicochemical characterization, molecular docking and antimicrobial testing. *Processes* **2020**, *8*, 1450. [CrossRef]
38. Hafeez, A.; Kazmi, I. Dacarbazine nanoparticle topical delivery system for the treatment of melanoma. *Sci. Rep.* **2017**, *7*, 16517. [CrossRef]
39. Mehling, A.; Kleber, M.; Hensen, H. Comparative studies on the ocular and dermal irritation potential of surfactants. *Food Chem. Toxicol.* **2007**, *45*, 747–758. [CrossRef]
40. Irimia, T.; Dinu-Pîrvu, C.E.; Ghica, M.V.; Lupuleasa, D.; Muntean, D.L.; Udeanu, D.I.; Popa, L. Chitosan-Based In Situ Gels for Ocular Delivery of Therapeutics: A State-of-the-Art Review. *Mar. Drugs* **2018**, *16*, 373. [CrossRef]
41. Zafar, A.; Alruwaili, N.K.; Imam, S.S.; Alotaibi, N.H.; Alharbi, K.S.; Afzal, M.; Ali, R.; Alshehri, S.; Alzarea, S.I.; Elmowafy, M.; et al. Bioactive Apigenin loaded oral nano bilosomes: Formulation optimization to preclinical assessment. *Saudi Pharm. J.* **2021**, *29*, 269–279. [CrossRef]
42. He, Y.; Luo, L.; Liang, S.; Long, M.; Xu, H. Influence of probe-sonication process on drug entrapment efficiency of liposomes loaded with a hydrophobic drug. *Int. J. Polym. Mater. Polym. Biomater.* **2019**, *68*, 193–197. [CrossRef]
43. Mavaddati, M.A.; Moztarzadeh, F.; Baghbani, F. Effect of Formulation and Processing Variables on Dexamethasone Entrapment and Release of Niosomes. *J. Clust. Sci.* **2015**, *26*, 2065–2078. [CrossRef]
44. Zhang, S.; Gao, H.; Bao, G. Physical Principles of Nanoparticle Cellular Endocytosis. *ACS Nano* **2015**, *9*, 8655–8671. [CrossRef]
45. Agarwal, S.; Mohamed, M.S.; Raveendran, S.; Rochani, A.K.; Maekawa, T.; Kumar, D.S. Formulation, characterization and evaluation of morusin loaded niosomes for potentiation of anticancer therapy. *RSC Adv.* **2018**, *8*, 32621–32636. [CrossRef]
46. Patel, D.K.; Kesharwani, R.; Kumar, V. Etodolac loaded solid lipid nanoparticle based topical gel for enhanced skin delivery. *Biocatal. Agric. Biotechnol.* **2020**, *29*, 101810. [CrossRef]
47. Riaz, A.; Hendrickx, S.; Elbrink, K.; Caljon, G.; Maes, L.; Ahmed, N.; Kiekens, F.; Khan, G.M. Preparation and Characterization of Nanostructured Lipid Carriers for Improved Topical Drug Delivery: Evaluation in Cutaneous Leishmaniasis and Vaginal Candidiasis Animal Models. *AAPS PharmSciTech* **2020**, *21*, 185. [CrossRef]
48. Koca, H.D.; Doganay, S.; Turgut, A.; Tavman, I.H.; Saidur, R.; Mahbubul, I.M. Effect of particle size on the viscosity of nanofluids: A review. *Renew. Sustain. Energy Rev.* **2018**, *82*, 1664–1674. [CrossRef]
49. Motawea, T.B.; El-Gawad, A. Topical phenytoin nanostructured lipid carriers: Design and development. *Drug Dev. Ind. Pharm.* **2018**, *44*, 144–151. [CrossRef]
50. Uprit, S.; Sahu, R.K.; Roy, A.; Pare, A. Preparation and characterization of minoxidil loaded nanostructured lipid carrier gel for effective treatment of alopecia. *Saudi Pharm. J.* **2013**, *21*, 379–385. [CrossRef]
51. Zeb, A.; Arif, S.T.; Malik, M.; Shah, F.A.; Din, F.U.; Qureshi, O.S.; Lee, E.S.; Lee, G.Y.; Kim, J.K. Potential of nanoparticulate carriers for improved drug delivery via skin. *J. Pharm. Investig.* **2019**, *49*, 485–517. [CrossRef]
52. Dudhipala, N.; Mohammed, R.P.; Youssef, A.A.A.; Banala, N. Effect of lipid and edge activator concentration on development of aceclofenac-loaded transfersomes gel for transdermal application: In vitro and *ex-vivo* skin permeation. *Drug Dev. Ind. Pharm.* **2020**, *46*, 1334–1344. [CrossRef] [PubMed]

53. Grande, F.; Ragno, G.; Muzzalupo, R.; Occhiuzzi, M.A.; Mazzotta, E.; De Luca, M.; Garofalo, A.; Ioele, G. Gel formulation of Nabumetone and a newly synthesized analog: Microemulsion as a photoprotective topical delivery system. *Pharmaceutics* **2020**, *12*, 423. [CrossRef] [PubMed]
54. Guo, Y.; Liu, Y.; Zhang, Z.; Chen, M.; Zhang, D.; Tian, C.; Liu, M.; Jiang, G. The Antibacterial Activity and Mechanism of Action of Luteolin Against *Trueperella pyogenes*. *Infect. Drug Resist.* **2020**, *13*, 1697–1711. [CrossRef]

pharmaceutics

MDPI

Article

M1 Macrophage-Derived Exosomes Loaded with Gemcitabine and Deferasirox against Chemoresistant Pancreatic Cancer

Yongmei Zhao [1], Yuanlin Zheng [1], Yan Zhu [1], Yi Zhang [2], Hongyan Zhu [1] and Tianqing Liu [3,*]

[1] School of Pharmacy, Nantong University, Nantong 226019, China; ymzhao@ntu.edu.cn (Y.Z.); 2019310034@stmail.ntu.edu.cn (Y.Z.); 2019320002@stmail.ntu.edu.cn (Y.Z.); amy600@ntu.edu.cn (H.Z.)
[2] School of Chemistry, University of Glasgow, Glasgow G12 8QQ, UK; 2449900z@student.gla.ac.uk
[3] NICM Health Research Institute, Western Sydney University, Sydney, NSW 2145, Australia
* Correspondence: M.Liu3@westernsydney.edu.au

Abstract: Pancreatic cancer is a malignant disease with high mortality and poor prognosis due to lack of early diagnosis and low treatment efficiency after diagnosis. Although Gemcitabine (GEM) is used as the first-line chemotherapeutic drug, chemoresistance is still the major problem that limits its therapeutic efficacy. Here in this study, we developed a specific M1 macrophage-derived exosome (M1Exo)-based drug delivery system against GEM resistance in pancreatic cancer. In addition to GEM, Deferasirox (DFX) was also loaded into drug carrier, M1Exo, in order to inhibit ribonucleotide reductase regulatory subunit M2 (RRM2) expression via depleting iron, and thus increase chemosensitivity of GEM. The M1Exo nanoformulations combining both GEM and DFX significantly enhanced the therapeutic efficacy on the GEM-resistant PANC-1/GEM cells and 3D tumor spheroids by inhibiting cancer cell proliferation, cell attachment and migration, and chemoresistance to GEM. These data demonstrated that M1Exo loaded with GEM and DFX offered an efficient therapeutic strategy for drug-resistant pancreatic cancer.

Keywords: exosomes; gemcitabine; deferasirox; pancreatic cancer; chemoresistance; RRM2

Citation: Zhao, Y.; Zheng, Y.; Zhu, Y.; Zhang, Y.; Zhu, H.; Liu, T. M1 Macrophage-Derived Exosomes Loaded with Gemcitabine and Deferasirox against Chemoresistant Pancreatic Cancer. *Pharmaceutics* **2021**, *13*, 1493. https://doi.org/10.3390/pharmaceutics13091493

Academic Editor: Mazzucchelli Serena

Received: 30 July 2021
Accepted: 14 September 2021
Published: 17 September 2021

1. Introduction

Pancreatic cancer is a lethal disease with poor survival rate and an increasing incidence due to lack of early diagnosis and low treatment efficiency after diagnosis [1]. As less than 20% of patients are suitable for resection, chemotherapy is still one of the main treatments for pancreatic cancer [2,3]. However, current treatment strategies are still unsatisfactory and have failed to significantly increase the overall survival time of patients with pancreatic cancer over the last decade.

Gemcitabine (GEM), a hydrophilic deoxycytidine analogue, is a first-line chemotherapeutic drug and is widely used in the treatment of unresectable pancreatic cancer. However, chemoresistance is still the major problem that limits the therapeutic efficacy of GEM. The positive response rate for standard GEM treatment in pancreatic cancer patients is only 6% [4]. Although not completely understood, the mechanism of GEM resistance is associated with regulation of drug transport, DNA damage and repair, and renewability of cancer stem cells [5–10]. Ribonucleotide reductase (RR) is an enzyme that catalyzes the formation of deoxyribonucleotides from ribonucleotides, which is essential for cell replication. This enzyme contains two subunits, M1 (RRM1) and M2 (RRM2). It has been reported that RRM2 plays an essential role in cancer cell proliferation and the development of resistance to GEM in pancreatic tumor cells [11,12]. Studies have demonstrated that overexpression of RRM2 enhanced DNA damage repair and replication, leading to decreased chemosensitivity of GEM [13]. In addition, clinical data demonstrated that patients who have elevated RRM2 expression had less response to gemcitabine-based chemotherapy, and high expression of RRM2 in pancreatic cancer is associated with a poor prognosis [14]. Therefore, RRM2 has become a potential target to overcome GEM resistance. Iron chelators have been identified as RRM2 inhibitors as they can interact with the essential diiron tyrosyl radical center and inhibit the enzyme activity. Deferasirox (DFX) is an oral iron chelator for the treatment of iron overload. Studies have demonstrated that DFX can significantly downregulate the expression of RRM2 and potentiate the therapeutic effects of GEM, resulting in improved anticancer efficacy [15]. However, the application against GEM-resistant cancer is limited due to variable bioavailability and low intratumoral distribution [16,17]. Therefore, there is an urgent need to develop advanced delivery strategies to overcome these issues.

Exosomes are nanosized (30–150 nm) extracellular vesicles secreted by cells for intercellular communications. They are promising natural drug carriers with advantages including excellent biocompatibility, long circulation, and low immunogenicity [18–20]. Moreover, exosomes as delivery vesicles can fuse with target cells and directly transport loaded drugs to receptor cells in order to overcome p-glycoprotein-involved drug resistance [21]. In addition, exosomes derived from different cell types contain cell-specific bioactive lipids, proteins, and genetic materials, which allows them to have the specific biological functions [22]. Macrophages are a key immune cell population involved in the tumor microenvironment and can be polarized into M1 or M2 phenotypes in response to different stimuli. M1 macrophages produce high levels of proinflammatory and immunostimulatory cytokines, including interleukin 12 (IL-12), interleukin 23 (IL-23), tumor necrosis factor alpha (TNFα) etc., leading to tumor suppression [23]. It has been reported that M1 macrophage-derived exosomes (M1Exo) can release pro-inflammatory signals and generate a stimulatory tumor immune-microenvironment, implying that they have great therapeutic potentials for anti-cancer therapy [24]. More recently, M1Exo has been used as drug carrier to deliver paclitaxel into the tumor tissue, and results demonstrated that M1Exo provided a pro-inflammatory environment which further enhanced the therapeutic efficacy of chemotherapy by activating the apoptosis pathway [25]. Similar findings were presented by Li and Wang et al., showing that anti-tumor efficiency of M1Exo loaded with cisplatin was significantly improved in vivo via upregulating Bcl-2-associated X protein (Bax) and caspase-3 in the apoptosis pathway [26]. Therefore, engineered M1Exo can be a promising approach for drug delivery in cancer therapy.

In this study, M1Exo was engineered as drug carrier to co-delivery DFX and GEM to overcome the chemoresistance of GEM and improve its therapeutic potential. Our aim is to achieve efficient delivery of DFX and at the same time sensitize GEM-resistant pancreatic cancer cells to this chemotherapy. Our results show that the M1Exo-based co-delivery of DFX and GEM can be used as a promising strategy for drug-resistant pancreatic cancer treatment via the inhibition of cancer cell proliferation, metastasis, and chemoresistance.

2. Materials and Methods

2.1. Cell Culture

Human monocyte THP-1 cells and human pancreatic cancer PANC-1 cells were purchased from National Collection of Authenticated Cell Cultures, China. THP-1 cells were maintained in Roswell Park Memorial Institute medium (RPMI 1640, Thermo Fisher Scientific, Waltham, MA, USA) supplemented with 10% fetal bovine serum (FBS, Thermo Fisher Scientific) and 1% penicillin-streptomycin (PS, Thermo Fisher Scientific). PANC-1 cells were cultured in Dulbecco's Modified Eagle Medium (DMEM, Thermo Fisher Scientific) supplemented with 10% FBS and 1% PS. THP-1 were differentiated to macrophage by culturing in serum-free RPMI1640 medium with 100 ng/mL phorbol ester (PMA, Sigma-Aldrich, New York, NY, USA) and 0.3% bovine Serum Albumin (BSA, Sigma-Aldrich) for 72 h. GEM-resistant pancreatic cancer cells (PANC-1/GEM) were induced by treatment with 18 µg/mL GEM (Sigma-Aldrich) for ten months. Both PANC-1 and PANC-1/GEM were cultured in a humidified incubator containing 5% CO_2 at 37 °C.

2.2. Exosome Isolation and Characterization

The TPH-1 differentiated macrophages (M0 phenotype) were seeded in 6-well plate at the density of 10^6 cells/mL. After 24 h incubation, cells were stimulated with 100 ng/mL lipopolysaccharide (LPS, Sigma-Aldrich) for 24 h to induce M1 macrophage polarization [27]. Next, the culture media of M1 macrophage were collected in order to obtain M1Exo. The media were centrifugated at 800× *g* (10 min), 3000× *g* (10 min), and 10,000× *g* (30 min) at 4 °C to remove cell debris and large extracellular vesicles. M1Exo were then harvested by ultracentrifugation at 100,000× *g* for 70 min at 4 °C using an ultracentrifuge (Optima XPN-100, Beckman Coulter, Indianapolis, IN, USA) [28].

The morphology of the exosomes was characterized using a transmission electron microscope (TEM, Talos F200X, Thermo Fisher Scientific, Waltham, MA, USA). In brief, a drop of the isolated exosomes in PBS solution was added onto carbon-coated copper grids (Sigma-Aldrich). After drying for 5 min at room temperature, the samples were stained with 1% uranium acetate (Sigma-Aldrich) for 1 min and the excess solution was removed via a filter paper. The samples were further dried for 20 min at room temperature and then imaged by TEM, and the TEM particle size data were converted directly to cumulative number-based distributions. Particle size and zeta potential of the exosomes were measured by dynamic light scattering (DLS, Nano-zs30, Malvern Panalytical, Malvern, UK).

2.3. Drug Loading and Quantification

The exosomes (100 µg) and drug mixture (100 µg) were mixed in 400 µL of PBS solution, and electroporated in 4mm path length electroporation cuvettes using a Bio-Rad electroporation instrument. The electroporation was performed at a voltage of 400 V and an electric capacity of 150 mF with 1 ms of discharging time. The mixture was then incubated at 37 °C for 30 min to recover the exosome membrane [29]. The un-encapsulated GEM and DFX (Sigma-Aldrich) was removed by passing through an amicon filter (100 kDa, Merk Millipore, Burlington, VT, USA) and centrifuged at 100,000× *g* for 60 min.

The drug loading efficiency in the exosomes was measured by dissolving the exosomes with methanol to completely release GEM and DFX, and the released GEM and DFX were quantified by a high-performance liquid chromatography with ultraviolet detection (HPLC-UV) (Agilent Scientific Instruments, Santa Clara, CA, USA) at wavelengths of 275 nm and 245 nm, respectively.

2.4. In Vitro Cell Viability

The cellular anti-cancer effects of M1Exo loaded with GEM and DFX (M1Exo-GEM-DFX) against both PANC-1 and PANC-1/GEM cells were evaluated by MTS (3-(4,5-dimethylthiazol-2-yl)-5-(3 carboxymethoxyphenyl)-2-(4-sulfophenyl)-2H-tetraz olium) assay (Sigma-Aldrich). Cells (1×10^4 cells/well) were seeded into 96-well plates. After 12 h incubation, the cells were treated with control, M1Exo, GEM, DFX, GEM&DFX, and M1Exo-GEM-DFX at corresponding concentrations of GEM (18 µg/mL) and DFX (18 µg/mL) for 48 h at 37 °C and 5% CO_2. After incubation, the culture medium in each well was replaced with 100 µL of MTS solution (20 µL CellTiter 96® AQueous One Solution Reagent and 80 µL tissue culture medium). After 1 h incubation, the absorbance was detected at 490 nm using a microplate reader (Thermo Fisher Scientific, Waltham, MA, USA) to investigate the cell viability.

2.5. Iron Removal Efficacy Study

Both PANC-1 and PANC-1/GEM cells were treated with PBS control, M1Exo, GEM, DFX, GEM&DFX, and M1Exo-GEM-DFX for 24 h. The cells were then harvested using Trypsin (Thermo Fisher Scientific), washed twice with PBS, and counted in a cell counting chamber. The final cell pellets were collected and digested, and their iron content was determined using inductively coupled plasma mass spectrometry (ICP-MS, Agilent 7700, Agilent Scientific Instruments, Santa Clara, CA, USA) under routine element operating conditions.

2.6. Multidrug Resistance (MDR) Study

Multidrug resistance study was performed by assessing drug efflux activity using the Vybrant Multidrug Resistance Assay (Thermo Fisher Scientific) following the manufacturer's instructions. PANC-1/GEM cells could express high levels of drug transporter, p-glycoprotein, that rapidly eliminates nonfluorescent calcein AM from the plasma membrane and reduces the intake of fluorescent calcein in the cytosol. Therefore, the p-glycoprotein activity can be quantitated by measuring the accumulation of intracellular calcein fluorescence. In brief, the cells were seeded on 12-well plates and treated with verapamil (Thermo Fisher Scientific), control calcein (Thermo Fisher Scientific), and other drug treatments including M1Exo, GEM, DFX, GEM&DFX, and M1Exo-GEM-DFX. The calcein retention was measured using a microplate reader, with calcein-specific fluorescence absorption maximum at 494 nm and the emission maximum at 517 nm.

2.7. Wound Healing Assays

Wound healing study was carried out using our method as previously described [30]. In brief, cells were seeded in 96-well plates and cultured until they reached more than 80% confluence and incubated in serum-free medium to eliminate the effect of cell proliferation. The wounds on the cell monolayers were generated using an Incucyte wound maker. The cells were then treated with fresh serum-free medium containing drug-loaded M1Exo or controls, their wound healing behaviors were monitored using Incucyte Zoom (Essen BioScience, Ann Arbor, MI, USA) and analyzed with IncuCyte Zoom software (IncuCyte®Scratch Wound Cell Migration Software Module, BioScience, Ann Arbor, MI, USA).

2.8. Cell Attachment Study

PANC-1/GEM cells were seeded onto 6-well plates and cultured until they reached more than 80% confluence. The cells were treated with drug-loaded M1Exo or controls for 24 h. After being washed with PBS and detached using versine, 5×10^4 cells per well were seeded onto gelatine precoated 96-well plates for 2 h and then fixed with 4% paraformaldehyde (PFA, Thermo Fisher Scientific) for 15 min. The attached cells were imaged using Incucyte Zoom (Essen) and the number of cells was quantified using IncuCyte Zoom software.

2.9. Western Blotting

The expression of both RRM2 and equilibrative nucleoside transporter-1 (hENT1) markers were assessed by Western blotting. Cells were treated with drug-loaded nanoparticles or controls for 24 h, and washed with PBS. Total protein was harvested and quantified using the bicinchoninic acid protein assay kit (Thermo Fisher Scientific). Protein samples were prepared at a concentration of 50 µg protein/20 µL. Samples in loading buffer were heated at 75 °C for 10 min, loaded into wells of 15% acrylamide gels (Thermo Fisher Scientific) with a protein ladder, and run on SDS-PAGE at 100 V. The proteins were subsequently transferred onto nitrocellulose membranes (Thermo Fisher Scientific) for 1 h and blocked at 4 °C overnight. Blots were treated with an anti-RRM2 antibody (1:1000, Abcam) or an anti-hENT1 antibody (1:1000, Abcam) in blocking buffer for 1 h at room temperature on a shaker, while an anti-β-actin monoclonal antibody was a loading control (1:1000, Abcam). After washing with Tris Buffered Saline with Tween (TBST, Thermo Fisher Scientific) 4 times, the blots were incubated with secondary antibodies (1:5000, Abcam) in the dark for 1 h. After washing 4 times and drying, the membranes were scanned with the Odyssey imaging system (LI-COR) (LI-COR Biosciences, Lincoln, Dearborn, MI, USA).

2.10. Inhibition of 3D PANC-1 and PANC-1/GEM Tumor Spheroids

PANC-1/GEM tumor spheroids were formed in microwell devices using our reported method [31–33]. The culture medium was changed every 2 days and the formation of the tumor spheroids was monitored using a light microscope (OLYMPUS, CKX53,Tokyo, Japan). When the size of the tumor spheroids reached 200 µm in diameter, the spheroids were treated with drug-loaded nanoparticles or controls at a concentration of 50 µg/mL of GEM or of 50 µg/mL of DFX, respectively, for 7 days. The growth of the spheroids was recorded using a light microscope and the roundness was calculated using the following formula: roundness (%) = 100 − (R − r)/R × 100 (R: represents the radius of the minimum circumscribed circle; r: represents the maximum inscribed concentric circle) and analyzed by imageJ [34]. In addition, the tumor spheroid volume was calculated with the following formula: $V = (\pi \times d_{max} \times d_{min})/6$ and the change ratio of the tumor spheroid volume was compared with initial volume of each group [35].

2.11. Statistical Analysis

Multiple group comparisons were carried out using t-tests using GraphPad Prism version 9 (GraphPad, San Diego, CA, USA). All results are shown as mean ± the standard deviations of at least three replicates. $p < 0.05$ was considered to be significantly different.

3. Results and Discussion

3.1. M1Exo Preparation, Characterization, and Drug Loading

TPH-1-derived macrophages were polarized using cytokines IFN-γ or LPS for 24 h until they turned into M1 phenotypes. M1Exo were collected and loaded with both DFX and GEM via electroporation. M1Exo with or without drug loading were characterized by morphology, hydrodynamic size, surface charge, and loading efficiency. The morphology of M1Exo and M1Exo-GEM-DFX was measured by TEM and showed that they were a uniform spherical shape with narrow size distribution (Figure 1). The hydrodynamic size of M1Exo and M1Exo-GEM-DFX was investigated by DLS analysis. The results showed that the size of M1Exo was 120.1 ± 0.5 nm, while the size increased to 150.9 ± 1.1 nm after encapsulation of GEM and DFX (Table 1). The loading efficiency was measured by HPLC via generating a standard curve at 275 nm and 245 nm for GEM and DFX, respectively. The encapsulation efficiency was around 6.5 ± 2.3 and 5.7 ± 1.4 for GEM and DFX, respectively, as shown in Table 1.

Table 1. Physical properties and drug loading efficiency of M1Exo formulations. [a] Determined by DLS.

Sample	Hydrodynamic Size D_h [a] (nm)	Zeta-Potential (mV)	Drug Loading (%)
M1Exo	120.1 ± 0.5	−36.32 ± 1.89	-
M1Exo-GEM-DFX	150.9 ± 0.3	−34.30 ± 3.25	6.5 ± 2.3 (GEM) 5.7 ± 1.4 (DFX)

3.2. In Vitro Cell Viability Study

Both PANC-1 and drug-resistant PANC-1/GEM cells were used to investigate the cytotoxicity effects of M1Exo-GEM-DFX. Based on the viability study using MTS assays, native empty M1Exo did not show significant cytotoxic effects on both PANC-1 and PANC-1/GEM cells (Figure 2). Although both free GEM and free DFX showed greater cytotoxicity to PANC-1 cells compared with the control group (Figure 2A), neither of them had an obvious cytotoxicity effect on PANC-1/GEM cells (Figure 2B). The cell viability of the GEM&DFX group against PANC-1 cells was around 33%, while higher cell viability (~55%) was observed when treated with PANC-1/GEM cells. These results indicated that free drugs including GEM, DFX, and GEM&DFX was able to inhibit the proliferation of PANC-1 cells, however, they had less or no cytotoxic effect on drug-resistant PANC-1/GEM cells. In contrast, M1Exo-GEM-DFX showed greatest inhibition of cell viability against both PANC-1 cells and PANC-1/GEM cells. Moreover, the cell viability of M1Exo-GEM-DFX against PANC-1/GEM cells was around 29%, which was significantly lower than other groups (Figure 2B), suggesting that M1Exo-GEM-DFX has great potential to reverse drug resistance for pancreatic cancer treatment.

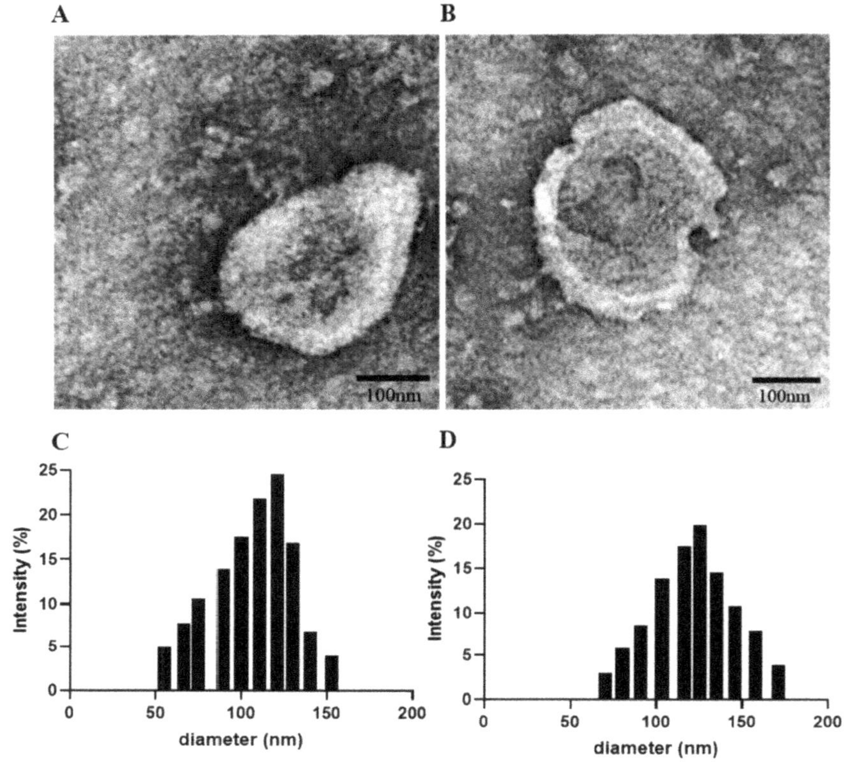

Figure 1. Characterization of M1Exo with or without drug loading: TEM images of (**A**) M1Exo and (**B**) M1Exo-GEM-DFX and the TEM particle size data which were converted directly to cumulative number-based distributions, (**C**) M1Exo, and (**D**) M1Exo-GEM-DFX.

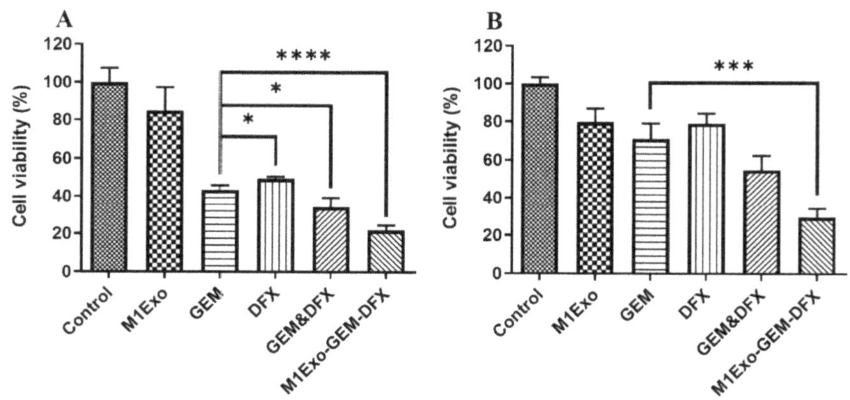

Figure 2. Cell viability study determined by the MTS assay with treatment of different drug formulations, including M1Exo, GEM, DFX, GEM&DFX, and M1Exo-GEM-DFX at a corresponding concentration of GEM and DFX in (**A**) PANC-1 cells and (**B**) PANC-1/GEM cells. (* $p < 0.05$, *** $p < 0.001$, **** $p < 0.0001$).

3.3. Iron Removal Efficacy

To understand whether M1Exo-GEM-DFX could efficiently remove iron from both PANC-1 and drug-resistant PANC-1/GEM cells, we measured the iron content after the treatment with the drug formulations. It was observed that native empty M1Exo did not show iron-removing ability on PANC-1 or PANC-1/GEM cells (Figure 3). GEM, DFX, and GEM&DFX slightly reduced the iron amount in PANC-1 cells compared with the control group (Figure 3A), whereas, in Figure 3B, their iron-removing ability decreased during the treatment of PANC-1/GEM cells. In addition, we found that the iron content of either PANC-1 or PANC-1/GEM cells treated with M1Exo-GEM-DFX was the lowest compared with other groups, indicating that M1Exo-GEM-DFX had the best iron removal efficacy, even in drug-resistant cells. This novel formulation can be an effective iron chelation strategy for cancer treatment.

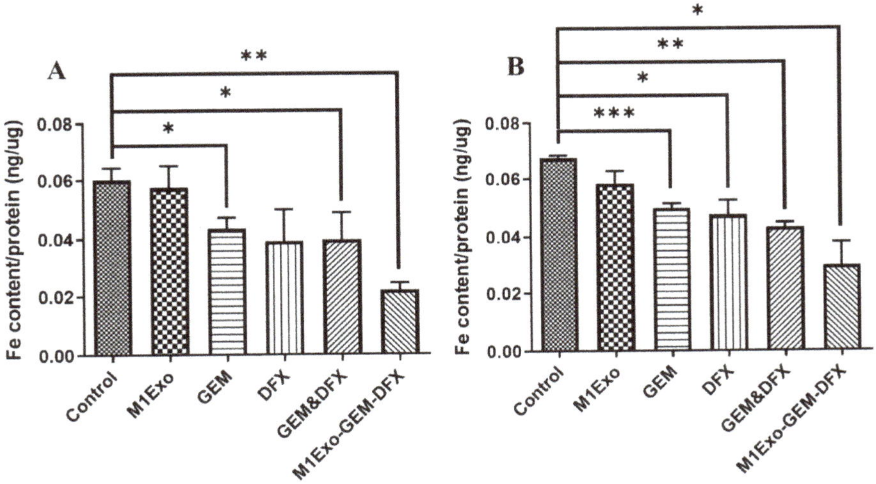

Figure 3. Iron removal efficacy of M1Exo, GEM, DFX, GEM&DFX, and M1Exo-GEM-DFX at a corresponding concentration of GEM and DFX in (**A**) PANC-1 cells and (**B**) PANC-1/GEM cells. (* $p < 0.05$, ** $p < 0.01$, *** $p < 0.001$).

3.4. Expression of Drug Resistance-Related Proteins

To investigate the mechanism and the degree of GEM resistance after the M1Exo-GEM-DFX treatments, the protein expression of RRM2 and hENT1 was detected using Western blot analysis. In Figure 4A,B, the band intensity of RRM2 of the M1Exo-GEM-DFX group was the lowest among all the treatment groups, suggesting the expression of the RRM2 protein was significantly downregulated in the cells treated with M1Exo-GEM-DFX. This reduced RRM2 expression is mainly related to the low iron supply induced by the iron chelator as hENT1 is the main transporter of GEM to penetrate the cell membrane [36]. We then measured the expression of hENT1 and found that the expression of hENT1 was significantly increased in the M1Exo-GEM-DFX group compared with other groups (Figure 4A,C). Taken together, these data demonstrated that M1Exo-GEM-DFX not only inhibited the RRM2 expression via iron depletion, but also promoted the hENT1 expression to efficiently transport the drug combinations into cells. Thus, M1Exo-GEM-DFX was able to overcome GEM resistance, which may lead to better therapeutic outcomes for patient with GEM-resistant pancreatic cancer.

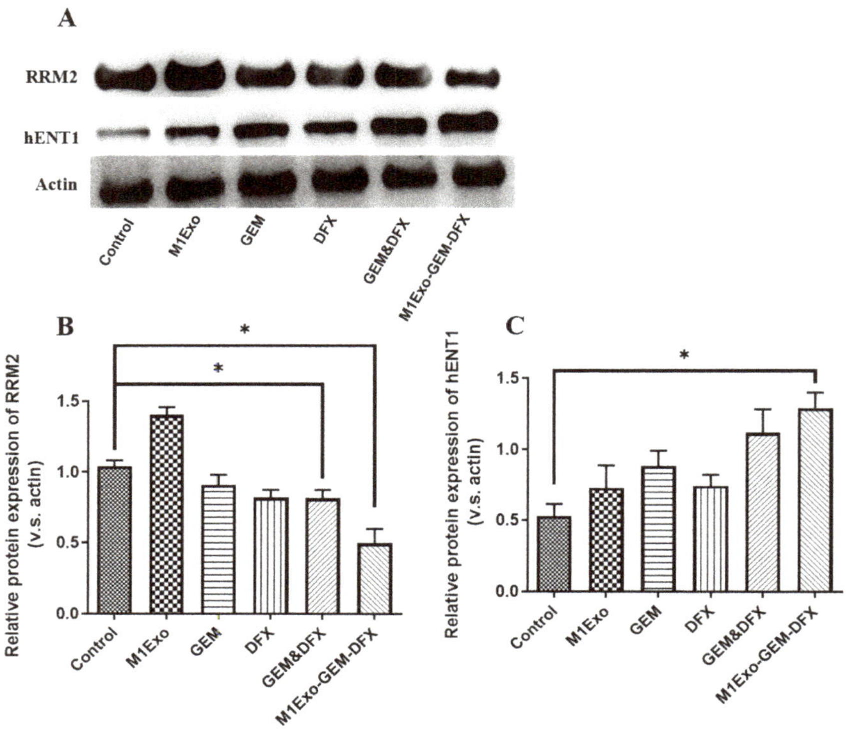

Figure 4. (**A**) Western blot of RRM2 and hENT1 expression in PANC-1/GEM cells at 24 h post-transplant. Figure 1. Exo, GEM, DFX, GEM&DFX, and M1Exo-GEM-DFX, while β-actin was used as internal control. (**B**) Relative expression level of RRM2. (**C**) Relative expression level of hENT1. (* $p < 0.05$).

3.5. Anti-MDR Effects

Overexpression of P-glycoprotein (p-gp) can lead to MDR in many cancers due to its ability to efflux intracellular anticancer drugs. Therefore, inhibiting p-gp is a promising strategy to improve the chemosensitivity of GEM in PANC-1/GEM cells. As shown in Figure 5A, both M1Exo and the blank control had relatively low level of calcein AM retention, while the calcein AM retention levels were increased for both GEM and DFX treatment groups. Meanwhile, when treated with M1Exo-GEM-DFX, the calcein AM retention level was over 80% compared to positive control verapamil, a classic P-gp inhibitor. These results indicated that M1Exo-GEM-DFX treatment efficiently suppressed the expression of P-gp in PANC-1/GEM cells, leading to increased accumulation of chemotherapeutic agents in the cytosol.

3.6. Cell Migration and Attachment

Cell migration and attachment plays a critical role in cancer cell invasion and tumor metastasis. Thus, we used wound-healing assay to investigate cell migration ability in vitro, which is a commonly used model to mimic cancer cell metastasis. As shown in Figure 5B,C, both wound length and rate of wound closure for PANC-1/GEM treated with M1Exo-GEM-DFX remain unchanged when compared to other treated groups. In Figure 5D, GEM, DFX, GEM&DFX treatment groups had slightly reduced cellular confluency compared to control and M1Exo groups. In contrast, PANC-1/GEM cells which received M1Exo-GEM-DFX treatment had the lowest cellular confluency during cell attachment assay due to its reduced adhesion ability of PANC-1/GEM cells. These results suggest that

M1Exo-GEM-DFX dramatically inhibited the invasiveness of GEM-resistant pancreatic cancer cells.

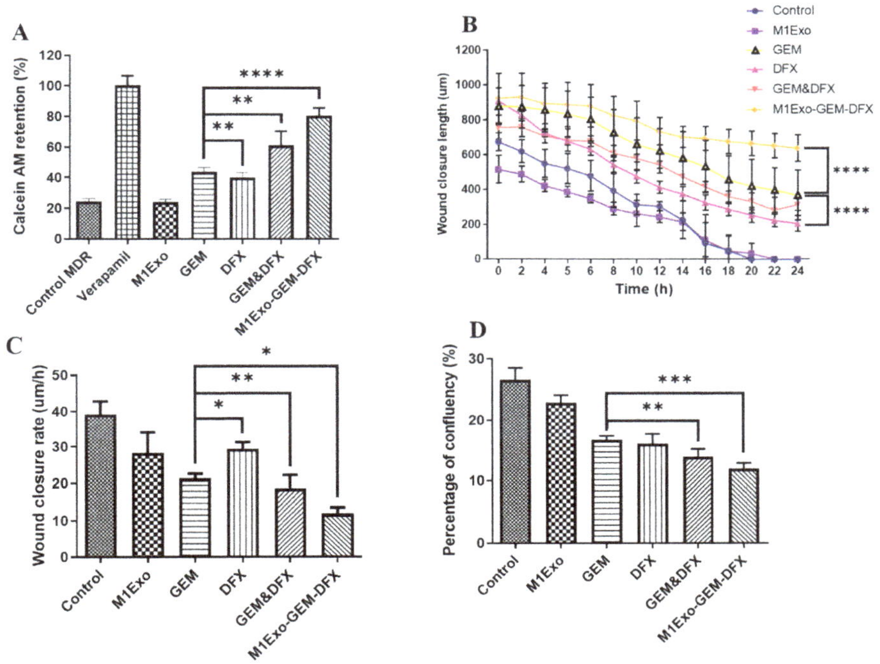

Figure 5. (**A**) Anti-MDR study of M1Exo, GEM, DFX, GEM&DFX, and M1Exo-GEM-DFX against PANC-1/GEM compared with blank and positive controls. Wound-healing study of M1Exo, GEM, DFX, GEM&DFX, and M1Exo-GEM-DFX against PANC-1/GEM: (**B**) Wound-healing assay of wound closure length. (**C**) Wound-closure rate. (**D**) Percentage of confluency measured by cell attachment assay. (* $p < 0.05$, ** $p < 0.01$, *** $p < 0.001$, **** $p < 0.0001$).

3.7. Tumor Spheroid Assay

The anticancer effect was further evaluated in a 3D tumor spheroid system which has been widely used for anticancer drug screening due to better mimicking the physiological properties of tumor tissue [37]. Here, PANC-1/GEM tumor spheroids were successfully cultured to evaluate the potential anti-tumor activity. Figure 6A represents the inhibitory effects of the applied formulations on the 3D PANC-1/GEM tumor spheroids. As shown in Figure 6B, native empty M1Exo had no effect on inhibiting the volume of the PANC-1/GEM tumor spheroids, as a solid cellular cluster structure was maintained after the treatment. Meanwhile, the tumor spheroid formation efficiency of both GEM&DFX and M1Exo-GEM-DFX decreased compared with other treatment groups in PANC-1/GEM tumor spheroids. In addition, the volume change ratio of tumor spheroids at day seven was 477.2 ± 17.2%, 448.5 ± 16.8%, 339.3 ± 17.3%, 300.6 ± 15.9%, 47.1 ± 14.1% and 20 ± 17.6% for control, M1Exo, GEM, DFX, GEM&DFX, and M1Exo-GEM-DFX, respectively. Among these formulations, M1Exo-GEM-DFX induced the strongest inhibitory effect on PANC-1/GEM tumor spheroid growth. The measurement of the spheroid roundness was compared, and we found that the tumor spheroids became granular and irregular on the periphery and finally broke into pieces when treated with M1Exo-GEM-DFX (Figure 6C). These results were consistent with the previous results of cell migration and attachment study. Our results indicated that M1Exo-GEM-DFX effectively inhibited the formation and growth of PANC-1/GEM tumor spheroids compared with free drugs. Thus, these data provided solid evidence that M1Exo-GEM-DFX can significantly enhance therapeutic efficacy toward drug-resistant PANC-1/GEM in vitro by inducing PANC-1/GEM cell death and increasing drug sensitivity to GEM.

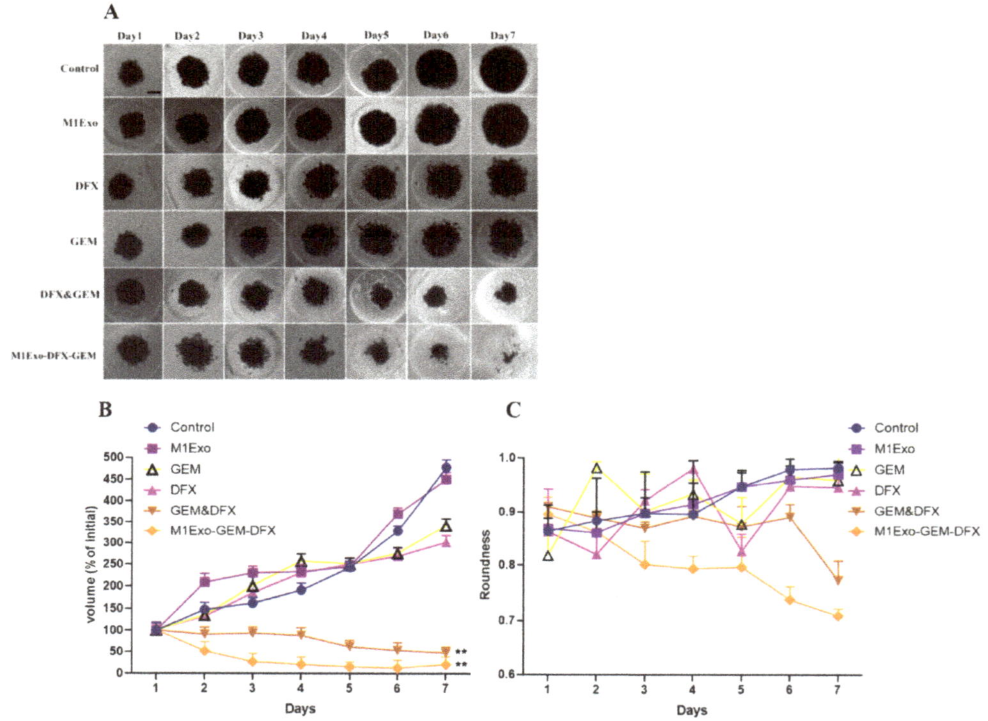

Figure 6. (**A**) Morphology of 3D PANC-1/GEM tumor spheroids treated with blank control, M1Exo, GEM, DFX, GEM&DFX, and M1Exo-GEM-DFX at the indicated concentration for 7 days. Scale bar: 200 µm. (**B**) Inhibitory effect on the growth of PANC-1/GEM tumor spheroids. (**C**) PANC-1/GEM tumor spheroid roundness after treatments, calculated by ImageJ. (** $p < 0.01$).

4. Conclusions

In summary, co-delivery of gemcitabine and Deferasirox using M1Exo offers an effective solution for treating drug-resistant pancreatic cancer. Our study revealed that M1Exo-GEM-DFX nanoformulation enhanced the cytotoxicity efficacy on the GEM-resistant PANC-1/GEM cell line. The mechanism of action was associated with increasing chemosensitivity of GEM in PANC-1/GEM cells by inhibition of RRM2 expression via depleting iron. We also investigated the anticancer therapeutic efficacy of the nanoformulations using a 3D tumor spheroid model. We demonstrated that M1Exo-GEM-DFX effectively inhibited the formation and growth of PANC-1/GEM tumor spheroids compared with free drugs. Overall, the present study suggested that M1Exo-GEM-DFX could be an efficient therapeutic strategy for the treatment of drug-resistant pancreatic tumors and provided insight into their mechanism of action.

Author Contributions: Conceptualization, T.L.; methodology, Y.Z. (Yongmei Zhao) and T.L., software, Y.Z. (Yi Zhang) and Y.Z. (Yuanlin Zheng); validation, Y.Z. (Yan Zhu) and Y.Z. (Yuanlin Zheng); formal analysis, Y.Z. (Yan Zhu) and Y.Z. (Yuanlin Zheng); investigation, Y.Z. (Yongmei Zhao), Y.Z. (Yan Zhu), Y.Z. (Yuanlin Zheng) and T.L.; writing—original draft preparation, Y.Z. (Yongmei Zhao);

writing—review and editing, T.L.; project administration, Y.Z. (Yongmei Zhao), H.Z. and T.L.; funding acquisition Y.Z. (Yongmei Zhao), H.Z. and T.L. All authors have read and agreed to the published version of the manuscript.

Funding: This research was funded by the National Health and Medical Research Council (NHMRC) Early Career Fellowship (Grant No. 1112258), WSU Vice-Chancellor's Research Fellowship, National Science Foundation of the Jiangsu Higher Education Institutions of China (Grant No. 19KJD350002), Natural Science Foundation of Jiangsu Province of China (BK20201443), Jiangsu's Mass Entrepreneurship and Innovation Program and the Large Instruments Open Foundation of Nantong University (KFJN2063).

Institutional Review Board Statement: Not applicable.

Informed Consent Statement: Not applicable.

Data Availability Statement: All data needed to validate the conclusion in this paper are present in the paper. Additional data related to this paper may be request from the authors.

Conflicts of Interest: The authors declare no conflict of interest.

References

1. Pourshams, A.; Sepanlou, S.G.; Ikuta, K.S.; Bisignano, C.; Safiri, S.; Roshandel, G.; Sharif, M.; Khatibian, M.; Fitzmaurice, C.; Nixon, M.R.; et al. The global, regional, and national burden of pancreatic cancer and its attributable risk factors in 195 countries and territories, 1990–2017: A systematic analysis for the Global Burden of Disease Study 2017. *Lancet Gastroenterol. Hepatol.* **2019**, *4*, 934–947. [CrossRef]
2. Lambert, A.; Schwarz, L.; Borbath, I.; Henry, A.; Van Laethem, J.L.; Malka, D.; Ducreux, M.; Conroy, T. An update on treatment options for pancreatic adenocarcinoma. *Ther. Adv. Med. Oncol.* **2019**, *11*. [CrossRef]
3. Hidalgo, M. Pancreatic cancer. *N. Engl. J. Med.* **2010**, *362*, 1605–1617. [CrossRef]
4. Ying, J.-E.; Zhu, L.-M.; Liu, B.-X. Developments in metastatic pancreatic cancer: Is gemcitabine still the standard? *World J. Gastroenterol.* **2012**, *18*, 736–745. [CrossRef]
5. Chen, M.; Xue, X.; Wang, F.; An, Y.; Tang, D.; Xu, Y.; Wang, H.; Yuan, Z.; Gao, W.; Wei, J.; et al. Expression and promoter methylation analysis of ATP-binding cassette genes in pancreatic cancer. *Oncol. Rep.* **2012**, *27*, 265–269. [CrossRef]
6. Zinzi, L.; Contino, M.; Cantore, M.; Capparelli, E.; Leopoldo, M.; Colabufo, N.A. ABC transporters in CSCs membranes as a novel target for treating tumor relapse. *Front. Pharmacol.* **2014**, *5*, 163. [CrossRef] [PubMed]
7. Quint, K.; Tonigold, M.; Di Fazio, P.; Montalbano, R.; Lingelbach, S.; Rückert, F.; Alinger, B.; Ocker, M.; Neureiter, D. Pancreatic cancer cells surviving gemcitabine treatment express markers of stem cell differentiation and epithelial-mesenchymal transition. *Int. J. Oncol.* **2012**, *41*, 2093–2102. [CrossRef] [PubMed]
8. Jia, Y.; Xie, J. Promising molecular mechanisms responsible for gemcitabine resistance in cancer. *Genes Dis.* **2015**, *2*, 299–306. [CrossRef]
9. Zeng, S.; Pöttler, M.; Lan, B.; Grützmann, R.; Pilarsky, C.; Yang, H. Chemoresistance in Pancreatic Cancer. *Int. J. Mol. Sci.* **2019**, *20*, 4504. [CrossRef] [PubMed]
10. Dauer, P.; Nomura, A.; Saluja, A.; Banerjee, S. Microenvironment in determining chemo-resistance in pancreatic cancer: Neighborhood matters. *Pancreatology* **2017**, *17*, 7–12. [CrossRef]
11. Itoi, T.; Sofuni, A.; Fukushima, N.; Itokawa, F.; Tsuchiya, T.; Kurihara, T.; Moriyasu, F.; Tsuchida, A.; Kasuya, K. Ribonucleotide reductase subunit M2 mRNA expression in pretreatment biopsies obtained from unresectable pancreatic carcinomas. *J. Gastroenterol.* **2007**, *42*, 389–394. [CrossRef]
12. Fisher, S.B.; Patel, S.H.; Bagci, P.; Kooby, D.A.; El-Rayes, B.F.; Staley, C.A., 3rd; Adsay, N.V.; Maithel, S.K. An analysis of human equilibrative nucleoside transporter-1, ribonucleoside reductase subunit M1, ribonucleoside reductase subunit M2, and excision repair cross-complementing gene-1 expression in patients with resected pancreas adenocarcinoma: Implications for adjuvant treatment. *Cancer* **2013**, *119*, 445–453. [CrossRef] [PubMed]
13. Zhan, Y.; Jiang, L.; Jin, X.; Ying, S.; Wu, Z.; Wang, L.; Yu, W.; Tong, J.; Zhang, L.; Lou, Y.; et al. Inhibiting RRM2 to enhance the anticancer activity of chemotherapy. *Biomed. Pharmacother.* **2021**, *133*, 110996. [CrossRef]
14. Fujita, H.; Ohuchida, K.; Mizumoto, K.; Itaba, S.; Ito, T.; Nakata, K.; Yu, J.; Kayashima, T.; Souzaki, R.; Tajiri, T.; et al. Gene Expression Levels as Predictive Markers of Outcome in Pancreatic Cancer after Gemcitabine-Based Adjuvant Chemotherapy. *Neoplasia* **2010**, *12*, 807-IN808. [CrossRef] [PubMed]
15. Shinoda, S.; Kaino, S.; Amano, S.; Harima, H.; Matsumoto, T.; Fujisawa, K.; Takami, T.; Yamamoto, N.; Yamasaki, T.; Sakaida, I. Deferasirox, an oral iron chelator, with gemcitabine synergistically inhibits pancreatic cancer cell growth in vitro and in vivo. *Oncotarget* **2018**, *9*, 28434–28444. [CrossRef]
16. Galanello, R.; Piga, A.; Cappellini, M.D.; Forni, G.L.; Zappu, A.; Origa, R.; Dutreix, C.; Belleli, R.; Ford, J.M.; Rivière, G.J.; et al. Effect of food, type of food, and time of food intake on deferasirox bioavailability: Recommendations for an optimal deferasirox administration regimen. *J. Clin. Pharmacol.* **2008**, *48*, 428–435. [CrossRef]

17. Theerasilp, M.; Chalermpanapun, P.; Ponlamuangdee, K.; Sukvanitvichai, D.; Nasongkla, N. Imidazole-modified deferasirox encapsulated polymeric micelles as pH-responsive iron-chelating nanocarrier for cancer chemotherapy. *RSC Adv.* **2017**, *7*, 11158–11169. [CrossRef]
18. Liao, W.; Du, Y.; Zhang, C.; Pan, F.; Yao, Y.; Zhang, T.; Peng, Q. Exosomes: The next generation of endogenous nanomaterials for advanced drug delivery and therapy. *Acta Biomater.* **2019**, *86*, 1–14. [CrossRef]
19. Johnsen, K.B.; Gudbergsson, J.M.; Skov, M.N.; Pilgaard, L.; Moos, T.; Duroux, M. A comprehensive overview of exosomes as drug delivery vehicles—endogenous nanocarriers for targeted cancer therapy. *Biochim. Biophys. Acta* **2014**, *1846*, 75–87. [CrossRef] [PubMed]
20. Li, J.; Cui, D.; Huang, J.; He, S.; Yang, Z.; Zhang, Y.; Luo, Y.; Pu, K. Organic Semiconducting Pro-nanostimulants for Near-Infrared Photoactivatable Cancer Immunotherapy. *Angew. Chem. Int. Ed. Engl.* **2019**, *58*, 12680–12687. [CrossRef]
21. Antimisiaris, S.G.; Mourtas, S.; Marazioti, A. Exosomes and Exosome-Inspired Vesicles for Targeted Drug Delivery. *Pharmaceutics* **2018**, *10*, 218. [CrossRef] [PubMed]
22. Zhang, M.; Zang, X.; Wang, M.; Li, Z.; Qiao, M.; Hu, H.; Chen, D. Exosome-based nanocarriers as bio-inspired and versatile vehicles for drug delivery: Recent advances and challenges. *J. Mater. Chem. B* **2019**, *7*, 2421–2433. [CrossRef] [PubMed]
23. Bowdish, D.M.E. Macrophage Activation and Polarization. In *Encyclopedia of Immunobiology*; Ratcliffe, M.J.H., Ed.; Academic Press: Oxford, UK, 2016; pp. 289–292. [CrossRef]
24. Li, P.; Gao, M.; Hu, Z.; Xu, T.; Chen, J.; Ma, Y.; Li, S.; Gu, Y. Synergistic ferroptosis and macrophage re-polarization using engineering exosome-mimic M1 nanovesicles for cancer metastasis suppression. *Chem. Eng. J.* **2021**, *409*, 128217. [CrossRef]
25. Wang, P.; Wang, H.; Huang, Q.; Peng, C.; Yao, L.; Chen, H.; Qiu, Z.; Wu, Y.; Wang, L.; Chen, W. Exosomes from M1-Polarized Macrophages Enhance Paclitaxel Antitumor Activity by Activating Macrophages-Mediated Inflammation. *Theranostics* **2019**, *9*, 1714–1727. [CrossRef]
26. Li, J.; Li, N.; Wang, J. M1 macrophage-derived exosome-encapsulated cisplatin can enhance its anti-lung cancer effect. *Minerva Med.* **2020**. [CrossRef]
27. Cheng, L.; Wang, Y.; Huang, L. Exosomes from M1-Polarized Macrophages Potentiate the Cancer Vaccine by Creating a Pro-inflammatory Microenvironment in the Lymph Node. *Mol. Ther.* **2017**, *25*, 1665–1675. [CrossRef]
28. Zhao, L.; Gu, C.; Gan, Y.; Shao, L.; Chen, H.; Zhu, H. Exosome-mediated siRNA delivery to suppress postoperative breast cancer metastasis. *J. Control Release* **2020**, *318*, 1–15. [CrossRef] [PubMed]
29. Tian, Y.; Li, S.; Song, J.; Ji, T.; Zhu, M.; Anderson, G.J.; Wei, J.; Nie, G. A doxorubicin delivery platform using engineered natural membrane vesicle exosomes for targeted tumor therapy. *Biomaterials* **2014**, *35*, 2383–2390. [CrossRef]
30. Zhao, Y.; Wang, K.; Zheng, Y.; Zeng, X.; Lim, Y.C.; Liu, T. Co-delivery of Salinomycin and Curcumin for Cancer Stem Cell Treatment by Inhibition of Cell Proliferation, Cell Cycle Arrest, and Epithelial-Mesenchymal Transition. *Front. Chem.* **2020**, *8*, 601649. [CrossRef]
31. Liu, T.; Winter, M.; Thierry, B. Quasi-spherical microwells on superhydrophobic substrates for long term culture of multicellular spheroids and high throughput assays. *Biomaterials* **2014**, *35*, 6060–6068. [CrossRef]
32. Liu, T.; Kempson, I.; de Jonge, M.; Howard, D.L.; Thierry, B. Quantitative synchrotron X-ray fluorescence study of the penetration of transferrin-conjugated gold nanoparticles inside model tumour tissues. *Nanoscale* **2014**, *6*, 9774–9782. [CrossRef]
33. Zhao, Y.; Fletcher, N.L.; Gemmell, A.; Houston, Z.H.; Howard, C.B.; Blakey, I.; Liu, T.; Thurecht, K.J. Investigation of the Therapeutic Potential of a Synergistic Delivery System through Dual Controlled Release of Camptothecin–Doxorubicin. *Adv. Ther.* **2020**, *3*, 1900202. [CrossRef]
34. Seo, J.; Kim, K.S.; Park, J.W.; Cho, J.Y.; Chang, H.; Fukuda, J.; Hong, K.Y.; Chun, Y.S. Metastasis-on-a-chip reveals adipocyte-derived lipids trigger cancer cell migration via HIF-1α activation in cancer cells. *Biomaterials* **2021**, *269*, 120622. [CrossRef] [PubMed]
35. Yao, H.J.; Zhang, Y.G.; Sun, L.; Liu, Y. The effect of hyaluronic acid functionalized carbon nanotubes loaded with salinomycin on gastric cancer stem cells. *Biomaterials* **2014**, *35*, 9208–9223. [CrossRef]
36. García-Manteiga, J.; Molina-Arcas, M.; Casado, F.J.; Mazo, A.; Pastor-Anglada, M. Nucleoside transporter profiles in human pancreatic cancer cells: Role of hCNT1 in 2′,2′-difluorodeoxycytidine- induced cytotoxicity. *Clin. Cancer Res. Off. J. Am. Assoc. Cancer Res.* **2003**, *9*, 5000–5008.
37. Gupta, P.B.; Onder, T.T.; Jiang, G.; Kai, T.; Kuperwasser, C.; Weinberg, R.A.; Lander, E.S. Identification of selective inhibitors of cancer stem cells by high-throughput screening. *Cell* **2009**, *138*, 645–659. [CrossRef] [PubMed]

pharmaceutics MDPI

Review

An Up-to-Date Review of Natural Nanoparticles for Cancer Management

Daniel Ion [1,2,†], Adelina-Gabriela Niculescu [3], Dan Nicolae Păduraru [1,2,†], Octavian Andronic [1,2,†], Florentina Mușat [1,2,†], Alexandru Mihai Grumezescu [3,4,5,*] and Alexandra Bolocan [1,2,†]

1 General Surgery Department, Faculty of Medicine, Carol Davila University of Medicine and Pharmacy, 050474 Bucharest, Romania; dr.daniel.ion@gmail.com (D.I.); dan.paduraru.nicolae@gmail.com (D.N.P.); andronicoctavian@gmail.com (O.A.); flori.musat94@gmail.com (F.M.); bolocan.alexa@gmail.com (A.B.)
2 3rd Clinic of General and Emergency Surgery, University Emergency Hospital of Bucharest, 050098 Bucharest, Romania
3 Department of Science and Engineering of Oxide Materials and Nanomaterials, Faculty of Applied Chemistry and Materials Science, Politehnica University of Bucharest, 011061 Bucharest, Romania; adelina.niculescu@upb.ro
4 Research Institute of the University of Bucharest—ICUB, University of Bucharest, 050657 Bucharest, Romania
5 Academy of Romanian Scientists, Ilfov No. 3, 050044 Bucharest, Romania
* Correspondence: grumezescu@yahoo.com or agrumezescu@upb.ro
† These authors contributed equally to this work.

Abstract: Cancer represents one of the leading causes of morbidity and mortality worldwide, imposing an urgent need to develop more efficient treatment alternatives. In this respect, much attention has been drawn from conventional cancer treatments to more modern approaches, such as the use of nanotechnology. Extensive research has been done for designing innovative nanoparticles able to specifically target tumor cells and ensure the controlled release of anticancer agents. To avoid the potential toxicity of synthetic materials, natural nanoparticles started to attract increasing scientific interest. In this context, this paper aims to review the most important natural nanoparticles used as active ingredients (e.g., polyphenols, polysaccharides, proteins, and sterol-like compounds) or as carriers (e.g., proteins, polysaccharides, viral nanoparticles, and exosomes) of various anticancer moieties, focusing on their recent applications in treating diverse malignancies.

Keywords: natural nanoparticles; natural anticancer compounds; natural cancer therapies; novel cancer treatment alternatives; natural nanocarriers; chemotherapeutic agents targeted delivery

Citation: Ion, D.; Niculescu, A.-G.; Păduraru, D.N.; Andronic, O.; Mușat, F.; Grumezescu, A.M.; Bolocan, A. An Up-to-Date Review of Natural Nanoparticles for Cancer Management. *Pharmaceutics* **2022**, *14*, 18. https://doi.org/10.3390/pharmaceutics14010018

Academic Editor: Mazzucchelli Serena

Received: 27 November 2021
Accepted: 20 December 2021
Published: 22 December 2021

1. Introduction

Cancer has long been a critical threat worldwide, imposing a global health and economic burden. Cancer cells can evade the immune system, multiply indefinitely, and perform angiogenesis, leading to challenging malignancies that directly damage human life [1–6].

The most frequently employed treatment option in fighting cancer is chemotherapy, which can be used either alone or in combinatorial approaches with radiotherapy, surgery, or adjuvant therapies (e.g., immunotherapy, hormone therapy, photothermal therapy, photodynamic therapy, and ablative techniques) to produce effective responses depending on the cancer stage [7–10]. However, the dissatisfying specificity coupled with poor aqueous solubility and short blood circulation of conventional anticancer drugs leads to low concentrations of drugs at the tumor site and the requirement of high doses [2,8,11–13]. In addition, the therapeutic efficacy of administered drugs diminishes over time due to the development of drug resistance [14,15]. Despite the growing number of anticancer agents developed in the last few decades, their severe toxicity, high production cost, and low patient compliance demand better antitumor alternatives [16].

To improve treatment outcomes, radiotherapy can be used complementarily to destroy cancer cells sensitized by chemotherapeutics. However, radiation also affects normal tissues, leading to side effects occurrence immediately or soon after radiotherapy treatment [17,18]. Thus, the lack of specificity of conventional therapies results in negative effects upon rapidly multiplying normal cells (e.g., bone marrow, gastrointestinal tract, and hair follicles) [13] or other healthy tissues, leading to multiple off-target adverse effects, including appetite loss, anemia, internal bleeding, fatigue, and hair loss [2,12,19]. Some of the most important disadvantages associated with the classic trio of cancer therapeutic options are summarized in Figure 1.

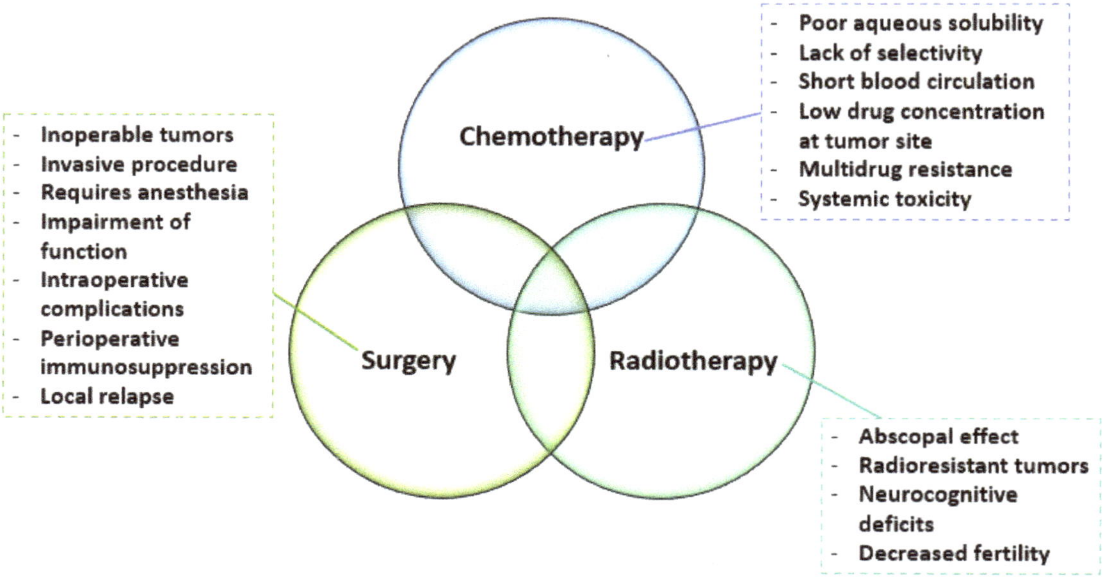

Figure 1. Limitations of conventional cancer treatment strategies. Created based on information from [13,20–25].

Nanotechnology has appeared as a promising solution to overcome the most pressing challenges of current cancer treatments [2,26,27]. The use of nanocarriers is a viable option for enhancing solubility and bioavailability of anticancer compounds of both natural and synthetic origin [28], delivering drugs across traditional biological barriers in the body [29], and combining therapeutic agents with imaging techniques towards achieving synergic results [12]. In more detail, anticancer drug delivery via nanoparticles (NPs) is influenced by biological barriers, counting tumor microenvironment (TME) and vasculature, reticuloendothelial system, blood–brain barrier (BBB), and kidney filtration [30]. Thus, special attention must be given to overcoming these barriers and ensuring tumor uptake of NPs.

Specifically, by taking advantage of the newly accumulated knowledge TME [15,31,32], multifunctional NPs can be designed to deliver bioactive agents directly to the tumor, reducing systemic side effects (Figure 2). This can be done either by surfaced functionalization with ligands specific for receptors expressed by tumor cells or other cells in TME or by functionalization with chemical groups that can respond to TME signals (e.g., secreted molecules, acidic pH, and hypoxic conditions) [11,29,33].

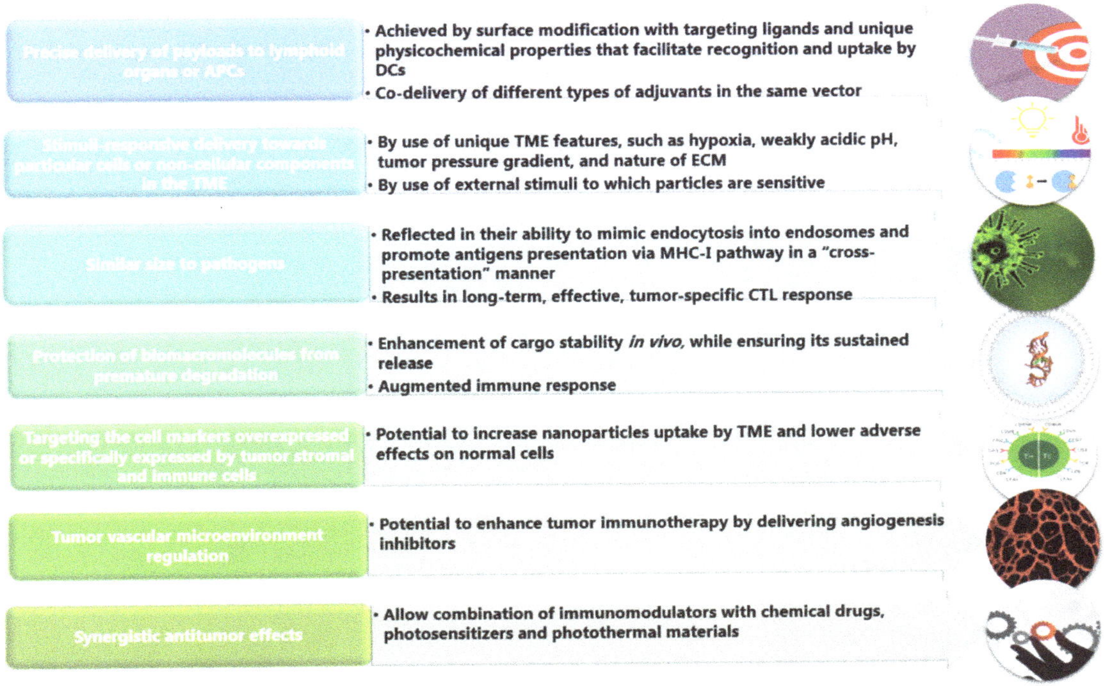

Figure 2. NPs roles in regulating TME and improving tumor immunotherapy. Created based on information from [1,30,34–36]. Abbreviations: APCs—antigen-presenting cells; DCs—dendritic cells; TME—tumor microenvironment; ECM—extracellular matrix; MHC—major histocompatibility complex; CTL—cytotoxic T lymphocyte.

To date, a wide array of materials have been investigated for producing effective anticancer NPs. Researchers developed different nanostructures of different shapes, sizes, architectures, and compositions using various materials, including lipids, proteins, polysaccharides, synthetic polymers, and inorganic materials [2,8,9,11,29].

In particular, the beneficial physicochemical and biological properties of natural materials have recently rendered these NPs among the most promising base materials for cancer therapy. As compared to most synthetic NPs, nanostructures of natural origin have been noticed to have better safety profiles, enhanced biocompatibility, biodegradability, and non-immunogenicity; they also present functional groups that facilitate their chemical modification towards obtaining even more performant formulations [8,11,37]. Moreover, as detailed toxicology assessments are fundamental for the clinical translation of nanoparticulate nanomedicines [38], the favorable biological behavior of natural NPs may represent an opportunity for their faster introduction into clinical trials and consequently into medical practice as compared to synthetic alternatives.

In this respect, the present paper further discusses the most recent advances in natural NPs for cancer management, including both natural molecules with inherent anticancer properties loaded into NPs and natural NPs used as carriers of various freights in cancer therapies. More specifically, in this review are considered "natural" the compounds and materials that can be obtained from biological sources.

2. Natural Compound-Based NPs with Intrinsic Anticancer Activity

Attempting to avoid the side effects and downsides of chemotherapeutic drugs, researchers have tackled the anticancer potential of natural anticancer agents from a variety of plants and organisms [39,40]. Some of the most relevant examples of nanoparticles of natural origin with intrinsic anticancer activity are further described in this section.

2.1. Polyphenols

Polyphenols are well recognized for their health benefits, showing biomedical potential in various diseases, such as tumors, inflammatory diseases, and cardiovascular diseases [41,42]. They can be easily included in the daily diet, as they can be found in diverse natural sources (e.g., tea, red wine, cocoa, fruits, and olive oil), or they can be extracted and processed for developing nutraceutical and pharmaceutical formulations with specific and enhanced activity [43].

Polyphenols represent a broad class of bioactive compounds, comprising a variety of chemical structures (Figure 3). Their molecular structures allow polyphenols to combine with other materials, such as proteins, metal ions, polymers, and nucleic acids, creating better delivery strategies [41]. In this respect, the following subsections discuss the newest approaches for polyphenols delivery that are relevant for alternative or complementary cancer therapies.

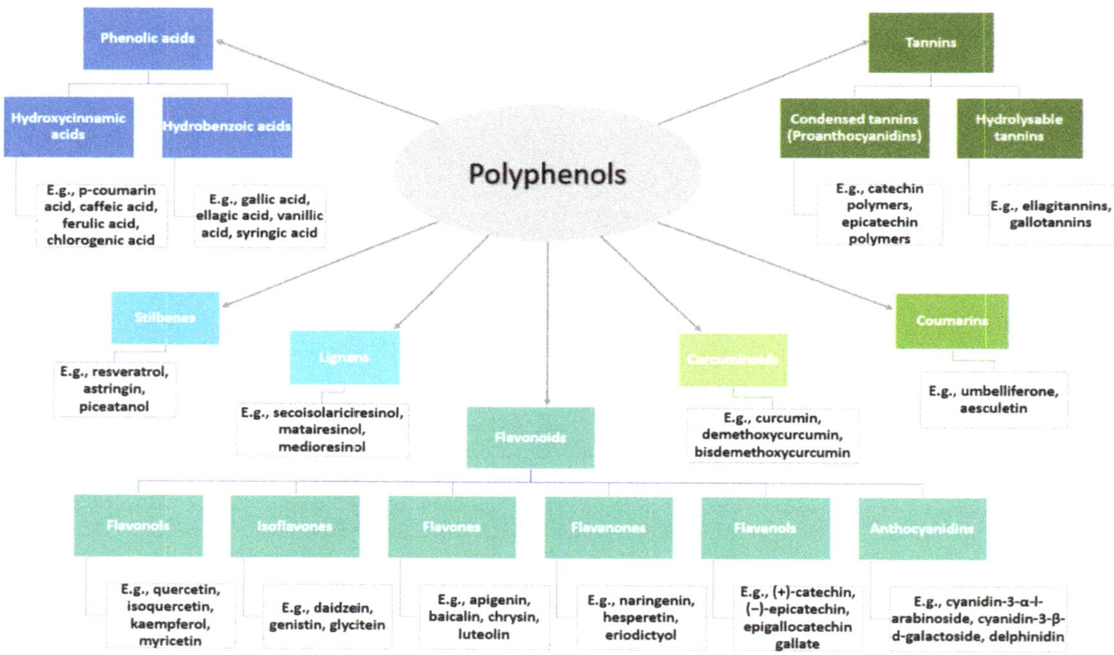

Figure 3. Polyphenols classification and examples. Created based on information from [44–46].

2.1.1. Flavonoids

Nanoparticles obtained from flavonoids have promising anticancer effects [28,47]. One such useful compound is chrysin, which has proven advantageous characteristics, including anti-inflammatory, antioxidant, antiallergic, and cancer chemopreventive properties [2,48,49]. Nonetheless, the poor water solubility and bioavailability of chrysin limit its use as an anticancer drug [48,50]. In this context, special attention has been drawn to investigate chrysin in nanoparticulate form. Mutha et al. [49] have developed chrysin NPs by probe sonication technique. The authors used small amounts of sodium dodecyl sulfate to enhance aqueous solubility of NPs and mannitol as a bulking agent. The as-such obtained NPs demonstrated potential cytotoxicity and significant control on the growth of MCF-7 breast cancer cell line. Another convenient strategy to overcome pure chrysin limitations is to encapsulate this compound in PLGA-PEG nanoparticles [48,50]. For instance, Mohammadian et al. [48] have evaluated chrysin-loaded PLGA-PEG NPs against a gastric cancer cell line. The researchers observed a significant decrease in IC_{50} value of nanocapsulated chrysin as compared to free flavonoid and a decline in miR-18a, miR-21, and miR-221 gene expression. Another example is offered by Tavakoli et al. [50], who have used PLGA-PEG NPs to co-deliver chrysin and curcumin. The scientists reported enhanced antiproliferative and anti-metastatic effects on melanoma cancer when encapsulating these polyphenols than when using them in free form.

Quercetin is another flavonoid with potential applications in cancer therapy as it exhibits a strong inhibitory effect on the growth of several cancer cell lines, including nasopharyngeal, lung, prostate, ovarian, breast, leukemic, skin, bone, and colon cancer cells [51–53]. Nonetheless, the application of quercetin in anticancer treatments is limited by low aqueous solubility, bioavailability, and chemical instability in neutral and alkaline media [54]. To overcome these drawbacks, quercetin can be delivered via different nanoplatforms. For instance, Rezaei-Sadabady et al. [55] have encapsulated quercetin into liposomes to enhance the hydrophilicity and deliverability of this flavonoid. The authors obtained promising results in terms of quercetin solubility and bioavailability, but the types of cancer most likely to benefit from this non-toxic therapy were yet to be determined. Another delivery possibility is proposed by Sadhukhan et al. [56], who have loaded quercetin into phenylboronic acid conjugated zinc oxide NPs. The scientists observed that their nanosystem was able to enhance oxidative stress and mitochondrial damage, leading to apoptotic cell death in human breast cancer cells. Moreover, tumor-associated toxicity in the liver, kidney, and spleen was reportedly reduced. Quercetin was also prepared in combination with other nanomaterials, including chitosan NPs [57,58], PEGylated-PLGA nanocapsules [59], pluronic-grafted gelatin copolymers [60], mesoporous silica NPs [61], and metal–organic frameworks [54].

Baicalein has also been extensively studied as an alternative to synthetic chemotherapeutic agents [2]. It has been proven as an anti-inflammatory, antioxidant, and antitumor agent for different types of cancers, including lung, breast, skin, and gastric cancers [62,63] (Figure 4). To overcome its poor bioavailability caused by its hydrophobic nature, scientists started to explore various baicalein-based nanoparticulate combinations. For instance, Wang et al. [62] developed self-assembled NPs containing dual-targeted ligands of folate and hyaluronic acid for co-delivery of baicalein and paclitaxel. The as-described system was proven efficient for targeted drug delivery, leading to synergistic anticancer effects and overcoming multidrug resistance in human lung cancer cells. A different strategy is offered by Joshi et al. [64], who have fabricated solid lipid nanoparticles of baicalein that had an increase of over ~300% in relative oral bioavailability as compared to free flavonoid administration. The researchers also reported better radioprotection to healthy cells and sensitized cancer cells to radiation-induced killing, concluding that these novel nanoparticles can be employed as an adjuvant in cancer radiotherapy.

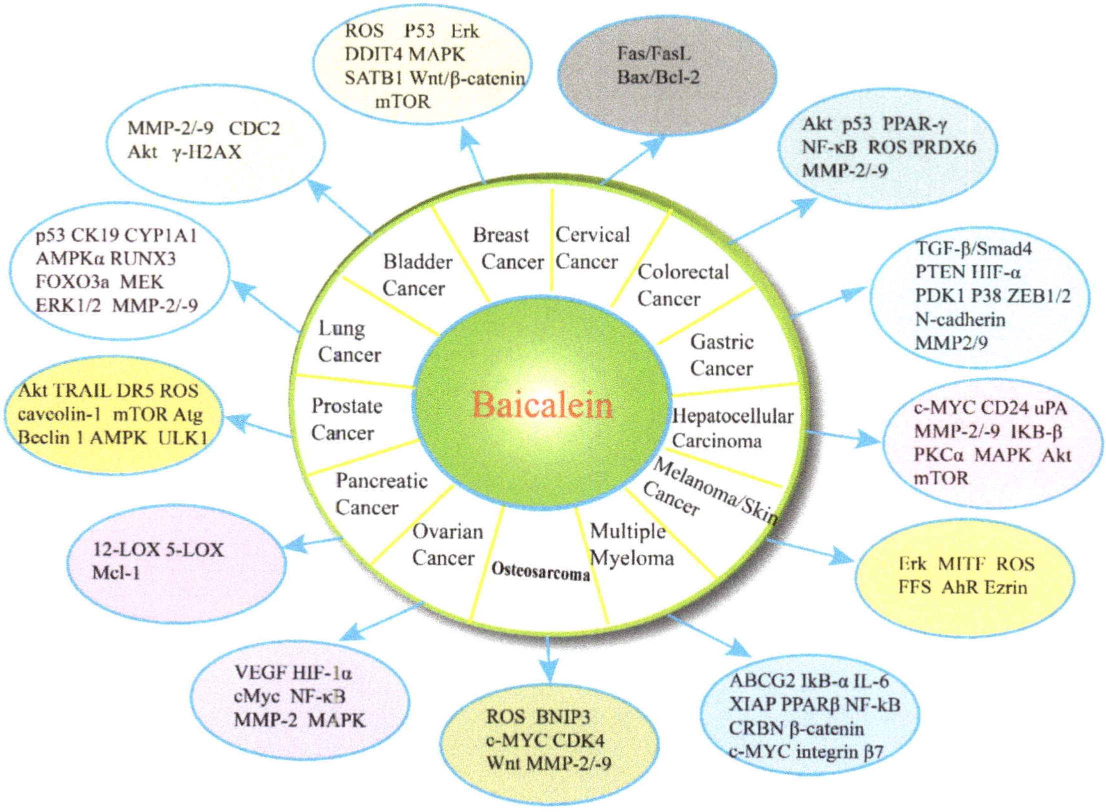

Figure 4. Baicalein anticancer activity by binding to and interacting with specific cellular targets. Reprinted from an open-access source [63].

Green tea catechins were also noted for their suppressive effects on cancer cell progression, metastasis, and angiogenesis [65]. The major bioactive constituent of green tea, (−)-epigallocatechin-3-gallate (EGCG), has been intensively studied for its chemopreventive and chemotherapeutic activity. However, its lack of target specificity, short half-life, low stability, and low bioavailability limit its free use and request for special delivery approaches [66,67] (Figure 5). EGCG-based nanosystems showed promising synergistic results by conjugation with gold NPs [68–70], encapsulation in synthetic [71–73] and natural [74–77] polymeric NPs, liposome delivery [78–81], and combination with various other anticancer agents [79,81–83].

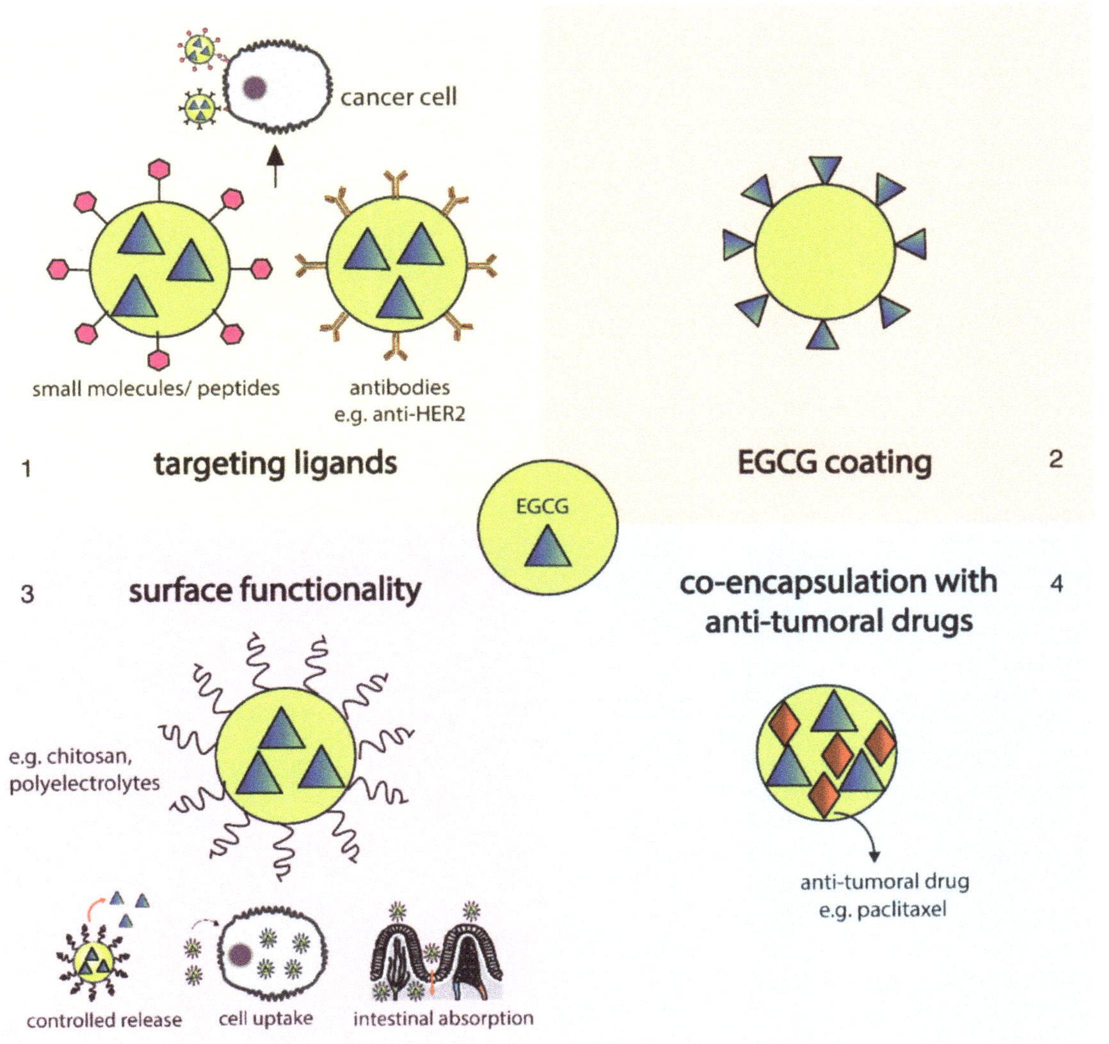

Figure 5. EGCG delivery possibilities for cancer therapy: (**1**) incorporation of ligands on NPs surface for specific targeting of cancer cell receptors or antigens; (**2**) EGCG used as a capping agent; (**3**) surface functionalization with polymers for improving drug release, cellular uptake, and intestinal absorption; (**4**) co-encapsulation with conventional chemotherapeutic agents. Reprinted from an open-access source [66].

2.1.2. Tannins

Condensed tannin extracts have also been remarked as promising anticancer candidates due to their antitumor activity and potential of inducing apoptosis in cancer cells via enzymes regulation [37,84,85]. Other anticancer mechanisms of tannins include negative regulation of transcription factors, growth factors, receptor kinases, and various oncogenic molecules [86].

Recently, scientists have moved from condensed tannins extracts to their nanoparticles' counterparts, as is the case of AlMalki et al. [37], who have synthesized NPs from a commercial product extracted from the bark of Pine trees. The researchers obtained potential anticancer effects against MCF-7 cells, concluding that tannin nanoparticles are promising candidates for treating breast cancer either alone or in combination with low doses of tamoxifen.

2.1.3. Resveratrol

Resveratrol is the most studied stilbene due to its large availability, antioxidant properties, and potential induction of cancer chemopreventive and therapeutic responses [87,88]. Nonetheless, similar to other polyphenols, resveratrol's direct use is hindered by its low bioavailability and rapid metabolism [88].

Thus, the development of nanoparticulated formulations has become a convenient solution for overcoming these drawbacks. Conjugating resveratrol to gold NPs was seen to improve its bioavailability, leading to optimal cellular uptake, and enhanced antitumor efficacy against breast, prostate, and pancreatic cancer cells [89]. Other studies emphasized the potent anticancer activity of resveratrol (RES) when loaded in solid lipid NPs (SLNs [90], functionalized mesoporous silica NPs (MSNs) [91], gelatin NPs [92,93], and more (Table 1).

Table 1. Examples of anticancer nanoparticle formulation based on resveratrol.

Nanoformulation	Physicochemical Characteristics	Type(s) of Cancer	Observations	Refs.
RES-conjugated gold NPs	• Shape: spherical • Average core size: 16.1 ± 5.0 nm • Zeta potential: −25 mV	Breast cancer Prostate cancer Pancreatic cancer	• Enhanced bioavailability of RES • Excellent loading of RES with subsequent efficient antitumor effects • Synergic anticancer activity due to dual action of gold and RES	[89]
RES-loaded SLNs	• Shape: spherical • Average diameter: 168 ± 10.7 nm • Zeta potential: −23.5 ± 1.6 mV • Loading capacity: 25.2 ± 1.7%	Breast cancer	• Enhanced bioavailability and anticancer activity of RES • Cell proliferation inhibited in a dose-dependent manner • Lower IC_{50} values for RES-SLNs than for free RES • Increased cell cycle arrest in the G0/G1 phase via CyclinD1 downregulation in cancer cells	[90]
Chitosan-coated-trans-RES and ferulic acid loaded SLNs conjugated with folic acid	• Shape: spherical • Average diameter: 174 ± 5 nm • Zeta potential: −25.9 mV	Colon cancer	• Good stability under acidic conditions • Effectively involved and increased cytotoxicity in cancer cells, resulting in apoptosis induction • Cancer cells specific delivery; induced cell death of HT-29 cell line but did not affect normal NIH 3T3 cells	[94]
RES-loaded in functionalized MSNs	• Shape: spherical • Average diameter: ~60 nm	Prostate cancer	• Significant control over RES release at 5.5 pH • Robust and dose-dependent sensitization of Docatexal in hypoxic cell environment • Enhanced antiproliferative potential	[91]

Table 1. *Cont.*

Nanoformulation	Physicochemical Characteristics	Type(s) of Cancer	Observations	Refs.
RES-loaded PLGA NPs	• Shape: spherical • Average diameter: 237.8 ± 4.93 nm • Encapsulation efficiency: 89.32 ± 3.51%	Prostate cancer	• Significant decrease cell viability with IC_{50} and IC_{90} of 15.6 ± 1.49 and 41.1 ± 2.19 μM, respectively • Significantly greater cytotoxicity than free RES • Anticancer effects mediated by apoptosis; confirmed by cell cycle arrest at G1-S transition phase, DNA nicking, loss of mitochondrial membrane potential, ROS generation, and externalization of phosphtidylserine	[95]
RES-loaded gelatin NPs	• Shape: spherical • Average diameter: 294 nm • Encapsulation efficiency: 93.6%	Non-small cell lung cancer	• Superior efficacy in NCI-H460 cells • Induced apoptosis via alteration in expression of p53, p21, caspase-3, Bax, Bcl-2 and NF-κB • Induced cell arrest in the G0/G1 phase of cell cycle	[93]
RES-cyclodextrin complex-loaded PLGA NPs	• Shape: spherical • Average diameter: 264.2 ± 3.4 nm	Non-small cell lung cancer	• Improved RES aqueous solubility by 66-folds • Intensified anticancer effects compared to free RES • Enhanced cellular uptake, cytotoxicity, and apoptosis • Very good aerosolization potential	[96]

2.1.4. Curcumin

Curcumin is one of the most studied herbal anticancer compounds, especially due to its multiple-fold action (Figure 6). Curcumin can inhibit carcinogenesis, angiogenesis, and tumor growth [97] by modulating or interacting with growth factors, interleukins, and protein kinases, thus being a promising alternative to conventional chemotherapeutic agents [39,40].

However, curcumin has low water solubility and is unstable in physiological conditions, drawbacks that further lead to poor bioavailability and unfavorable biodistribution [39,98]. To overcome these challenges and enhance anticancer bioactivity, scientists employed curcumin (CUR) in the development of composite nanoparticles by conjugation with human serum albumin (HSA) [99], encapsulation in chitosan [40,100], silk fibroin [101], soybean polysaccharide [102], and more (Table 2).

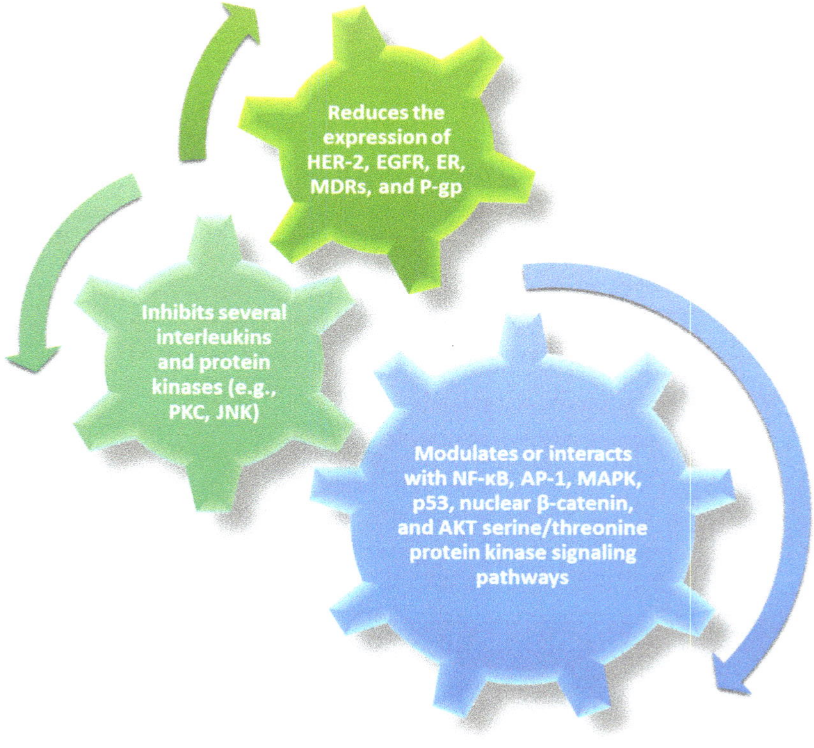

Figure 6. Anticancer effects of curcumin. Created based on information from [39,97].

Table 2. Examples of anticancer nanoparticle formulations based on curcumin.

Nanoformulation	Physicochemical Characteristics	Type(s) of Cancer	Observations	Refs.
CUR-conjugated with HSA	• Shape: spherical • Average diameter: 180 ± 2nm • Zeta potential: • −7mV • Loading capacity: 12% • Encapsulation efficiency: 70%	Breast cancer	• Enhanced stability of CUR both in physiological and acidic conditions • Significant increase in CUR cytotoxicity on cancer cells without increasing the toxicity on healthy cells	[99]
CUR and liquid fluorocarbon perfluorohexane (PHF) co-loaded in ferritin nanocages conjugated with folic acid	• Shape: spherical • Average diameter: 47 nm • Zeta potential: −37mV • CUR loading ratio: 125.8 ± 2.1%	Ovarian cancer	• Significant tumor treatment effects • Under low-intensity focused ultrasound (LIFU) and 5.0 pH, the nanoplatform released 53.2% of encapsulated drugs in 24 h • After 4 min of LIFU at 5.0 pH, the system provided contrast-enhanced ultrasound imaging capabilities	[103]

Table 2. *Cont.*

Nanoformulation	Physicochemical Characteristics	Type(s) of Cancer	Observations	Refs.
CUR-loaded chitosan NPs	• Shape: spherical • Average diameter: 115 ± 21 nm • Zeta potential: 33.8mV • Loading capacity: 21.6% • Encapsulation efficiency: 83.8%	Colon cancer Lung cancer	• CUR was mostly released in the first 5 h then gradually released up to 90 h • Higher release in pH 5.2 than in pH 7 • Time-dependent decrement of cancer cells viability • After 96 h of exposure 67.6% HCT-116 cells and 73.8% A-546 cells were dead	[40]
CUR-loaded chitosan NPs	• Shape: spherical • Average diameter: 415.30 ± 9.03 nm • Zeta potential: 33.37 ± 0.21 mV • Encapsulation efficiency: 73.56 ± 6.01%	Lung cancer	• Effective and precisely controllable NPs induced cytotoxicity only upon irradiation with 457 nm LED light NPs • Upon photoactivation, CUR induced chromatin condensation and DNA fragmentation leading to cancer cells destruction	[100]
CUR-loaded silk fibroin NPs	• Shape: spherical • Average diameter: 155–175 nm • Zeta potential: −45 mV	Hepatocellular carcinoma Neuroblastoma	• Local long-term sustained drug delivery • Cytotoxicity against cancer cells, while no decreasing viability reported for healthy cells • Higher efficacy against neuroblastoma cells than against hepatocellular carcinoma cells	[101]
CUR-loaded soybean polysaccharide nanocapsules	• Shape: spherical • Average diameter: 200–300 nm • Encapsulation efficiency: ~90%	Colon cancer Mammary adenocarcinoma	• No significant difference in the viability of HCT 116 and MCF-7 cells challenged with DMSO-dissolved and nanoencapsulated CUR • Most of antiproliferative activity of the nanosystem manifested after sim, ulated gastric and intestinal digestions	[102]
CUR-loaded PEGylated MSNs	• Shape: spherical • Average diameter: 197 nm • Loading capacity: 8.1% • Encapsulation efficiency: 89.1%	Cervical cancer	• Significantly increased solubility and enhanced bioavailability of CUR for photodynamic therapy • Smooth and steady release at physiological pH, while at 5.0 pH the release rate was slightly speeded up	[104]
CUR-loaded poloxamer188-*b*-PCL NPs	• Shape: spherical • Average diameter: 100 nm	Esophageal squamous carcinoma	• Improved in vitro antioxidant activity compared to crude CUR powder • Particles could biodistribute into liver, kidney, and lung tissues, acting as protection agents in cancer radiotherapy	[105]

Table 2. *Cont.*

Nanoformulation	Physicochemical Characteristics	Type(s) of Cancer	Observations	Refs.
CUR-loaded therapeutic lipid NPs	• Shape: spherical • Average diameter: 19.8 ± 4.2 nm	Nasopharyngeal carcinoma (NPC)	• Effective targeting ability, suppressed cellular proliferation, and induced apoptosis in vitro • Enhanced inhibitory effect on NPC tumor growth and metastasis in vivo	[106]
CUR-loaded in niosomal NPs	• Shape: spherical • Average diameter: ~60 nm • Zeta potential: −35 mV	Glioblastoma	• Dose-dependent decrease in cell proliferation and viability of glioblastoma stem-like cells (GSC) • Higher effects on GSC viability, apoptosis, cell cycle arrest, and expression of Bax than free CUR • Significantly impaired GSC migration	[107]

2.2. *Proteins*

Increasing interest has been directed towards using ovalbumin (OVA) protein antigen in developing novel anticancer nanoparticles. This non-toxic, temperature- and pH-sensitive, and economical material is convenient for use in cancer immunotherapy as it elicits cellular and humoral immune responses [108,109]. Habibi et al. [108] have created different OVA-based NPs by chemically linking individual OVA molecules via PEG units. The scientists noticed that by controlling the PEG/OVA ratio, the physicochemical characteristics of the protein NPs, such as size, elasticity, and network structure, can be tailored. The authors reported better results as the PEG/OVA ratio decreased, i.e., increased uptake by lymph node macrophages, dendritic cells, and B cells, more effective processing by dendritic cells, enhanced lymphatic drainage, higher anti-OVA antibody titers in vivo, and overall improved humoral responses. Moreover, when compared to solute OVA, NPs significantly increased the median survival rate in a mouse model of B16F10-OVA melanoma.

Silk sericin is another protein of interest for improving cancer treatment strategies. Sericin's biocompatibility, non-immunogenicity, and antioxidant properties are the main factors contributing to its research for anticancer purposes. It has been proven that this natural material can reduce oxidative stress or suppress cancer cytokines for skin and colon cancer. Moreover, silk sericin has a self-assembling capacity and unique chemistry that favors surface modifications, thus also being a promising nanocarrier for anticancer drugs [110,111].

Keratin may also be employed in designing NPs for cancer therapies. This protein is particularly appealing due to its unique amino acid sequences that can specifically bind vitronectin integrin receptors overexpressed by several cancer cells [112]. Nonetheless, this inherent anticancer potential has been mostly exploited for delivery purposes, to create stimuli-responsive nanocarriers for chemotherapeutics [112–115], or to enhance the effects of phototherapies [116–119].

2.3. *Polysaccharides*

Fucoidan is a marine polysaccharide that can be extracted from different species of brown algae. As this polymer exhibited great promise in treating several types of cancer, including colon, bladder, liver, lymphoma, and gastric cancers, researchers started to investigate its effects in the nanoscale form [120,121]. For instance, Etman et al. [120] have developed fucoidan-based NPs by polyelectrolyte interaction with lactoferrin targeted ligand. The researchers have reported cytotoxic properties against pancreatic cancer cells, enhanced ability to prevent tumor cells migration and invasion, with a 2.3-fold decreased IC_{50} value for fucoidan NPs than for fucoidan solution.

Chitosan, a polymeric biomaterial found in shellfish exoskeletons, can also be used in developing novel anticancer formulations. This polysaccharide manifests its antitumor activity by affecting cell digestion and inhibiting cell development. However, most of the studies approach chitosan as an assistant agent and nanocarrier rather than for its intrinsic anticancer properties [122].

2.4. Sterol-Like Compounds

Sterol-like natural compounds have also been shown to have promising antitumor activity, gaining interest in developing innovative nanoparticulate cancer therapeutics. For instance, Qiu et al. [123] have loaded 20(S)-ginsenoside (Rg3) into pH-sensitive polymeric NPs that can target cancer cells and prolong circulation time. It was reported that Rg3 could be released rapidly at the tumor site, significantly inhibiting tumor proliferation. Specifically, the sterol-like compound decreased the expressions of proliferating cell nuclear antigen, producing the apoptosis of colorectal cancer cells through the increased expressions of caspase-3.

A different approach was used by Kim et al. [124], who have prepared flower-shaped nanocomposites based on zinc oxide and hyaluronic acid, which they functionalized with ginsenoside Rh2. This complex nanosystem proved successful against lung, colon, and breast cancer cells, exerting its anticancer activity through several mechanisms (e.g., ROS production, upregulation of p53 and BAX, downregulation of BCL2, induction of morphological changes in the nucleus of tumor cells).

3. Natural Nanoparticles for Anticancer Drug and Gene Delivery

A lot of research effort has been put into developing biocompatible targeted delivery vehicles from materials of natural origin. Numerous studies have been published evaluating the anticancer action of various natural-based nanosystems able to effectively and efficiently deliver different cargos to the tumor site. Several recently developed nanoformulations based on proteins, polysaccharides, viral NPs, exosomes, and other natural materials are further presented to make an overview of current progress in the field and emphasize the versatility of these nanoparticles for cancer management.

3.1. Protein-Based NPs

3.1.1. Albumin

Human serum albumin (HSA) is an appealing material for developing novel nanomedicines as it has a high drug loading ability, self-assembling properties, long half-life, and preferential uptake by tumor and inflamed tissues [11,125]. Moreover, increasing interest has been drawn to using HSA-based NPs in cancer management after the FDA approval of paclitaxel-bound albumin NPs (Abraxane^{TM}) [7,126]. Albumin NPs also show a good affinity for other cancer drugs, such as doxorubicin, curcumin, and tacrolimus [109].

For instance, Chaiwaree et al. [127] have successfully encapsulated doxorubicin in HSA submicron particles. The researchers reported excellent A549 cell uptake (up to 98%) and localization of drug nanoformulation within the cell lysosomal compartment. These observations were also reflected in the reduction of cancer cell metabolic activities after 72 h and less than 1% drug release within 5 h at physiological pH. Thus, it can be concluded that this delivery system is a potential candidate for cancer therapy.

Another promising strategy was developed by Yu et al. [128], who have fabricated albumin NPs loaded with docetaxel and functionalized with nucleolin-targeted aptamers. Their targeted drug delivery system was preferentially ingested by nucleolin-expressing CT26 colon cancer cells, leading to enhanced antitumor efficacy, low systemic toxicity, and prolonged survival of CT26-bearing mice.

3.1.2. Keratin

The targeting potential and ease of functionalization of keratin nanoparticles have been exploited by Avancini et al. [116], who have used them as nanocarriers for salinomycin, chlorin e6 photosensitizer, and vitamin E acetate. The researchers tested this novel drug delivery system in vitro on breast cancer cell lines and cancer stem cell (CSC)-enriched mammospheres, reporting synergistic cell killing, limited self-renewal capacity, and eradication of CSCs. Further in vivo tests on zebrafish embryos confirmed the results and revealed that keratin encapsulation of the drug does not alter its CSC-specific cytotoxicity.

Lu et al. [117] have also investigated keratin-based NPs for photodynamic therapy against breast cancer. The complete nanosystem comprised keratin as nitric oxide generator, phenylboronic acid (PBA)-modified d-α-tocopherol polyethylene glycol 1000 succinate (TPGS) as targeting ligand, and methylene blue as photosensitizer. In vivo studies showed that the developed constructs induced extensive cell apoptosis, leading to significant inhibition of in situ tumor growth and lung metastases.

3.1.3. Silk Sericin and Fibroin

The two proteins of silk (i.e., sericin and fibroin) can be used for engineering silk-based nanoparticles suitable for drug delivery and cancer treatment [129–131]. For instance, Huang et al. [111] have fabricated folate-conjugated sericin nanoparticles for tumor targeting and pH-responsive delivery of doxorubicin. The as-described NPs possessed good cytotoxicity and hemocompatibility, specifically releasing the encapsulated drug freight into the lysosomes of folate receptor-rich KB cells.

Silk sericin has also found use as a photosensitizer carrier. More specifically, Gao et al. [132] have used this protein to create NPs for chlorin e6 delivery. These nanosystems achieved superior accumulation in tumor sites compared to with free photosensitizer agents, suppressing tumor growth and avoiding side effects occurrence.

Pandey et al. [133] have developed silk fibroin NPs coated with hydrophilic stabilizers to allow longer circulation times and facilitate their uptake by low-density lipoprotein receptors. These nanocarriers loaded with doxorubicin showed a proinflammatory response, sustained drug release, and better cytotoxicity than the free drug, being a potential candidate for glioblastoma treatment.

Silk fibroin was also proven useful in the delivery of drugs for colorectal cancer treatment. Particularly, Hudita et al. [134] have designed silk fibroin-based NPs for the delivery of 5-fluorouracil (5-FU). The scientists reported great antitumor efficacy in vitro and promising results in vivo, as the proposed system was able to ameliorate mucositis induced during 5-FU treatment.

3.1.4. Ferritin

Ferritin is a convenient protein for drug encapsulation as it self-assemblies naturally into a hollow nanocage with an inner diameter of 7–8 nm. This structure is composed of 24 subunits of either heavy chain ferritin (HFn), light chain ferritin (LFn), or a mix between the two types. In particular, HFn has been noted to bind human cells by interacting with transferrin receptor 1 (TfR1), which is highly expressed on cancer cells and is commonly used as a targeting marker for tumor diagnosis and therapy [135,136].

In this respect, Liang et al. [136] have investigated HFn nanocages for the delivery of doxorubicin. The researchers demonstrated that these ferritin nanostructures were able to transport high drug doses for tumor-specific targeting and killing without requiring any additional functionalization or property modulation. Doxorubicin-loaded HFn internalized into tumor cells via interaction with overexpressed TfR1, releasing the drug in lysosomes and significantly inhibiting tumor growth after a single dose injection with minimum healthy organ drug exposure. A similar approach was followed by Jiang et al. [137], who have used doxorubicin-loaded ferritin nanosystems against hepatocellular carcinoma. The scientists also used a targeted ligand (i.e., GRP78-targeted peptide SP94) to improve release selectivity. The as-such designed platform was able to encapsulate a high amount of drug, ensuring a lower dosage of carrier and fewer adverse effects. Moreover, a better therapeutic effect was observed as compared to currently reported nanomedicines.

A different strategy is offered by Sitia et al. [138], who have encapsulated navitoclax and functionalized HFn nanocages with fibroblast activation protein (FAP) antibody fragments aiming to create cancer-associated fibroblast (CAF)-targeted drug delivery agents. The obtained results are promising as there were reported significantly higher drug release only in FAP$^+$ cells, and considerably higher cytotoxicity than non-functionalized systems. The researchers concluded it would be interesting to study if this nanoplatform is also able to reduce metastases formation.

3.2. *Polysaccharide-Based NPs*

3.2.1. Chitosan

Chitosan has many advantageous properties recommending it as a suitable delivery vehicle for a variety of biomolecules [139]. Being non-toxic, biodegradable, and biocompatible, chitosan became the material of choice for developing many nanoparticulate formulations for biomedical applications in general, and cancer management in particular [40].

For instance, chitosan has been demonstrated successful in improving the efficacy of photodynamic therapy. Ding et al. [140] have prepared chitosan NPs encapsulated with chlorin e6 photosensitizer by a nonsolvent-aided counterion complexation method, noting enhanced biocompatibility and dramatically increased therapy efficiency compared with free photosensitizer administration.

Chitosan NPs have also been remarked as a promising anticancer drug delivery platform [141]. Gounden et al. [142] have developed silver NPs conjugated with chitosan and loaded with cisplatin. The authors reported specificity towards breast cancer cells with minimal cytotoxicity towards normal cells, as the drug was efficiently and rapidly released from the nanosystem at low pH.

3.2.2. Fucoidan

Etman et al. [143] have used fucoidan extracted from *Undaria pinnatifida* to encapsulate quinacrine. Fucoidan acted as both delivery vehicle and active targeting ligand, while for some particles lactoferrin was added as a second active targeting ligand. The researchers obtained promising results for both single- and dual-targeting particles against pancreatic cancer, with a higher animal survival rate and no hepatotoxicity. In particular, dual-targeted particles were reported to enhance quinacrine activity 5.7-fold compared to free drug solution, having a higher ability in inhibiting cancer migration and invasion.

Jafari et al. [144] have also taken advantage of fucoidan's targeting ability. The authors used fucoidan-conjugated doxorubicin nanoparticles to target P-selectin overexpressed by malignant cells, obtaining enhanced cellular uptake and cytotoxicity against the MDA-MB-231 cell line.

Alternatively, Coutinho et al. [145] have combined fucoidan and chitosan to create an oral delivery system for methotrexate. The nanoplatform was reported safe towards fibroblasts but hindered lung cancer cell proliferation via an apoptotic process, being 7-times more effective than the free drug.

3.2.3. Alginate

Alginate has attracted research interest especially due to its biocompatibility, biodegradability, ease of production, and functionalization [146]. This natural polymer has been used as a base material for several novel anticancer formulations. As an example, Pourjavadi et al. [147] have employed this polysaccharide in the development of multifunctional nanocarriers. The researchers fabricated a magnetic core nanocarrier with a lipophilic surface based on oleic acid chains onto which paclitaxel and doxorubicin were adsorbed and covered by a smart pH-sensitive alginate shell. The system exhibited increased stability, enhanced biocompatibility, faster drug release in the acidic medium than at physiological pH, and even higher toxicity toward MCF-7 and HeLa cells than the free drugs.

3.2.4. Hyaluronic Acid

Hyaluronic acid (HA) is another polysaccharide [148] of interest for cancer drug delivery especially due to its ability to specifically bind CD44 receptor overexpressed by cancer cells [11]. For instance, Gaio et al. [149] have proposed the co-delivery of docetaxel and meso-tetraphenyl chlorine disulfonate ($TPCS_{2a}$) photosensitizer via HA-coated NPs. Combining chemotherapy and photodynamic therapy, their nanosystem demonstrated superior efficacy over monotherapies in lowering self-renewal capacity and inhibiting the growth of breast cancer CSCs.

Moreover, thiolated HA (HA-SS) is sensitive to glutathione [19], and is thus a potential targeting agent for TME-responsive delivery. In this respect, Debele et al. [150] have created a dual-sensitive HA-SS conjugated with 6-mercaptopurine for doxorubicin-targeted delivery to parental colon cancer and colon cancer CSCs. The synthesized nanoformulation was uptaken via CD44 receptor, accumulating more in the tumor region than in any other organ.

3.3. *Viral NPs*

The use of plant viruses in humans is considered a safe and promising alternative for intravital imaging and drug delivery [151]. Particularly, cowpea mosaic virus (CPMV) was reported advantageous due to its capsid's icosahedral shape, which allows enhanced multifunctional group display and endows the viral NP with the ability to carry specific cargos. In what concerns cancer therapies, CPMV was noted to have enhanced permeability and retention effect, allowing these viral NPs to preferentially extravasate from tumor neovasculature and efficiently penetrate tumors [152].

For instance, Lam et al. [153] have used CPMV for the delivery of mitoxantrone (MTO) antineoplastic chemotherapeutic to treat glioblastoma multiforme (GBM). The researchers reported CPMV-MTO uptake in glioma cells and significant in vitro cytotoxic effects both as a solo therapy and in combination with tumor necrosis factor-related apoptosis-inducing ligand, concluding that these plant virus-based NPs are promising platforms for GBM treatment.

Cowpea chlorotic mottle virus (CCMV) has also been exploited in developing innovative cancer treatments. Cai et al. [154] have employed this plant virus in the targeted delivery of oligodeoxynucleotides (ODN) with CpG motifs to tumor-associated macrophages (TAMs). This nanoformulation promoted ODN uptake by TAMS, enhancing their phagocytic activity. Moreover, the direct injection of these engineered NPs into tumor tissues induced a robust antitumor response increasing the phagocytic activity in the TME.

Another plant-based virus reported for anticancer drug delivery is tobacco mosaic virus (TMV). Franke et al. [155] used TMV as a nanocarrier of cisplatin as a potential treatment for platinum-resistant ovarian cancer. The scientists observed more efficient cell uptake, superior cytotoxicity and DNA double-strand breakage in both platinum-sensitive and platinum-resistant cancer cells than for free cisplatin, concluding that their newly developed nanoplatform may be a powerful tool in combating ovarian cancer.

Other viruses reported for cancer applications include potato virus X, red clover necrotic mosaic virus, papaya mosaic virus and physalis mottle virus [151].

3.4. *Exosomes*

Exosomes have recently emerged as potential nanocarriers of anticancer therapeutics due to their biocompatibility, low immunogenicity, long circulation time, and high loading capacity. These nanoscale extracellular vesicles benefit from an excellent tumor cell uptake and high specificity to tumor-associated cells [19,29].

Aqil et al. [156] have used bovine milk exosomes as nanocarriers for siRNA delivery. The scientists reported a dose-dependent antiproliferative activity against A549 cells treated with folic acid-functionalized exosomes loaded with $siKRAS^{G12S}$ gene, concluding that this nanocarrier is suitable for siRNA delivery and effective for tumor growth inhibition.

Another application of bovine milk exosomes is offered by Li et al. [157], who have encapsulated doxorubicin in these vesicles and decorated them with HA as a targeting ligand for the selective delivery into overexpressed CD44 tumor cells. Munagala et al. [158] have also developed drug-loaded exosomes for cancer therapy. The authors loaded various chemopreventive and chemotherapeutic agents (i.e., withaferin A, bilberry-derived anthocyanidins, curcumin, paclitaxel, and docetaxel) into exosomes and used folic acid to achieve tumor targeting, reporting enhanced biological efficacy, improved specificity, and elimination of off-target side effects of encapsulated drugs.

Furthermore, cancer cell-derived exosomes have unique characteristics that can be exploited in early cancer diagnosis, detecting highly metastatic cancer cells, and assessing cancer heterogeneity [159].

3.5. Other Natural NPs

Studies have also evaluated drug delivery nanoplatforms that do not fit under any of the above-presented categories. One such example is proposed by Carvalho et al. [160], who have encapsulated *Solanum lycocarpum* alkaloidic extract in natural lipid-based NPs with the aim of creating a better bladder cancer treatment strategy. The authors reported a sustained release profile 36 h after administration, antitumor activity in targeted cancer cells, and high antiproliferative activity, with a 5.4 times lower cell viability than with free extract.

Olive oil nanocapsules have also been described in the literature as efficient delivery vehicles for anticancer drugs. For instance, Galisteo-Gonzalez et al. [161] have developed nanocarriers with an olive oil core covered by a cross-linked HSA shell loaded with curcumin. The as described nanoplatforms had a similar IC_{50} value to that of free curcumin, but also avoided issues associated with excipients and displayed an excellent uptake performance (entered human breast cancer cells massively in less than one minute). Another approach of using olive oil nanocapsules was taken by Navarro-Marchal et al. [162], who have surrounded them with a different shell containing phospholipids, a nonionic surfactant and deoxycholic acid molecules, further coated with an αCD44 antibody. This complex nanosystem was loaded with paclitaxel, leading to high targeted uptake and increased antitumor efficacy (up to four times compared to free drug in pancreatic CSCs).

As it has been recently suggested that hydroxyapatite NPs have selective anticancer activity for lung cancer cells, this natural inorganic material can also be employed in designing targeted delivery vehicles [163]. Thus, Chen et al. [164] have developed hafnium-doped hydroxyapatite NPs able to enhance photodynamic therapy efficiency and tumor growth when bombarded with ionizing radiation. The authors concluded that their newly developed nanoplatform is suitable for palliative treatment after lung surgical intervention.

Recent progress has been reported in developing DNA nanocages for delivery purposes. For instance, Tam et al. [113] have designed self-assembled DNA nanocages functionalized with or without blood–brain barrier (BBB)-targeting ligands. Their nanocarriers were able to carry anticancer drugs and penetrate BBB to inhibit the tumor growth in a U-87 MG xenograft mouse model, being a safe and cost-effective targeted delivery platform for brain tumors. DNA nanocages were also proposed as multifunctional vehicles, such as platforms for the co-delivery of anti-miR-21 and doxorubicin [165], or fluorometric detection of human 8-oxoG DNA glycosylase 1 and doxorubicin delivery [166].

Another highly researched recent delivery concept in cancer therapy is the use of bio-inspired nanoparticles mimicking natural components in the body [7,167]. More specifically, cell membrane coating technology has emerged as a promising strategy to deliver drugs into tumors as it endows nanomaterials with functions and properties that are inherent to source cells [29,168]. In this respect, researchers started to disguise nanoparticles using the plasma membranes of various cell types, including erythrocytes [168–171], leukocytes [172,173], platelets [174,175], macrophages [176,177], bacterial cells [178], stem cells [179,180], cancer cells [181,182], and different hybrid cell coatings [183–186].

3.6. Summative Discussion

A variety of natural materials can be employed in developing nanoconstructs for safe and efficient anticancer therapies. Incorporating a wide range of either conventional drugs or natural bioactive molecules, natural nanocarriers have been studied against diverse cancer cell lines in vitro or as in vivo therapeutic approaches in animal models. Summarizing the above-discussed research studies, Figure 7 overviews the four main categories of nanocarriers in a clear and concise manner in relation to their delivered moieties and cancer types for which they have been investigated.

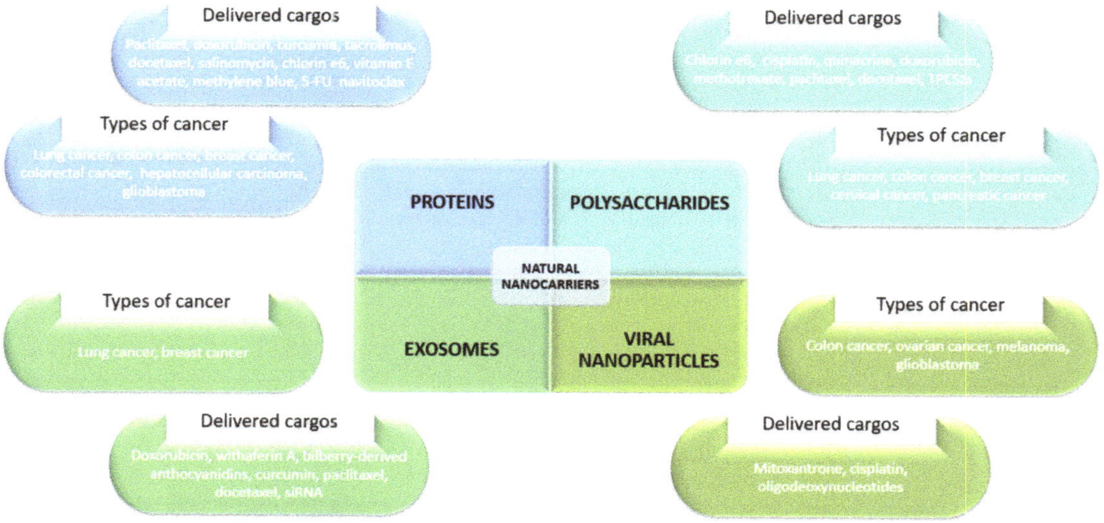

Figure 7. Overview of the main discussed categories of natural nanocarriers for cancer management.

Proteins and polysaccharides can be considered dual-role biomaterials as some of them can act both as delivery vehicles and active ingredients. Moreover, several exponents of these categories may be concomitantly used as active targeting ligands (e.g., fucoidan, hyaluronic acid, ferritin, and keratin). Nonetheless, additional research is needed for improving the yields of synthesis methods, reducing inter-batch variability, and reducing the costs implied by the design of precisely tailored NPs.

Viral NPs represent an interesting alternative as the use of plant viruses is considered harmless for drug delivery in humans. Moreover, these nanoplatforms have been reported successful against drug-resistant tumors. Nevertheless, their in vivo behavior is still difficult to predict, demanding an individual investigation of each viral-based delivery system.

Exosomes and biomimetic NPs have recently gained interest in the scientific world, being highly promising natural carriers. However, these nanostructures are still in their infancy in cancer management, requiring more research, especially concerning optimizing their preparation processes.

In addition, comparative data on the performance and treatment efficiency of various natural-origin-based NPs are still scarce in the literature [187]. In this context, further research should also be focused on testing several types of nanocarriers in the same experimental conditions in order to create a basis of understanding of which is the best material and design option for a given type of tumor.

In an attempt to compare at least from a qualitative point of view the discussed biomaterials, Table 3 summarizes the advantages and disadvantages of each class of the presented nanocarriers.

Table 3. Advantages and disadvantages of main discussed natural nanocarriers for cancer management.

Nanocarrier Type	Examples	Advantages	Disadvantages	Refs.
Proteins	Albumin, keratin, silk fibroin, silk sericin, ferritin	• Inherent targeting potential • Increased cellular uptake • Good affinity to anticancer drugs • High drug-binding capacity • High stability • Possibility of self-assembly • Ease of functionalization • Possibility of use in photodynamic therapy	• Batch-to-batch variations • Low yields • High costs • Some may cause in vivo inflammation	[11,109,116,117,132,133,135,136,187–189]
Polysaccharides	Chitosan, fucoidan, alginate, hyaluronic acid	• Wide availability • Ease of production • Ease of functionalization • Possibility of developing multifunctional carriers • Some may act as targeting agents • High stability • Possibility of use in photodynamic therapy	• Complicated manufacturing process • Unclear metabolism in the body • Poor solubility in common solvents	[11,109,140,143,144,146,147,190,191]
Viral NPs	Cowpea mosaic virus, cowpea chlorotic mosaic virus, tobacco mosaic virus, potato virus X, red clover necrotic mosaic virus, papaya mosaic virus, physalis mottle virus	• Enhanced multifunctional group display • Some offer the possibility to treat platinum-resistant cancers • Possibility of use in photodynamic therapy • Can be used for intravital imaging, drug and gene delivery, and immunotherapy	• Difficult to predict in vivo behavior; each viral-based delivery system must be evaluated on a case-by-case basis prior to clinical testing	[151,192,193]
Exosomes	Bovine milk exosomes, cancer cell-derived exosomes	• High loading capacity for drugs and genes • High stability • Inherent targeting potential • Excellent tumor cell uptake • High specificity to tumor-associated cells • Enhanced permeability and retention effect • Can be employed in cancer diagnosis	• Lack of consensus on the best method for obtaining high yields of pure exosomes • Challenging to load cargos and targeting agents without corrupting exosomes	[19,29,159,194]
Biomimetic NPs	NPs coated with the membranes of red blood cells, white blood cells, platelets, macrophages, bacterial cells, stem cells, cancer cells	• Unique construction • Long lifespan • Excellent targeting ability • Enhanced drug retention • Retained cytotoxicity to cancer cells	• Some may raise immunogenicity concerns • Challenging translation from bench to bedside • Complex preparation process • Low yields	[29,167,168,195]

Moreover, natural NPs are considered attractive for developing theranostic systems. Through an interdisciplinary approach, multifunctional and multivalent nanostructures can be created towards generating innovative and performant cancer therapies [7,152]. Specifically, several of the discussed nanocarriers have been reported able to encapsulate and co-deliver imaging moieties, drugs, and genes, and even detect cancer cells by binding to specific receptors. Thus, complex natural-based nanoparticulated formulations may hold the answer for effective cancer management.

4. Conclusions

To summarize, cancers remain a global health concern, requiring urgently improved therapeutic strategies. As the current treatment options are limited by invasiveness, severe adverse effects, and poor patient compliance, increasing scientific interest has been shifted to developing carrier systems able to ensure targeted delivery and controlled drug release. Numerous studies have revolved around various synthetic and natural nanomaterials as delivery platforms, yet, in recent years, the latter started to gain more ground. Alternatively, several natural anticancer compounds have been reported to be fabricated into nanoparticles, eliminating the need for synthetic chemotherapeutic agents.

Thus, through an interdisciplinary approach, better treatment strategies can be envisaged in the foreseeable future. Nonetheless, before applying the described nanoformulations in the clinic, more detailed research has to be performed concerning their efficacy and safety in human use. Moreover, further progress in drug encapsulation optimization, ligand conjugation efficiency, and high reproducibility fabrication with low costs is needed prior to undergoing mass production.

Author Contributions: All authors participated in the review, writing, and revision. All authors have read and agreed to the published version of the manuscript.

Funding: This research received no external funding.

Institutional Review Board Statement: Not applicable.

Informed Consent Statement: Not applicable.

Data Availability Statement: Not applicable.

Acknowledgments: Not applicable.

Conflicts of Interest: The authors declare no conflict of interest.

References

1. Gao, S.; Yang, D.; Fang, Y.; Lin, X.; Jin, X.; Wang, Q.; Wang, X.; Ke, L.; Shi, K. Engineering Nanoparticles for Targeted Remodeling of the Tumor Microenvironment to Improve Cancer Immunotherapy. *Theranostics* **2019**, *9*, 126–151. [CrossRef]
2. Davatgaran-Taghipour, Y.; Masoomzadeh, S.; Farzaei, M.H.; Bahramsoltani, R.; Karimi-Soureh, Z.; Rahimi, R.; Abdollahi, M. Polyphenol nanoformulations for cancer therapy: Experimental evidence and clinical perspective. *Int. J. Nanomed.* **2017**, *12*, 2689–2702. [CrossRef] [PubMed]
3. Sayiner, M.; Golabi, P.; Younossi, Z.M. Disease Burden of Hepatocellular Carcinoma: A Global Perspective. *Dig. Dis. Sci.* **2019**, *64*, 910–917. [CrossRef] [PubMed]
4. Bray, F.; Ferlay, J.; Soerjomataram, I.; Siegel, R.L.; Torre, L.A.; Jemal, A. Global cancer statistics 2018: GLOBOCAN estimates of incidence and mortality worldwide for 36 cancers in 185 countries. *CA Cancer J. Clin.* **2018**, *68*, 394–424. [CrossRef]
5. Arifin, M.Z.; Parikesit, A.A.; Agustriawan, D. Molecular simulation oF MDM2 and E6AP proteins as P53 regulator in cervical cancer. *Biointerface Res. Appl. Chem.* **2020**, *10*, 5875–5879. [CrossRef]
6. Naeimi, M.; Mollaamin, F.; Monajjemi, M. Removing skin-cancer damaging based on destroying thymine dimer complexes. *Biointerface Res. Appl. Chem.* **2020**, *10*, 5696–5703. [CrossRef]
7. Madamsetty, V.S.; Mukherjee, A.; Mukherjee, S. Recent trends of the bio-inspired nanoparticles in cancer theranostics. *Front. Pharmacol.* **2019**, *10*, 1264. [CrossRef]
8. Choi, D.G.; Venkatesan, J.; Shim, M.S. Selective Anticancer Therapy Using Pro-Oxidant Drug-Loaded Chitosan–Fucoidan Nanoparticles. *Int. J. Mol. Sci.* **2019**, *20*, 3220. [CrossRef] [PubMed]
9. Abdelaziz, H.M.; Gaber, M.; Abd-Elwakil, M.M.; Mabrouk, M.T.; Elgohary, M.M.; Kamel, N.M.; Kabary, D.M.; Freag, M.S.; Samaha, M.W.; Mortada, S.M.; et al. Inhalable particulate drug delivery systems for lung cancer therapy: Nanoparticles, microparticles, nanocomposites and nanoaggregates. *J. Control. Release* **2018**, *269*, 374–392. [CrossRef]
10. Yagawa, Y.; Tanigawa, K.; Kobayashi, Y.; Yamamoto, M. Cancer immunity and therapy using hyperthermia with immunotherapy, radiotherapy, chemotherapy, and surgery. *J. Cancer Metastasis Treat.* **2017**, *3*, 218–230. [CrossRef]
11. Curcio, M.; Diaz-Gomez, L.; Cirillo, G.; Nicoletta, F.P.; Leggio, A.; Iemma, F. Dual-Targeted Hyaluronic Acid/Albumin Micelle-Like Nanoparticles for the Vectorization of Doxorubicin. *Pharmaceutics* **2021**, *13*, 304. [CrossRef]

12. Khan, H.; Ullah, H.; Martorell, M.; Valdes, S.E.; Belwal, T.; Tejada, S.; Sureda, A.; Kamal, M.A. Flavonoids nanoparticles in cancer: Treatment, prevention and clinical prospects. *Semin. Cancer Biol.* **2021**, *69*, 200–211. [CrossRef]
13. Amjad, M.T.; Kasi, A. Cancer Chemotherapy. StatPearls. 2020. Available online: https://www.ncbi.nlm.nih.gov/books/NBK564367/ (accessed on 10 November 2021).
14. Bukowski, K.; Kciuk, M.; Kontek, R. Mechanisms of Multidrug Resistance in Cancer Chemotherapy. *Int. J. Mol. Sci.* **2020**, *21*, 3233. [CrossRef]
15. Mansoori, B.; Mohammadi, A.; Davudian, S.; Shirjang, S.; Baradaran, B. The Different Mechanisms of Cancer Drug Resistance: A Brief Review. *Adv. Pharm. Bull.* **2017**, *7*, 339–348. [CrossRef] [PubMed]
16. Alshehri, A.A.; Almughem, F.A.; Aldossary, A.M.; Tawfik, E.A.; Al-Fahad, A.J.; Alyahya, S.; Alomary, M.N. Microbial Nanoparticles for Cancer Treatment. In *Microbial Nanotechnology: Green Synthesis and Applications*; Ansari, M.A., Rehman, S., Eds.; Springer: Singapore, 2021; pp. 217–235. [CrossRef]
17. Barazzuol, L.; Coppes, R.P.; van Luijk, P. Prevention and treatment of radiotherapy-induced side effects. *Mol. Oncol.* **2020**, *14*, 1538–1554. [CrossRef] [PubMed]
18. Biondi, A.; Lirosi, M.C.; D'Ugo, D.; Fico, V.; Ricci, R.; Santullo, F.; Rizzuto, A.; Cananzi, F.C.; Persiani, R. Neo-adjuvant chemo(radio)therapy in gastric cancer: Current status and future perspectives. *World J. Gastrointest. Oncol.* **2015**, *7*, 389–400. [CrossRef]
19. Ertas, Y.N.; Abedi Dorcheh, K.; Akbari, A.; Jabbari, E. Nanoparticles for Targeted Drug Delivery to Cancer Stem Cells: A Review of Recent Advances. *Nanomaterials* **2021**, *11*, 1755. [CrossRef] [PubMed]
20. Chidambaram, M.; Manavalan, R.; Kathiresan, K. Nanotherapeutics to overcome conventional cancer chemotherapy limitations. *J. Pharm. Pharm. Sci.* **2011**, *14*, 67–77. [CrossRef]
21. Powell, J.W.; Dexter, E.; Scalzetti, E.M.; Bogart, J.A. Treatment advances for medically inoperable non-small-cell lung cancer: Emphasis on prospective trials. *Lancet Oncol.* **2009**, *10*, 885–894. [CrossRef]
22. Kareliotis, G.; Tremi, I.; Kaitatzi, M.; Drakaki, E.; Serafetinides, A.A.; Makropoulou, M.; Georgakilas, A.G. Combined radiation strategies for novel and enhanced cancer treatment. *Int. J. Radiat. Biol.* **2020**, *96*, 1087–1103. [CrossRef] [PubMed]
23. Monje, M.; Dietrich, J. Cognitive side effects of cancer therapy demonstrate a functional role for adult neurogenesis. *Behav. Brain Res.* **2012**, *227*, 376–379. [CrossRef]
24. Vallejo, R.; Hord, E.D.; Barna, S.A.; Santiago-Palma, J.; Ahmed, S. Perioperative immunosuppression in cancer patients. *J. Environ. Pathol. Toxicol. Oncol.* **2003**, *22*, 8–46. [CrossRef]
25. Wo, J.Y.; Viswanathan, A.N. Impact of Radiotherapy on Fertility, Pregnancy, and Neonatal Outcomes in Female Cancer Patients. *Int. J. Radiat. Oncol.* **2009**, *73*, 1304–1312. [CrossRef] [PubMed]
26. Alphandéry, E. Natural Metallic Nanoparticles for Application in Nano-Oncology. *Int. J. Mol. Sci.* **2020**, *21*, 4412. [CrossRef] [PubMed]
27. Kanchi, S.; Inamuddin; Khan, A. Biogenic Synthesis of Selenium Nanoparticles with Edible Mushroom Extract: Evaluation of Cytotoxicity on Prostate Cancer Cell Lines and Their Antioxidant, and Antibacterial Activity. *Biointerface Res. Appl. Chem.* **2020**, *10*, 6629–6639. [CrossRef]
28. Aiello, P.; Consalvi, S.; Poce, G.; Raguzzini, A.; Toti, E.; Palmery, M.; Biava, M.; Bernardi, M.; Kamal, M.A.; Perry, G. Dietary flavonoids: Nano delivery and nanoparticles for cancer therapy. *Semin. Cancer Biol.* **2021**, *69*, 150–165. [CrossRef]
29. De la Torre, P.; Pérez-Lorenzo, M.J.; Alcázar-Garrido, Á.; Flores, A.I. Cell-based nanoparticles delivery systems for targeted cancer therapy: Lessons from anti-angiogenesis treatments. *Molecules* **2020**, *25*, 715. [CrossRef] [PubMed]
30. Lungu, I.I.; Grumezescu, A.M.; Volceanov, A.; Andronescu, E. Nanobiomaterials Used in Cancer Therapy: An Up-To-Date Overview. *Molecules* **2019**, *24*, 3547. [CrossRef]
31. Baghban, R.; Roshangar, L.; Jahanban-Esfahlan, R.; Seidi, K.; Ebrahimi-Kalan, A.; Jaymand, M.; Kolahian, S.; Javaheri, T.; Zare, P. Tumor microenvironment complexity and therapeutic implications at a glance. *Cell Commun. Signal.* **2020**, *18*, 59. [CrossRef]
32. Jin, M.-Z.; Jin, W.-L. The updated landscape of tumor microenvironment and drug repurposing. *Signal Transduct. Target. Ther.* **2020**, *5*, 166. [CrossRef] [PubMed]
33. Cheng, Z.; Li, M.; Dey, R.; Chen, Y. Nanomaterials for cancer therapy: Current progress and perspectives. *J. Hematol. Oncol.* **2021**, *14*, 85. [CrossRef]
34. Pham, S.H.; Choi, Y.; Choi, J. Stimuli-Responsive Nanomaterials for Application in Antitumor Therapy and Drug Delivery. *Pharmaceutics* **2020**, *12*, 630. [CrossRef] [PubMed]
35. Zhou, L.; Wang, H.; Li, Y. Stimuli-Responsive Nanomedicines for Overcoming Cancer Multidrug Resistance. *Theranostics* **2018**, *8*, 1059–1074. [CrossRef]
36. Saha, R.; Verbanic, S.; Chen, I.A. Lipid vesicles chaperone an encapsulated RNA aptamer. *Nat. Commun.* **2018**, *9*, 2313. [CrossRef]
37. AlMalki, F.A.; Hassan, A.M.; Klaab, Z.M.; Abdulla, S.; Pizzi, A. Tannin Nanoparticles (NP99) Enhances the Anticancer Effect of Tamoxifen on ER [supercript]+[/supercript] Breast Cancer Cells. *J. Renew. Mater.* **2021**, *9*, 2077. [CrossRef]
38. Hua, S.; de Matos, M.B.C.; Metselaar, J.M.; Storm, G. Current Trends and Challenges in the Clinical Translation of Nanoparticulate Nanomedicines: Pathways for Translational Development and Commercialization. *Front. Pharmacol.* **2018**, *9*, 790. [CrossRef] [PubMed]
39. Saw, P.E.; Lee, S.; Jon, S. Naturally Occurring Bioactive Compound-Derived Nanoparticles for Biomedical Applications. *Adv. Ther.* **2019**, *2*, 1800146. [CrossRef]

40. Almutairi, F.M.; El Rabey, H.A.; Tayel, A.A.; Alalawy, A.I.; Al-Duais, M.A.; Sakran, M.I.; Zidan, N.S. Augmented anticancer activity of curcumin loaded fungal chitosan nanoparticles. *Int. J. Biol. Macromol.* **2020**, *155*, 861–867. [CrossRef]
41. Chen, Z.; Farag, M.A.; Zhong, Z.; Zhang, C.; Yang, Y.; Wang, S.; Wang, Y. Multifaceted role of phyto-derived polyphenols in nanodrug delivery systems. *Adv. Drug Deliv. Rev.* **2021**, *176*, 113870. [CrossRef]
42. Noormand, F.; Kermani, A.S.; Raviz, E.K.; Esmaeilpour, K.; Golshani, M.; Bashiri, H.; Kalantaripour, T.P.; Asadi-Shekaari, M. Investigating the neuroprotective effects of Resveratrol on encephalopathy induced by bile duct ligation in male rats. *Biointerface Res. Appl. Chem.* **2020**, *10*, 5512–5515. [CrossRef]
43. Abbas, M.; Saeed, F.; Anjum, F.M.; Afzaal, M.; Tufail, T.; Bashir, M.S.; Ishtiaq, A.; Hussain, S.; Suleria, H.A.R. Natural polyphenols: An overview. *Int. J. Food Prop.* **2017**, *20*, 1689–1699. [CrossRef]
44. Basli, A.; Belkacem, N.; Amrani, I. Health benefits of phenolic compounds against cancers. In *Phenolic Compounds–Biological activity*; IntechOpen: London, UK, 2017; pp. 193–210.
45. Beconcini, D.; Felice, F.; Fabiano, A.; Sarmento, B.; Zambito, Y.; Di Stefano, R. Antioxidant and Anti-Inflammatory Properties of Cherry Extract: Nanosystems-Based Strategies to Improve Endothelial Function and Intestinal Absorption. *Foods* **2020**, *9*, 207. [CrossRef]
46. Belščak-Cvitanović, A.; Durgo, K.; Huđek, A.; Bačun-Družina, V.; Komes, D. 1—Overview of polyphenols and their properties. In *Polyphenols: Properties, Recovery, and Applications*; Galanakis, C.M., Ed.; Woodhead Publishing: Duxford, UK, 2018; pp. 3–44. [CrossRef]
47. Dobrzynska, M.; Napierala, M.; Florek, E. Flavonoid Nanoparticles: A Promising Approach for Cancer Therapy. *Biomolecules* **2020**, *10*, 1268. [CrossRef] [PubMed]
48. Mohammadian, F.; Pilehvar-Soltanahmadi, Y.; Mofarrah, M.; Dastani-Habashi, M.; Zarghami, N. Down regulation of miR-18a, miR-21 and miR-221 genes in gastric cancer cell line by chrysin-loaded PLGA-PEG nanoparticles. *Artif. Cells Nanomed. Biotechnol.* **2016**, *44*, 1972–1978. [CrossRef]
49. Mutha, R.E.; Surana, S.J. Ultrasonic frequency based development of chrysin nanoparticles: Assessment of bioavailability, anti-cancer activity and stability. *Mater. Technol.* **2018**, *33*, 495–505. [CrossRef]
50. Tavakoli, F.; Jahanban-Esfahlan, R.; Seidi, K.; Jabbari, M.; Behzadi, R.; Pilehvar-Soltanahmadi, Y.; Zarghami, N. Effects of nano-encapsulated curcumin-chrysin on telomerase, MMPs and TIMPs gene expression in mouse B16F10 melanoma tumour model. *Artif. Cells Nanomed. Biotechnol.* **2018**, *46*, 75–86. [CrossRef] [PubMed]
51. Kakran, M.; Sahoo, N.G.; Li, L.; Judeh, Z. Fabrication of quercetin nanoparticles by anti-solvent precipitation method for enhanced dissolution. *Powder Technol.* **2012**, *223*, 59–64. [CrossRef]
52. Salehi, B.; Machin, L.; Monzote, L.; Sharifi-Rad, J.; Ezzat, S.M.; Salem, M.A.; Merghany, R.M.; El Mahdy, N.M.; Kılıç, C.S.; Sytar, O.; et al. Therapeutic Potential of Quercetin: New Insights and Perspectives for Human Health. *ACS Omega* **2020**, *5*, 11849–11872. [CrossRef] [PubMed]
53. Aljadaan, S.A.N.; Elias, R.S.; Al-Anssari, R.A. Investigation of the Antioxidant and Antibacterial Activity of Novel Quercetin Derivatives. *Biointerface Res. Appl. Chem.* **2020**, *10*, 7329–7336. [CrossRef]
54. Jiang, W.; Zhang, H.; Wu, J.; Zhai, G.; Li, Z.; Luan, Y.; Garg, S. CuS@MOF-Based Well-Designed Quercetin Delivery System for Chemo–Photothermal Therapy. *ACS Appl. Mater. Interfaces* **2018**, *10*, 34513–34523. [CrossRef]
55. Rezaei-Sadabady, R.; Eidi, A.; Zarghami, N.; Barzegar, A. Intracellular ROS protection efficiency and free radical-scavenging activity of quercetin and quercetin-encapsulated liposomes. *Artif. Cells Nanomed. Biotechnol.* **2016**, *44*, 128–134. [CrossRef] [PubMed]
56. Sadhukhan, P.; Kundu, M.; Chatterjee, S.; Ghosh, N.; Manna, P.; Das, J.; Sil, P.C. Targeted delivery of quercetin via pH-responsive zinc oxide nanoparticles for breast cancer therapy. *Mater. Sci. Eng. C* **2019**, *100*, 129–140. [CrossRef]
57. Wang, Y.; Yu, H.; Wang, S.; Gai, C.; Cui, X.; Xu, Z.; Li, W.; Zhang, W. Targeted delivery of quercetin by nanoparticles based on chitosan sensitizing paclitaxel-resistant lung cancer cells to paclitaxel. *Mater. Sci. Eng.* **2021**, *119*, 111442. [CrossRef]
58. De Oliveira Pedro, R.; Hoffmann, S.; Pereira, S.; Goycoolea, F.M.; Schmitt, C.C.; Neumann, M.G. Self-assembled amphiphilic chitosan nanoparticles for quercetin delivery to breast cancer cells. *Eur. J. Pharm. Biopharm.* **2018**, *131*, 203–210. [CrossRef] [PubMed]
59. Shitole, A.A.; Sharma, N.; Giram, P.; Khandwekar, A.; Baruah, M.; Garnaik, B.; Koratkar, S. LHRH-conjugated, PEGylated, poly-lactide-co-glycolide nanocapsules for targeted delivery of combinational chemotherapeutic drugs Docetaxel and Quercetin for prostate cancer. *Mater. Sci. Eng. C* **2020**, *114*, 111035. [CrossRef]
60. Van Thoai, D.; Nguyen, D.T.; Dang, L.H.; Nguyen, N.H.; Nguyen, V.T.; Doan, P.; Nguyen, B.T.; Le Van, T.; Tung, N.N.; Quyen, T.N. Lipophilic effect of various pluronic-grafted gelatin copolymers on the quercetin delivery efficiency in these self-assembly nanogels. *J. Polym. Res.* **2020**, *27*, 369 [CrossRef]
61. Fang, J.; Zhang, S.; Xue, X.; Zhu, X.; Song, S.; Wang, B.; Jiang, L.; Qin, M.; Liang, H.; Gao, L. Quercetin and doxorubicin co-delivery using mesoporous silica nanoparticles enhance the efficacy of gastric carcinoma chemotherapy. *Int. J. Nanomed.* **2018**, *13*, 5113–5126. [CrossRef]
62. Wang, W.; Xi, M.; Duan, X.; Wang, Y.; Kong, F. Delivery of baicalein and paclitaxel using self-assembled nanoparticles: Synergistic antitumor effect In Vitro and In Vivo *Int. J. Nanomed.* **2015**, *10*, 3737–3750. [CrossRef]
63. Liu, H.; Dong, Y.; Gao, Y.; Du, Z.; Wang, Y.; Cheng, P.; Chen, A.; Huang, H. The Fascinating Effects of Baicalein on Cancer: A Review. *Int. J. Mol. Sci.* **2016**, *17*, 1681. [CrossRef] [PubMed]

64. Joshi, H.A.; Patwardhan, R.S.; Sharma, D.; Sandur, S.K.; Devarajan, P.V. Pre-clinical evaluation of an innovative oral nano-formulation of baicalein for modulation of radiation responses. *Int. J. Pharm.* **2021**, *595*, 120181. [CrossRef] [PubMed]
65. Cheng, Z.; Zhang, Z.; Han, Y.; Wang, J.; Wang, Y.; Chen, X.; Shao, Y.; Cheng, Y.; Zhou, W.; Lu, X.; et al. A review on anti-cancer effect of green tea catechins. *J. Funct. Foods* **2020**, *74*, 104172. [CrossRef]
66. Granja, A.; Pinheiro, M.; Reis, S. Epigallocatechin Gallate Nanodelivery Systems for Cancer Therapy. *Nutrients* **2016**, *8*, 307. [CrossRef] [PubMed]
67. Dai, W.; Ruan, C.; Zhang, Y.; Wang, J.; Han, J.; Shao, Z.; Sun, Y.; Liang, J. Bioavailability enhancement of EGCG by structural modification and nano-delivery: A review. *J. Funct. Foods* **2020**, *65*, 103732. [CrossRef]
68. Mostafa, S.M.; Gamal-Eldeen, A.M.; Maksoud, N.A.E.; Fahmi, A.A. Epigallocatechin gallate-capped gold nanoparticles enhanced the tumor suppressors let-7a and miR-34a in hepatocellular carcinoma cells. *An. Da Acad. Bras. De Ciências* **2020**, *92*, e20200574. [CrossRef] [PubMed]
69. Chavva, S.R.; Deshmukh, S.K.; Kanchanapally, R.; Tyagi, N.; Coym, J.W.; Singh, A.P.; Singh, S. *EGCG-Gold Nanoparticles Exhibit Greater Anti-Tumor Activity over Conventional Gold Nanoparticles or EGCG Due to Potential Synergistic Interactions*; AACR: Atlanta, GA, USA, 2019.
70. Mukherjee, S.; Ghosh, S.; Das, D.K.; Chakraborty, P.; Choudhury, S.; Gupta, P.; Adhikary, A.; Dey, S.; Chattopadhyay, S. Gold-conjugated green tea nanoparticles for enhanced anti-tumor activities and hepatoprotection—Synthesis, characterization and in vitro evaluation. *J. Nutr. Biochem.* **2015**, *26*, 1283–1297. [CrossRef] [PubMed]
71. Zhang, L.; Chen, W.; Tu, G.; Chen, X.; Lu, Y.; Wu, L.; Zheng, D. Enhanced Chemotherapeutic Efficacy of PLGA-Encapsulated Epigallocatechin Gallate (EGCG) Against Human Lung Cancer. *Int. J. Nanomed.* **2020**, *15*, 4417–4429. [CrossRef]
72. Kazi, J.; Sen, R.; Ganguly, S.; Jha, T.; Ganguly, S.; Chatterjee Debnath, M. Folate decorated epigallocatechin-3-gallate (EGCG) loaded PLGA nanoparticles; in-vitro and in-vivo targeting efficacy against MDA-MB-231 tumor xenograft. *Int. J. Pharm.* **2020**, *585*, 119449. [CrossRef]
73. Sanna, V.; Singh, C.K.; Jashari, R.; Adhami, V.M.; Chamcheu, J.C.; Rady, I.; Sechi, M.; Mukhtar, H.; Siddiqui, I.A. Targeted nanoparticles encapsulating (−)-epigallocatechin-3-gallate for prostate cancer prevention and therapy. *Sci. Rep.* **2017**, *7*, 41573. [CrossRef]
74. Shutava, T.G.; Balkundi, S.S.; Vangala, P.; Steffan, J.J.; Bigelow, R.L.; Cardelli, J.A.; O'Neal, D.P.; Lvov, Y.M. Layer-by-layer-coated gelatin nanoparticles as a vehicle for delivery of natural polyphenols. *ACS Nano* **2009**, *3*, 1877–1885. [CrossRef] [PubMed]
75. Kulandaivelu, K.; Mandal, A.K.A. Improved bioavailability and pharmacokinetics of tea polyphenols by encapsulation into gelatin nanoparticles. *IET Nanobiotechnol.* **2017**, *11*, 469–476. [CrossRef]
76. Safer, A.-M.; Leporatti, S.; Jose, J.; Soliman, M.S. Conjugation Of EGCG And Chitosan NPs As A Novel Nano-Drug Delivery System. *Int. J. Nanomed.* **2019**, *14*, 8033–8046. [CrossRef] [PubMed]
77. Liang, J.; Yan, H.; Puligundla, P.; Gao, X.; Zhou, Y.; Wan, X. Applications of chitosan nanoparticles to enhance absorption and bioavailability of tea polyphenols: A review. *Food Hydrocoll.* **2017**, *69*, 286–292. [CrossRef]
78. De Pace, R.C.; Liu, X.; Sun, M.; Nie, S.; Zhang, J.; Cai, Q.; Gao, W.; Pan, X.; Fan, Z.; Wang, S. Anticancer activities of (-)-epigallocatechin-3-gallate encapsulated nanoliposomes in MCF7 breast cancer cells. *J. Liposome Res.* **2013**, *23*, 187–196. [CrossRef]
79. Ramadass, S.K.; Anantharaman, N.V.; Subramanian, S.; Sivasubramanian, S.; Madhan, B. Paclitaxel/epigallocatechin gallate coloaded liposome: A synergistic delivery to control the invasiveness of MDA-MB-231 breast cancer cells. *Colloids Surfaces B Biointerfaces* **2015**, *125*, 65–72. [CrossRef] [PubMed]
80. Marwah, M.; Perrie, Y.; Badhan, R.K.S.; Lowry, D. Intracellular uptake of EGCG-loaded deformable controlled release liposomes for skin cancer. *J. Liposome Res.* **2020**, *30*, 136–149. [CrossRef]
81. Zununi Vahed, S.; Salehi, R.; Davaran, S.; Sharifi, S. Liposome-based drug co-delivery systems in cancer cells. *Mater. Sci. Eng. C* **2017**, *71*, 1327–1341. [CrossRef]
82. Chu, P.-Y.; Tsai, S.-C.; Ko, H.-Y.; Wu, C.-C.; Lin, Y.-H. Co-Delivery of Natural Compounds with a Dual-Targeted Nanoparticle Delivery System for Improving Synergistic Therapy in an Orthotopic Tumor Model. *ACS Appl. Mater. Interfaces* **2019**, *11*, 23880–23892. [CrossRef] [PubMed]
83. Mi, F.-L.; Wang, L.-F.; Chu, P.-Y.; Peng, S.-L.; Feng, C.-L.; Lai, Y.-J.; Li, J.-N.; Lin, Y.-H. Active Tumor-Targeted co-Delivery of Epigallocatechin Gallate and Doxorubicin in Nanoparticles for Combination Gastric Cancer Therapy. *ACS Biomater. Sci. Eng.* **2018**, *4*, 2847–2859. [CrossRef] [PubMed]
84. Nguyen, V.B.; Ton, T.Q.; Nguyen, D.N.; Nguyen, T.T.; Ngu, T.N.; Nguyen, T.H.; Doan, C.T.; Tran, T.N.; Nguyen, M.T.; Ho, N.D.; et al. Reclamation of beneficial bioactivities of herbal antioxidant condensed tannin extracted from Euonymus laxiflorus. *Res. Chem. Intermed.* **2020**, *46*, 4751–4766. [CrossRef]
85. Ke, H.; Wang, X.; Zhou, Z.; Ai, W.; Wu, Z.; Zhang, Y. Effect of weimaining on apoptosis and Caspase-3 expression in a breast cancer mouse model. *J. Ethnopharmacol.* **2021**, *264*, 113363. [CrossRef]
86. Rajasekar, N.; Sivanantham, A.; Ravikumar, V.; Rajasekaran, S. An overview on the role of plant-derived tannins for the treatment of lung cancer. *Phytochemistry* **2021**, *188*, 112799. [CrossRef] [PubMed]
87. Vervandier-Fasseur, D.; Latruffe, N. The Potential Use of Resveratrol for Cancer Prevention. *Molecules* **2019**, *24*, 4506. [CrossRef] [PubMed]
88. Ndiaye, M.; Kumar, R.; Ahmad, N. Resveratrol in cancer management: Where are we and where we go from here? *Ann. N. Y. Acad. Sci.* **2011**, *1215*, 144–149. [CrossRef] [PubMed]

89. Thipe, V.C.; Panjtan Amiri, K.; Bloebaum, P.; Raphael Karikachery, A.; Khoobchandani, M.; Katti, K.K.; Jurisson, S.S.; Katti, K.V. Development of resveratrol-conjugated gold nanoparticles: Interrelationship of increased resveratrol corona on anti-tumor efficacy against breast, pancreatic and prostate cancers. *Int. J. Nanomed.* **2019**, *14*, 4413–4428. [CrossRef] [PubMed]
90. Wang, W.; Zhang, L.; Chen, T.; Guo, W.; Bao, X.; Wang, D.; Ren, B.; Wang, H.; Li, Y.; Wang, Y.; et al. Anticancer Effects of Resveratrol-Loaded Solid Lipid Nanoparticles on Human Breast Cancer Cells. *Molecules* **2017**, *22*, 1814. [CrossRef] [PubMed]
91. Chaudhary, Z.; Subramaniam, S.; Khan, G.M.; Abeer, M.M.; Qu, Z.; Janjua, T.; Kumeria, T.; Batra, J.; Popat, A. Encapsulation and Controlled Release of Resveratrol Within Functionalized Mesoporous Silica Nanoparticles for Prostate Cancer Therapy. *Front. Bioeng. Biotechnol.* **2019**, *7*, 225. [CrossRef]
92. Karthikeyan, S.; Rajendra Prasad, N.; Ganamani, A.; Balamurugan, E. Anticancer activity of resveratrol-loaded gelatin nanoparticles on NCI-H460 non-small cell lung cancer cells. *Biomed. Prev. Nutr.* **2013**, *3*, 64–73. [CrossRef]
93. Karthikeyan, S.; Hoti, S.L.; Prasad, N.R. Resveratrol loaded gelatin nanoparticles synergistically inhibits cell cycle progression and constitutive NF-kappaB activation, and induces apoptosis in non-small cell lung cancer cells. *Biomed. Pharmacother.* **2015**, *70*, 274–282. [CrossRef] [PubMed]
94. Senthil Kumar, C.; Thangam, R.; Mary, S.A.; Kannan, P.R.; Arun, G.; Madhan, B. Targeted delivery and apoptosis induction of trans-resveratrol-ferulic acid loaded chitosan coated folic acid conjugate solid lipid nanoparticles in colon cancer cells. *Carbohydr. Polym.* **2020**, *231*, 115682. [CrossRef] [PubMed]
95. Nassir, A.M.; Shahzad, N.; Ibrahim, I.A.A.; Ahmad, I.; Md, S.; Ain, M.R. Resveratrol-loaded PLGA nanoparticles mediated programmed cell death in prostate cancer cells. *Saudi Pharm. J.* **2018**, *26*, 876–885. [CrossRef] [PubMed]
96. Wang, X.; Parvathaneni, V.; Shukla, S.K.; Kulkarni, N.S.; Muth, A.; Kunda, N.K.; Gupta, V. Inhalable resveratrol-cyclodextrin complex loaded biodegradable nanoparticles for enhanced efficacy against non-small cell lung cancer. *Int. J. Biol. Macromol.* **2020**, *164*, 638–650. [CrossRef]
97. Mansouri, K.; Rasoulpoor, S.; Daneshkhah, A.; Abolfathi, S.; Salari, N.; Mohammadi, M.; Rasoulpoor, S.; Shabani, S. Clinical effects of curcumin in enhancing cancer therapy: A systematic review. *BMC Cancer* **2020**, *20*, 791. [CrossRef] [PubMed]
98. Arzani, H.; Adabi, M.; Mosafer, J.; Dorkoosh, F.; Khosravani, M.; Maleki, H.; Nekounam, H.; Kamali, M. Preparation of curcumin-loaded PLGA nanoparticles and investigation of its cytotoxicity effects on human glioblastoma U87MG cells. *Biointerface Res. Appl. Chem.* **2019**, *9*, 4225–4231.
99. Matloubi, Z.; Hassan, Z. HSA-curcumin nanoparticles: A promising substitution for Curcumin as a Cancer chemoprevention and therapy. *DARU J. Pharm. Sci.* **2020**, *28*, 209–219. [CrossRef]
100. Duse, L.; Baghdan, E.; Pinnapireddy, S.R.; Engelhardt, K.H.; Jedelská, J.; Schaefer, J.; Quendt, P.; Bakowsky, U. Preparation and Characterization of Curcumin Loaded Chitosan Nanoparticles for Photodynamic Therapy. *Phys. Status Solidi* **2018**, *215*, 1700709. [CrossRef]
101. Montalbán, M.G.; Coburn, J.M.; Lozano-Pérez, A.A.; Cenis, J.L.; Víllora, G.; Kaplan, D.L. Production of Curcumin-Loaded Silk Fibroin Nanoparticles for Cancer Therapy. *Nanomaterials* **2018**, *8*, 126. [CrossRef]
102. Pan, K.; Chen, H.; Baek, S.J.; Zhong, Q. Self-assembled curcumin-soluble soybean polysaccharide nanoparticles: Physicochemical properties and in vitro anti-proliferation activity against cancer cells. *Food Chem.* **2018**, *246*, 82–89. [CrossRef] [PubMed]
103. Guo, X.; Mei, J.; Jing, Y.; Wang, S. Curcumin-Loaded Nanoparticles with Low-Intensity Focused Ultrasound-Induced Phase Transformation as Tumor-Targeted and pH-Sensitive Theranostic Nanoplatform of Ovarian Cancer. *Nanoscale Res. Lett.* **2020**, *15*, 73. [CrossRef] [PubMed]
104. Kuang, G.; Zhang, Q.; He, S.; Liu, Y. Curcumin-loaded PEGylated mesoporous silica nanoparticles for effective photodynamic therapy. *RSC Adv.* **2020**, *10*, 24624–24630. [CrossRef]
105. Lin, X.; Shi, Y.; Yu, S.; Li, S.; Li, W.; Li, M.; Chen, S.; Wang, Y.; Cong, M. Preparation of Poloxamer188-b-PCL and Study on in vitro Radioprotection Activity of Curcumin-Loaded Nanoparticles. *Front. Chem.* **2020**, *8*, 212. [CrossRef] [PubMed]
106. Luo, H.; Lu, L.; Liu, N.; Li, Q.; Yang, X.; Zhang, Z. Curcumin loaded sub-30 nm targeting therapeutic lipid nanoparticles for synergistically blocking nasopharyngeal cancer growth and metastasis. *J. Nanobiotechnol.* **2021**, *19*, 224. [CrossRef]
107. Sahab-Negah, S.; Ariakia, F.; Jalili-Nik, M.; Afshari, A.R.; Salehi, S.; Samini, F.; Rajabzadeh, G.; Gorji, A. Curcumin loaded in niosomal nanoparticles improved the anti-tumor effects of free curcumin on glioblastoma stem-like cells: An in vitro study. *Mol. Neurobiol.* **2020**, *57*, 3391–3411. [CrossRef]
108. Habibi, N.; Christau, S.; Ochyl, L.J.; Fan, Z.; Hassani Najafabadi, A.; Kuehnhammer, M.; Zhang, M.; Helgeson, M.; von Klitzing, R.; Moon, J.J.; et al. Engineered Ovalbumin Nanoparticles for Cancer Immunotherapy. *Adv. Ther.* **2020**, *3*, 2000100. [CrossRef]
109. Idrees, H.; Zaidi, S.Z.; Sabir, A.; Khan, R.U.; Zhang, X.; Hassan, S.-U. A Review of Biodegradable Natural Polymer-Based Nanoparticles for Drug Delivery Applications. *Nanomaterials* **2020**, *10*, 1970. [CrossRef]
110. Radu, I.-C.; Zaharia, C.; Hudiță, A.; Tanasă, E.; Ginghină, O.; Marin, M.; Gălățeanu, B.; Costache, M. In Vitro Interaction of Doxorubicin-Loaded Silk Sericin Nanocarriers with MCF-7 Breast Cancer Cells Leads to DNA Damage. *Polymers* **2021**, *13*, 2047. [CrossRef]
111. Huang, L.; Tao, K.; Liu, J.; Qi, C.; Xu, L.; Chang, P.; Gao, J.; Shuai, X.; Wang, G.; Wang, Z.; et al. Design and Fabrication of Multifunctional Sericin Nanoparticles for Tumor Targeting and pH-Responsive Subcellular Delivery of Cancer Chemotherapy Drugs. *ACS Appl. Mater. Interfaces* **2016**, *8*, 6577–6585. [CrossRef]
112. Ferroni, C.; Varchi, G. Keratin-Based Nanoparticles as Drug Delivery Carriers. *Appl. Sci.* **2021**, *11*, 9417. [CrossRef]

113. Tam, D.Y.; Ho, J.W.-T.; Chan, M.S.; Lau, C.H.; Chang, T.J.H.; Leung, H.M.; Liu, L.S.; Wang, F.; Chan, L.L.H.; Tin, C.; et al. Penetrating the Blood–Brain Barrier by Self-Assembled 3D DNA Nanocages as Drug Delivery Vehicles for Brain Cancer Therapy. *ACS Appl. Mater. Interfaces* **2020**, *12*, 28928–28940. [CrossRef]
114. Wang, L.; Du, J.; Han, X.; Dou, J.; Shen, J.; Yuan, J. Self-crosslinked keratin nanoparticles for pH and GSH dual responsive drug carriers. *J. Biomater. Sci. Polym. Ed.* **2020**, *31*, 1994–2006. [CrossRef] [PubMed]
115. Du, J.; Wang, L.; Han, X.; Dou, J.; Jiang, X.; Yuan, J. Polydopamine/keratin complexes as gatekeepers of mesoporous silica nanoparticles for pH and GSH dual responsive drug delivery. *Mater. Lett.* **2021**, *293*, 129676. [CrossRef]
116. Avancini, G.; Guerrini, A.; Ferroni, C.; Tedesco, D.; Ballestri, M.; Columbaro, M.; Menilli, L.; Reddi, E.; Costa, R.; Leanza, L.; et al. Keratin nanoparticles and photodynamic therapy enhance the anticancer stem cells activity of salinomycin. *Mater. Sci. Eng. C* **2021**, *122*, 111899. [CrossRef] [PubMed]
117. Lu, T.-Y.; Lu, W.-F.; Wang, Y.-H.; Liao, M.-Y.; Wei, Y.; Fan, Y.-J.; Chuang, E.-Y.; Yu, J. Keratin-Based Nanoparticles with Tumor-Targeting and Cascade Catalytic Capabilities for the Combinational Oxidation Phototherapy of Breast Cancer. *ACS Appl. Mater. Interfaces* **2021**, *13*, 38074–38089. [CrossRef]
118. Guglielmelli, A.; Rosa, P.; Contardi, M.; Prato, M.; Mangino, G.; Miglietta, S.; Petrozza, V.; Pani, R.; Calogero, A.; Athanassiou, A. Biomimetic keratin gold nanoparticle-mediated in vitro photothermal therapy on glioblastoma multiforme. *Nanomedicine* **2020**, *16*, 121–138. [CrossRef] [PubMed]
119. Frantellizzi, V.; Pontico, M.; De Vincentis, G.; Civitelli, L.; Petronella, F.; Pani, R.; Calogero, A.; Perotto, G.; De Sio, L. *99mTc-labeled Keratin-Coated Gold Nanoparticles for Selective Anticancer Photothermal Therapy*; Optical Society of America: Washington, DC, USA, 2021; p. OTu1E-5.
120. Etman, S.M.; Abdallah, O.Y.; Elnaggar, Y.S.R. Novel fucoidan based bioactive targeted nanoparticles from Undaria Pinnatifida for treatment of pancreatic cancer. *Int. J. Biol. Macromol.* **2020**, *145*, 390–401. [CrossRef]
121. Etman, S.M.; Elnaggar, Y.S.R.; Abdallah, O.Y. Fucoidan, a natural biopolymer in cancer combating: From edible algae to nanocarrier tailoring. *Int. J. Biol. Macromol.* **2020**, *147*, 799–808. [CrossRef]
122. Saeed, A.F.U.H.; Su, J.; Ouyang, S. Marine-derived drugs: Recent advances in cancer therapy and immune signaling. *Biomed. Pharmacother.* **2021**, *134*, 111091. [CrossRef]
123. Qiu, R.; Qian, F.; Wang, X.; Li, H.; Wang, L. Targeted delivery of 20(S)-ginsenoside Rg3-based polypeptide nanoparticles to treat colon cancer. *Biomed. Microdevices* **2019**, *21*, 18. [CrossRef] [PubMed]
124. Kim, Y.J.; Perumalsamy, H.; Castro-Aceituno, V.; Kim, D.; Markus, J.; Lee, S.; Kim, S.; Liu, Y.; Yang, D.C. Photoluminescent And Self-Assembled Hyaluronic Acid-Zinc Oxide-Ginsenoside Rh2 Nanoparticles And Their Potential Caspase-9 Apoptotic Mechanism Towards Cancer Cell Lines. *Int. J. Nanomed.* **2019**, *14*, 8195–8208. [CrossRef] [PubMed]
125. Abd Halim, A.A.; Ridzwan, F.W.; Bin Mohamad, S.; Tayyab, S. Targeting the Nalidixic Acid Binding Site on Human Serum Albumin Through Computational Approach: A Re-Investigation. *Biointerface Res. Appl. Chem.* **2022**, *12*, 1520–1525. [CrossRef]
126. Iqbal, H.; Yang, T.; Li, T.; Zhang, M.; Ke, H.; Ding, D.; Deng, Y.; Chen, H. Serum protein-based nanoparticles for cancer diagnosis and treatment. *J. Control. Release* **2021**, *329*, 997–1022. [CrossRef]
127. Chaiwaree, S.; Prapan, A.; Suwannasom, N.; Laporte, T.; Neumann, T.; Pruß, A.; Georgieva, R.; Bäumler, H. Doxorubicin–Loaded Human Serum Albumin Submicron Particles: Preparation, Characterization and In Vitro Cellular Uptake. *Pharmaceutics* **2020**, *12*, 224. [CrossRef]
128. Yu, Z.; Li, X.; Duan, J.; Yang, X.-D. Targeted Treatment of Colon Cancer with Aptamer-Guided Albumin Nanoparticles Loaded with Docetaxel. *Int. J. Nanomed.* **2020**, *15*, 6737–6748. [CrossRef]
129. Elahi, M.; Ali, S.; Tahir, H.M.; Mushtaq, R.; Bhatti, M.F. Sericin and fibroin nanoparticles—natural product for cancer therapy: A comprehensive review. *Int. J. Polym. Mater. Polym. Biomater.* **2021**, *70*, 256–269. [CrossRef]
130. Ali, G.W.; Abd Ellatif, M.A.; Abdel-Fattah, W.I. Extraction of Natural Cellulose and Zein Protein from Corn Silk: Physico-Chemical and Biological Characterization. *Biointerface Res. Appl. Chem.* **2021**, *11*, 10614–10619. [CrossRef]
131. El-Khatib, E.M.; Ali, N.F.; Nassar, S.H.; El-Shemy, N.S. Functionalization of Natural Fibers Properties by using TiO(2)Nanoparticles to Improve its Antimicrobial Activity. *Biointerface Res. Appl. Chem.* **2022**, *12*, 4177–4191. [CrossRef]
132. Gao, Y.-E.; Hou, S.; Cheng, J.; Li, X.; Wu, Y.; Tang, Y.; Li, Y.; Xue, P.; Kang, Y.; Xu, Z.; et al. Silk Sericin-Based Nanoparticle as the Photosensitizer Chlorin e6 Carrier for Enhanced Cancer Photodynamic Therapy. *ACS Sustain. Chem. Eng.* **2021**, *9*, 3213–3222. [CrossRef]
133. Pandey, V.; Haider, T.; Chandak, A.R.; Chakraborty, A.; Banerjee, S.; Soni, V. Surface modified silk fibroin nanoparticles for improved delivery of doxorubicin: Development, characterization, in-vitro studies. *Int. J. Biol. Macromol.* **2020**, *164*, 2018–2027. [CrossRef] [PubMed]
134. Hudita, A.; Radu, I.C.; Galateanu, B.; Ginghina, O.; Herman, H.; Balta, C.; Rosu, M.; Zaharia, C.; Costache, M.; Tanasa, E.; et al. Bioinspired silk fibroin nano-delivery systems protect against 5-FU induced gastrointestinal mucositis in a mouse model and display antitumor effects on HT-29 colorectal cancer cells in vitro. *Nanotoxicology* **2021**, *15*, 973–994. [CrossRef] [PubMed]
135. Zhang, L.; Li, L.; Di Penta, A.; Carmona, U.; Yang, F.; Schöps, R.; Brandsch, M.; Zugaza, J.L.; Knez, M. H-Chain Ferritin: A Natural Nuclei Targeting and Bioactive Delivery Nanovector. *Adv. Healthc. Mater.* **2015**, *4*, 1305–1310. [CrossRef] [PubMed]
136. Liang, M.; Fan, K.; Zhou, M.; Duan, D.; Zheng, J.; Yang, D.; Feng, J.; Yan, X. H-ferritin-nanocaged doxorubicin nanoparticles specifically target and kill tumors with a single-dose injection. *Proc. Natl. Acad. Sci. USA* **2014**, *111*, 14900–14905. [CrossRef]

137. Jiang, B.; Zhang, R.; Zhang, J.; Hou. Y.; Chen, X.; Zhou, M.; Tian, X.; Hao, C.; Fan, K.; Yan, X. GRP78-targeted ferritin nanocaged ultra-high dose of doxorubicin for hepatocellular carcinoma therapy. *Theranostics* **2019**, *9*, 2167–2182. [CrossRef] [PubMed]
138. Sitia, L.; Bonizzi, A.; Mazzucchelli, S.; Negri, S.; Sottani, C.; Grignani, E.; Rizzuto, M.A.; Prosperi, D.; Sorrentino, L.; Morasso, C.; et al. Selective Targeting of Cancer-Associated Fibroblasts by Engineered H-Ferritin Nanocages Loaded with Navitoclax. *Cells* **2021**, *10*, 328. [CrossRef] [PubMed]
139. Bahrami, M.K.; Movafeghi, A.; Mahdavinia, G.R.; Hassanpouraghdam, M.B.; Gohari, G. Effects of bare and chitosan-coated Fe3O4 magnetic nanoparticles on seed germination and seedling growth of Capsicum annuum L. *Biointerface Res. Appl. Chem.* **2018**, *8*, 3552–3559.
140. Ding, Y.-F.; Li, S.; Liang, L.; Huang, Q.; Yuwen, L.; Yang, W.; Wang, R.; Wang, L.-H. Highly biocompatible chlorin e6-loaded chitosan nanoparticles for improved photodynamic cancer therapy. *ACS Appl. Mater. Interfaces* **2018**, *10*, 9980–9987. [CrossRef]
141. Sabir, F.; Barani, M.; Randar, A.; Bilal, M.; Zafar, M.N.; Bungau, S.; Kyzas, G.Z. How to Face Skin Cancer with Nanomaterials: A Review. *Biointerface Res. Appl. Chem.* **2021**, *11*, 11931–11955. [CrossRef]
142. Gounden, S.; Daniels, A.; Singh, M. Chitosan-Modified Silver Nanoparticles Enhance Cisplatin Activity in Breast Cancer Cells. *Biointerface Res. Appl. Chem.* **2020**, *11*, 10572–10584.
143. Etman, S.M.; Mehanna, R.A.; Bary, A.A.; Elnaggar, Y.S.R.; Abdallah, O.Y. Undaria pinnatifida fucoidan nanoparticles loaded with quinacrine attenuate growth and metastasis of pancreatic cancer. *Int. J. Biol. Macromol.* **2021**, *170*, 284–297. [CrossRef]
144. Jafari, M.; Sriram, V.; Xu, Z.; Harris, G.M.; Lee, J.-Y. Fucoidan-Doxorubicin Nanoparticles Targeting P-Selectin for Effective Breast Cancer Therapy. *Carbohydr. Polym.* **2020**, *249*, 116837. [CrossRef]
145. Coutinho, A.J.; Costa Lima, S.A.; Afonso, C.M.M.; Reis, S. Mucoadhesive and pH responsive fucoidan-chitosan nanoparticles for the oral delivery of methotrexate. *Int. J. Biol. Macromol.* **2020**, *158*, 180–188. [CrossRef] [PubMed]
146. Bakil, S.N.A.; Kamal, H.; Abdullah, H.Z.; Idris, M.I. Sodium alginate-zinc oxide nanocomposite film for antibacterial wound healing applications. *Biointerface Res. Appl. Chem* **2020**, *10*, 6289–6296.
147. Pourjavadi, A.; Amin, S.S.; Hosseini, S.H. Delivery of Hydrophobic Anticancer Drugs by Hydrophobically Modified Alginate Based Magnetic Nanocarrier. *Ind. Eng. Chem. Res.* **2018**, *57*, 822–832. [CrossRef]
148. Khalil, A.M. Interpenetrating polymeric hydrogels as favorable materials for hygienic applications. *Biointerface Res. Appl. Chem.* **2020**, *10*, 5011–5020. [CrossRef]
149. Gaio, E.; Conte, C.; Esposito, D.; Reddi, E.; Quaglia, F.; Moret, F. CD44 Targeting Mediated by Polymeric Nanoparticles and Combination of Chlorine TPCS2a-PDT and Docetaxel-Chemotherapy for Efficient Killing of Breast Differentiated and Stem Cancer Cells In Vitro. *Cancers* **2020**, *12*, 278. [CrossRef]
150. Debele, T.A.; Yu, L.-Y.; Yang, C.-S.; Shen, Y.-A.; Lo, C.-L. pH- and GSH-Sensitive Hyaluronic Acid-MP Conjugate Micelles for Intracellular Delivery of Doxorubicin to Colon Cancer Cells and Cancer Stem Cells. *Biomacromolecules* **2018**, *19*, 3725–3737. [CrossRef] [PubMed]
151. Chung, Y.H.; Cai, H.; Steinmetz, N.F. Viral nanoparticles for drug delivery, imaging, immunotherapy, and theranostic applications. *Adv. Drug Deliv. Rev.* **2020**, *156*, 214–235. [CrossRef] [PubMed]
152. Beatty, P.H.; Lewis, J.D. Cowpea mosaic virus nanoparticles for cancer imaging and therapy. *Adv. Drug Deliv. Rev.* **2019**, *145*, 130–144. [CrossRef]
153. Lam, P.; Lin, R.D.; Steinmetz, N.F. Delivery of mitoxantrone using a plant virus-based nanoparticle for the treatment of glioblastomas. *J. Mater. Chem. B* **2018**, *6*, 5888–5895. [CrossRef]
154. Cai, H.; Shukla, S.; Steinmetz, N.F. The Antitumor Efficacy of CpG Oligonucleotides is Improved by Encapsulation in Plant Virus-Like Particles. *Adv. Funct. Mater.* **2020**, *30*, 1908743. [CrossRef] [PubMed]
155. Franke, C.E.; Czapar, A.E.; Patel, R.B.; Steinmetz, N.F. Tobacco Mosaic Virus-Delivered Cisplatin Restores Efficacy in Platinum-Resistant Ovarian Cancer Cells. *Mol. Pharm.* **2017**, *15*, 2922–2931. [CrossRef]
156. Aqil, F.; Munagala, R.; Jeyabalan, J.; Agrawal, A.K.; Kyakulaga, A.-H.; Wilcher, S.A.; Gupta, R.C. Milk exosomes—Natural nanoparticles for siRNA delivery. *Cancer Lett.* **2019**, *449*, 186–195. [CrossRef] [PubMed]
157. Li, D.; Yao, S.; Zhou, Z.; Shi, J.; Huang, Z.; Wu, Z. Hyaluronan decoration of milk exosomes directs tumor-specific delivery of doxorubicin. *Carbohydr. Res.* **2020**, *493*, 108032. [CrossRef] [PubMed]
158. Munagala, R.; Aqil, F.; Jeyabalan, J.; Gupta, R.C. Bovine milk-derived exosomes for drug delivery. *Cancer Lett.* **2016**, *371*, 48–61. [CrossRef] [PubMed]
159. Ale Ebrahim, S.; Ashtari, A.; Zamani Pedram, M.; Ale Ebrahim, N.; Sanati-Nezhad, A. Publication Trends in Exosomes Nanoparticles for Cancer Detection. *Int. J. Nanomed.* **2020**, *15*, 4453–4470. [CrossRef]
160. Carvalho, I.P.S.; Miranda, M.A.; Silva, L.B.; Chrysostomo-Massaro, T.N.; Paschoal, J.A.R.; Bastos, J.K.; Marcato, P.D. IN VITRO Anticancer Activity and Physicochemical Properties of SOLANUM LYCOCARPUM Alkaloidic Extract Loaded in Natural Lipid-Based Nanoparticles. *Colloid Interface Sci. Commun.* **2019**, *28*, 5–14. [CrossRef]
161. Galisteo-González, F.; Molina-Bolívar, J.A.; Navarro, S.A.; Boulaiz, H.; Aguilera-Garrido, A.; Ramírez, A.; Marchal, J.A. Albumin-covered lipid nanocapsules exhibit enhanced uptake performance by breast-tumor cells. *Colloids Surf. B* **2018**, *165*, 103–110. [CrossRef]
162. Navarro-Marchal, S.A.; Griñán-Lisón, C.; Entrena, J.-M.; Ruiz-Alcalá, G.; Tristán-Manzano, M.; Martin, F.; Pérez-Victoria, I.; Peula-García, J.M.; Marchal, J.A. Anti-CD44-Conjugated Olive Oil Liquid Nanocapsules for Targeting Pancreatic Cancer Stem Cells. *Biomacromolecules* **2021**, *22*, 1374–1388. [CrossRef] [PubMed]

163. Gomes Silva, B.; Pereira da Silva, W.F.; Santos Soares, J.C.; Lima Cavalcanti, I.D.; De Souza, I.A.; Cavalcanti, L. Nanoparticles in the Use of Natural Products for the Treatment of Lung Cancer. *Ars Pharm* **2019**, *60*, 185–192.
164. Chen, M.H.; Hanagata, N.; Ikoma, T.; Huang, J.Y.; Li, K.Y.; Lin, C.P.; Lin, F.H. Hafnium-doped hydroxyapatite nanoparticles with ionizing radiation for lung cancer treatment. *Acta Biom.* **2016**, *37*, 165–173. [CrossRef]
165. Raniolo, S.; Unida, V.; Vindigni, G.; Stolfi, C.; Iacovelli, F.; Desideri, A.; Biocca, S. Combined and selective miR-21 silencing and doxorubicin delivery in cancer cells using tailored DNA nanostructures. *Cell Death Dis.* **2021**, *12*, 7. [CrossRef]
166. Jie, G.; Gao, X.; Ge, J.; Li, C. Multifunctional DNA nanocage with CdTe quantum dots for fluorescence detection of human 8-oxoG DNA glycosylase 1 and doxorubicin delivery to cancer cells. *Microchim. Acta* **2019**, *186*, 85. [CrossRef]
167. Li, R.; He, Y.; Zhang, S.; Qin, J.; Wang, J. Cell membrane-based nanoparticles: A new biomimetic platform for tumor diagnosis and treatment. *Acta Pharm. Sin. B* **2018**, *8*, 14–22. [CrossRef]
168. Jiang, Q.; Liu, Y.; Guo, R.; Yao, X.; Sung, S.; Pang, Z.; Yang, W. Erythrocyte-cancer hybrid membrane-camouflaged melanin nanoparticles for enhancing photothermal therapy efficacy in tumors. *Biomaterials* **2019**, *192*, 292–308. [CrossRef]
169. Jiang, Q.; Luo, Z.; Men, Y.; Yang, P.; Peng, H.; Guo, R.; Tian, Y.; Pang, Z.; Yang, W. Red blood cell membrane-camouflaged melanin nanoparticles for enhanced photothermal therapy. *Biomaterials* **2017**, *143*, 29–45. [CrossRef]
170. Zhu, D.-M.; Xie, W.; Xiao, Y.-S.; Suo, M.; Zan, M.-H.; Liao, Q.-Q.; Hu, X.-J.; Chen, L.-B.; Chen, B.; Wu, W.-T. Erythrocyte membrane-coated gold nanocages for targeted photothermal and chemical cancer therapy. *Nanotechnology* **2018**, *29*, 084002. [CrossRef]
171. Peng, J.; Yang, Q.; Li, W.; Tan, L.; Xiao, Y.; Chen, L.; Hao, Y.; Qian, Z. Erythrocyte-membrane-coated prussian blue/manganese dioxide nanoparticles as H_2O_2-responsive oxygen generators to enhance cancer chemotherapy/photothermal therapy. *ACS Appl. Mater. Interfaces* **2017**, *9*, 44410–44422. [CrossRef]
172. Wang, D.; Gao, C.; Zhou, C.; Lin, Z.; He, Q. Leukocyte membrane-coated liquid metal nanoswimmers for actively targeted delivery and synergistic chemophotothermal therapy. *Research* **2020**, *2020*, 676954. [CrossRef] [PubMed]
173. Fan, M.; Jiang, M. Core-shell nanotherapeutics with leukocyte membrane camouflage for biomedical applications. *J. Drug Target.* **2020**, *28*, 873–881. [CrossRef] [PubMed]
174. Wu, L.; Xie, W.; Zan, H.-M.; Liu, Z.; Wang, G.; Wang, Y.; Liu, W.; Dong, W. Platelet membrane-coated nanoparticles for targeted drug delivery and local chemo-photothermal therapy of orthotopic hepatocellular carcinoma. *J. Mater. Chem. B* **2020**, *8*, 4648–4659. [CrossRef] [PubMed]
175. Zhou, M.; Lai, W.; Li, G.; Wang, F.; Liu, W.; Liao, J.; Yang, H.; Liu, Y.; Zhang, Q.; Tang, Q.; et al. Platelet Membrane-Coated and VAR2CSA Malaria Protein-Functionalized Nanoparticles for Targeted Treatment of Primary and Metastatic Cancer. *ACS Appl. Mater. Interfaces* **2021**, *13*, 25635–25648. [CrossRef]
176. Zhang, Y.; Cai, K.; Li, C.; Guo, Q.; Chen, Q.; He, X.; Liu, L.; Zhang, Y.; Lu, Y.; Chen, X. Macrophage-membrane-coated nanoparticles for tumor-targeted chemotherapy. *Nano Lett.* **2018**, *18*, 1908–1915. [CrossRef] [PubMed]
177. Meng, Q.-F.; Rao, L.; Zan, M.; Chen, M.; Yu, G.-T.; Wei, X.; Wu, Z.; Sun, Y.; Guo, S.-S.; Zhao, X.-Z. Macrophage membrane-coated iron oxide nanoparticles for enhanced photothermal tumor therapy. *Nanotechnology* **2018**, *29*, 134004. [CrossRef] [PubMed]
178. Patel, R.B.; Ye, M.; Carlson, P.M.; Jaquish, A.; Zangl, L.; Ma, B.; Wang, Y.; Arthur, I.; Xie, R.; Brown, R.J.; et al. Development of an In Situ Cancer Vaccine via Combinational Radiation and Bacterial-Membrane-Coated Nanoparticles. *Adv. Mater.* **2019**, *31*, 1902626. [CrossRef]
179. Gao, C.; Lin, Z.; Jurado-Sánchez, B.; Lin, X.; Wu, Z.; He, Q. Stem cell membrane-coated nanogels for highly efficient In Vivo tumor targeted drug delivery. *Small* **2016**, *12*, 4056–4062. [CrossRef] [PubMed]
180. Mu, X.; Li, J.; Yan, S.; Zhang, H.; Zhang, W.; Zhang, F.; Jiang, J. siRNA delivery with stem cell membrane-coated magnetic nanoparticles for imaging-guided photothermal therapy and gene therapy. *ACS Biomater. Sci. Eng.* **2018**, *4*, 3895–3905. [CrossRef]
181. Rao, L.; Yu, G.-T.; Meng, Q.-F.; Bu, L.-L.; Tian, R.; Lin, L.-S.; Deng, H.; Yang, W.; Zan, M.; Ding, J.; et al. Cancer Cell Membrane-Coated Nanoparticles for Personalized Therapy in Patient-Derived Xenograft Models. *Adv. Funct. Mater.* **2019**, *29*, 1905671. [CrossRef]
182. Jiang, Y. Natural and Engineered Cancer Cell Membrane-Coated Nanoparticles for Antitumor Immunotherapy. Ph.D. Thesis, University of California San Diego, La Jolla, CA, USA, 2021.
183. Rao, L.; Meng, Q.F.; Huang, Q.; Wang, Z.; Yu, G.T.; Li, A.; Ma, W.; Zhang, N.; Guo, S.S.; Zhao, X.Z. Platelet–leukocyte hybrid membrane-coated immunomagnetic beads for highly efficient and highly specific isolation of circulating tumor cells. *Adv. Funct. Mater.* **2018**, *28*, 1803531. [CrossRef]
184. Wang, D.; Liu, C.; You, S.; Zhang, K.; Li, M.; Cao, Y.; Wang, C.; Dong, H.; Zhang, X. Bacterial Vesicle-Cancer Cell Hybrid Membrane-Coated Nanoparticles for Tumor Specific Immune Activation and Photothermal Therapy. *ACS Appl. Mater. Interfaces* **2020**, *12*, 41138–41147. [CrossRef]
185. Gong, C.; Yu, X.; You, B.; Wu, Y.; Wang, R.; Han, L.; Wang, Y.; Gao, S.; Yuan, Y. Macrophage-cancer hybrid membrane-coated nanoparticles for targeting lung metastasis in breast cancer therapy. *J. Nanobiotechnol.* **2020**, *18*, 92. [CrossRef]
186. Xiong, J.; Wu, M.; Chen, J.; Liu, Y.; Chen, Y.; Fan, G.; Liu, Y.; Cheng, J.; Wang, Z.; Wang, S.; et al. Cancer-Erythrocyte Hybrid Membrane-Camouflaged Magnetic Nanoparticles with Enhanced Photothermal-Immunotherapy for Ovarian Cancer. *ACS Nano* **2021**. [CrossRef]
187. Hong, S.; Choi, D.W.; Kim, H.N.; Park, C.G.; Lee, W.; Park, H.H. Protein-Based Nanoparticles as Drug Delivery Systems. *Pharmaceutics* **2020**, *12*, 604. [CrossRef]

188. Jain, A.; Singh, S.K.; Arya, S.K.; Kundu, S.C.; Kapoor, S. Protein Nanoparticles: Promising Platforms for Drug Delivery Applications. *ACS Biomater. Sci. Eng.* **2018**, *4*, 3939–3961. [CrossRef]
189. Kianfar, E. Protein nanoparticles in drug delivery: Animal protein, plant proteins and protein cages, albumin nanoparticles. *J. Nanobiotechnol.* **2021**, *19*, 159. [CrossRef]
190. Sun, Y.; Ma, X.; Hu, H. Marine Polysaccharides as a Versatile Biomass for the Construction of Nano Drug Delivery Systems. *Mar. Drugs* **2021**, *19*, 345. [CrossRef]
191. Ghassami, E.; Varshosaz, J.; Taymouri, S. Redox Sensitive Polysaccharide Based Nanoparticles for Improved Cancer Treatment: A Comprehensive Review. *Curr. Pharm. Des.* **2018**, *24*, 3303–3319. [CrossRef]
192. Czapar, A.E.; Steinmetz, N.F. Plant viruses and bacteriophages for drug delivery in medicine and biotechnology. *Curr. Opin. Chem. Biol.* **2017**, *38*, 108–116. [CrossRef]
193. Steinmetz, N.F. Viral nanoparticles as platforms for next-generation therapeutics and imaging devices. *Nanomed. Nanotechnol. Biol. Med.* **2010**, *6*, 634–641. [CrossRef] [PubMed]
194. Jiang, X.-C.; Gao, J.-Q. Exosomes as novel bio-carriers for gene and drug delivery. *Int. J. Pharm.* **2017**, *521*, 167–175. [CrossRef] [PubMed]
195. Zhang, M.; Du, Y.; Wang, S.; Chen, B. A Review of Biomimetic Nanoparticle Drug Delivery Systems Based on Cell Membranes. *Drug Des. Dev. Ther.* **2020**, *14*, 5495–5503. [CrossRef] [PubMed]

pharmaceutics MDPI

Review

Protein-Based Nanoparticles for the Imaging and Treatment of Solid Tumors: The Case of Ferritin Nanocages, a Narrative Review

Francesco Mainini [1], Arianna Bonizzi [1], Marta Sevieri [1], Leopoldo Sitia [1], Marta Truffi [2], Fabio Corsi [1,2,*] and Serena Mazzucchelli [1,*]

[1] Dipartimento di Scienze Biomediche e Cliniche "L. Sacco", Università di Milano, 20157 Milano, Italy; francesco.mainini@unimi.it (F.M.); arianna.bonizzi@unimi.it (A.B.); marta.sevieri@unimi.it (M.S.); leopoldo.sitia@unimi.it (L.S.)
[2] Istituti Clinici Scientifici Maugeri IRCCS, 27100 Pavia, Italy; marta.truffi@icsmaugeri.it
* Correspondence: fabio.corsi@unimi.it (F.C.); serena.mazzucchelli@unimi.it (S.M.)

Abstract: Protein nanocages have been studied extensively, due to their unique architecture, exceptional biocompatibility and highly customization capabilities. In particular, ferritin nanocages (FNs) have been employed for the delivery of a vast array of molecules, ranging from chemotherapeutics to imaging agents, among others. One of the main favorable characteristics of FNs is their intrinsic targeting efficiency toward the Transferrin Receptor 1, which is overexpressed in many tumors. Furthermore, genetic manipulation can be employed to introduce novel variants that are able to improve the loading capacity, targeting capabilities and bio-availability of this versatile drug delivery system. In this review, we discuss the main characteristics of FN and the most recent applications of this promising nanotechnology in the field of oncology with a particular emphasis on the imaging and treatment of solid tumors.

Keywords: ferritin; cancer; tumor targeting; drug delivery; imaging

Citation: Mainini, F.; Bonizzi, A.; Sevieri, M.; Sitia, L.; Truffi, M.; Corsi, F.; Mazzucchelli, S. Protein-Based Nanoparticles for the Imaging and Treatment of Solid Tumors: The Case of Ferritin Nanocages, a Narrative Review. *Pharmaceutics* **2021**, *13*, 2000. https://doi.org/10.3390/pharmaceutics13122000

Academic Editor: Xiangyang Shi

Received: 2 November 2021
Accepted: 19 November 2021
Published: 25 November 2021

1. Introduction

Nanoparticle-based drug delivery systems have the capacity to enhance the physicochemical properties of a wide variety of drugs used in oncology to limit off-site side effects and improve their therapeutic efficacy [1–3]. The ideal nanocarrier should be bio-compatible and be able to avoid recognition by the reticuloendothelial system (RES), composed of tissue-resident macrophages and phagocytes in the bloodstream, capable of efficiently clearing exogenous nanoparticles (NP) from the circulation [4]. Natural proteins nanocages have a distinctive advantage in this regard, in comparison to synthetic NP (liposomes, polymeric NP, micelles and dendrimers), since they are virtually invisible to the immune system and display great biocompatibility coupled with minimal toxicity. An exception is represented by virus-like particles (VLPs), which are also composed of self-assembled proteins that are, in some cases, highly immunogenic [5].

Endogenous self-assembled NPs can be synthesized by many cell types and are primarily used to store and/or distribute to different tissues a wide variety of molecules, such as nutrients and biochemical signals. NPs of this kind are quite diverse in terms of size and physiological activity. Some examples are ferritin nanocages (FNs), heat-shock protein cages, vault ribonucleoparticles, the E2 protein of the pyruvate dehydrogenase multienzyme complex, chaperones, carboxysomes and other enzyme complexes [6]. Unfortunately, most of these protein-based NP are understudied and have, so far, limited applications in the field of oncology. However, FNs have been studied extensively due to their intrinsic targeting capabilities toward the Transferrin Receptor 1 (TfR1), which is highly expressed in many tumors, making them very appealing for drug-delivery applications in oncology. Furthermore, the small size and high customization potential make them ideal candidates for the development of novel nanomedicines able to deliver a wide variety of drugs to the tumor microenvironment (TME). This review describes the structure and function of FN, modifications of the nanocages by chemical or genetic manipulation (Figure 1) and novel applications of this nanotechnology for the imaging and treatment of solid tumors (Figure 2).

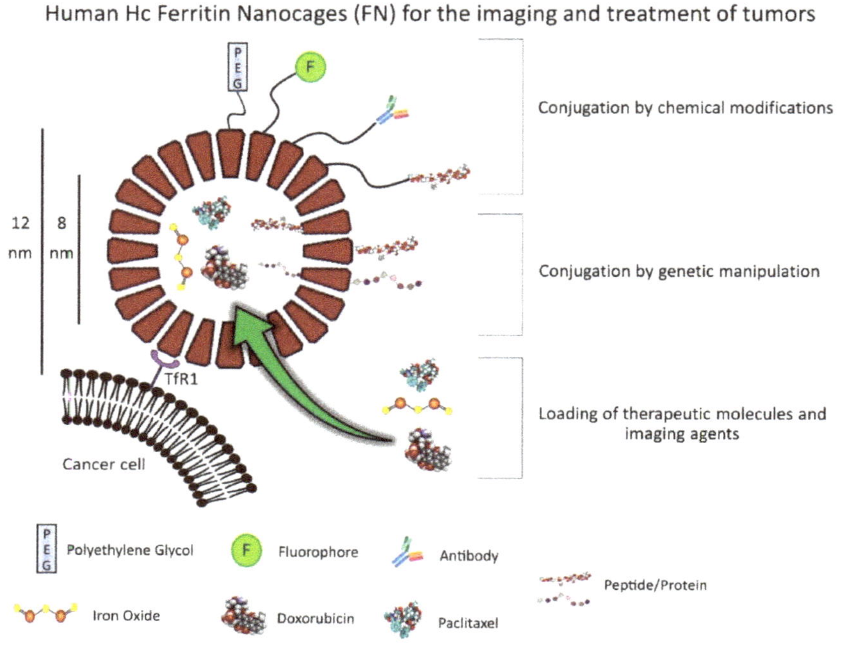

Figure 1. FN as a protein-based delivery system for oncological therapeutics and imaging agents. FNs are composed of 24 Hc subunits that can be chemically or genetically modified to couple a large variety of molecules (antibodies, peptides, fluorophores, polyethylene glycol (PEG) and others) to their surface (N-terminus) or internal cavity (C-terminus). Furthermore, FNs can be loaded with different drugs and imaging agents and have intrinsic targeting capabilities toward the receptor TfR1, which is overexpressed in many tumors. This cartoon was created by using BioRender (https://biorender.com/, accessed on 2 November 2021).

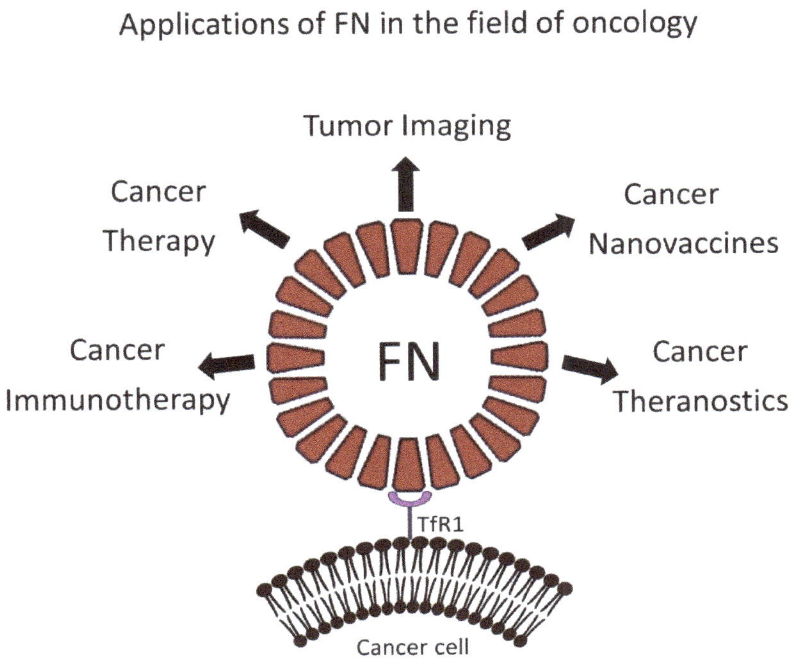

Figure 2. Applications of FNs in the field of oncology. FNs are versatile drug delivery systems. They can be loaded simultaneously with anticancer and imaging agents to provide effective antitumor therapy that can be monitored by different imaging modalities. In addition, FNs can be loaded with immunomodulatory drugs to remodel the TME or can be developed to incorporate tumor associated antigens to induce specific adaptive immune responses against cancer cells, in the case of nanovaccines. This cartoon was created by using BioRender (https://biorender.com/, accessed on 2 November 2021).

2. FN Structure and Properties

Ferritin self-assembles in hollow icosahedral-shaped nanocages with inner and outer dimensions of 8 and 12 nm, respectively [7]. In mammalian cells, ferritin is composed of heavy-chain (Hc, 21 kDa) and light-chain (Lc, 19 kDa) subunits (24 in total between the two) which are structurally similar. FN employed as a delivery device in cancer application are mostly constituted only by Hc subunits of human ferritin. Ferritin and FNs are remarkably stable in biological fluids and are resistant to denaturants, including high temperatures (>80 °C) [8]. Each subunit is composed of four long helixes, a short helix and a long loop [9]. The C-terminal of each subunit folds into the inner cavity, while the N-terminal is exposed on the outer surface of the nanocage. The ratio between Hc and Lc subunits is determined by ferritin's primary role in tissues. For example, in the hearth and brain, the Hc is more abundant, while, in the liver and spleen, the Lc is predominant. The Hc subunit contains a dinuclear ferroxidase site that is located within the four-helix bundle, while the Lc provides efficient sites for iron nucleation and mineralization [10]. Ferritin and FN carry six C4 channels and eight C3 channels. The C3 channels have hydrophilic properties and allow the passage of Fe(II) ions and water molecules in and out of the protein cage. On the other hand, the C4 channels allow the passage of small hydrophobic molecules [8].

Ferritin in the bloodstream is mainly composed of Lc subunits [8], which seem to be secreted primarily by macrophages [11]; however, their role in the serum is still highly debated. Nonetheless, high ferritin levels have been linked with ongoing infections and chronic inflammation, while its reduced levels have been correlated with iron deficiency [12–14]. Interestingly, it can be localized in cells both in the cytoplasm and in the nucleus. Iron stored in ferritin can be utilized by the cell in a process mediated by autophagy, where it is transported to the lysosomes and iron is released in a pH-dependent manner [15]. On the other hand, in conditions of oxidative stress, ferritin can convert DNA damaging Fe(II) to harmless Fe(III), thus limiting DNA damage mediated by the formation of hydroxyl radicals through the Fenton reaction [16–18]. Two proteins, poly(rC)-binding protein 1 (PCBP1) and nuclear receptor coactivator 4 (NCOA4), are involved in the transport of iron inside and outside ferritin [19]. Furthermore, it has been proposed that O-glycosylation of the Hc could be involved in the nuclear translocation of Ferritin, which maintains its intact structure during this process [20,21]. FNs share all structural features and properties with their physiological form, and they often have been demonstrated to be managed by the cells and the tissue as natural ferritin [21,22].

Strategies for Loading FN

Molecular cargoes can be loaded into the inner core of FN by different methodologies (Figure 3). Extreme pH (2 or 13) is used to transiently disassemble the protein nanocage into monomers that can reassemble by adjusting the pH toward neutrality. By employing this methodology, FN can be loaded with different chemotherapeutic drugs. Interestingly, only minor differences in the loading efficiency between doxorubicin (DOX), epirubicin (EPI), daunorubicin (DAU) and idarubicin (IDA) were seen, despite their differences in terms of hydrophobicity [23]. In addition, high concentrations of guanidine hydrochloride (GuHCl) or urea are able to disrupt the non-covalent forces which support FNs' structure, leading to their disassembly. This process can be reversed by dialysis to remove the excess of chaotropic agents, leading to the recovery of the original nanostructure with consequent loading of molecular cargoes in the inner cavity [24]. More recently, atmospheric cold plasma (ACP) technology was implemented to reduce the α-helix/β-sheet contents and thermal stability of FN to allow disassembly at a pH of 4. This technique can be utilized to load molecules which are susceptible to extreme pH conditions and could be degraded during the loading procedure [25]. In addition, our group has showed that the loading of molecules sensitive to low pH can be achieved during the reassembly phase by adding the molecule of interest to the ferritin-containing solution after the adjustment of the pH toward neutrality [26]. In another report, Jiang and colleagues developed a methodology that is able to provide high loading of DOX and high recovery of FN by incubating DOX with FN at 60 °C for 4 h [27]. This loading methodology enables the opening of FN's channel to introduce DOX without disrupting FN's structure. Lastly, by taking advantage of the natural capacity of FNs to encapsulate iron in their cavity, several metal ions can be coupled to molecules of interest that can then be loaded into FNs. In this case, the final loading efficiency of the chosen drug depends on its binding affinity for the metal ion, the FN species used and preparation conditions. For example, Zhen and colleagues suggested that, between Cu(II), Mn(II), Zn(II) and Fe(III) metal ions, the use of copper resulted in the highest loading rate of DOX into FN [28]. A more detailed comparison of loading methodologies for DOX into FN has been reviewed by He and colleagues [29], while Zhang and colleagues provide specific protocols regarding the loading of various drugs into FNs by using different methodologies [30].

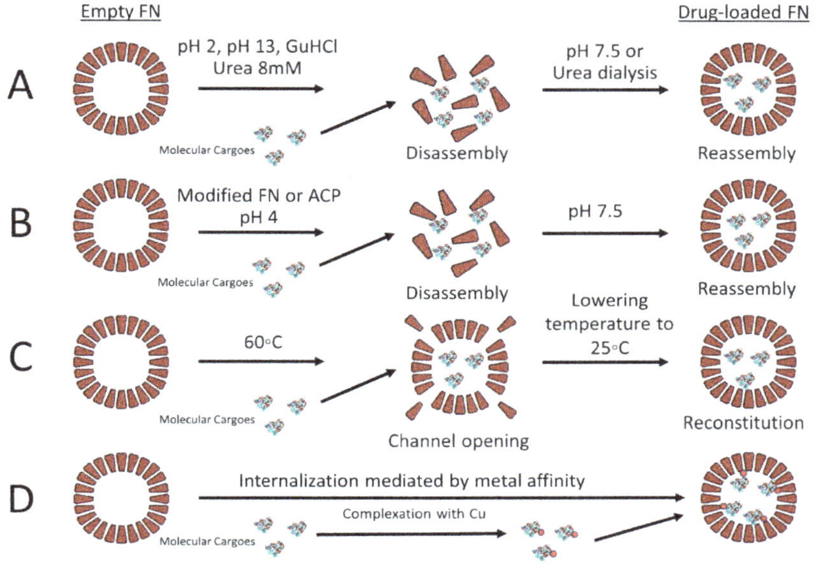

Figure 3. Biochemical strategies used to load different cargoes into FNs: (**A**) pH or urea-mediated disassembly–reassembly methodology; (**B**) modified ferritins or ACP can be utilized to disassemble FN at pH 4, and then pH 7.5 is used to reassemble FN; (**C**) high temperatures partially destabilize FN to allow channel openings with consequent drug loading, and then lowering the temperature can slowly reconstitute the natural conformation of FN; (**D**) molecular cargoes can be complexed with Cu(II) or other metal ions which have a high affinity for the internal cavity of FN. This methodology permits the loading of hydrophobic molecules with some limitations.

Overall, FNs' physicochemical properties (small size and negative Z potential), together with their intrinsic capacity to avoid recognition by RES and targeting capability toward TfR1-expressing tumor cells, make them an ideal candidate for the development of drug delivery systems for nanobiotechnological applications in the field of oncology (Table 1). Moreover, drug loading and targeting efficiency could be enhanced by chemical and genetic manipulations of FNs, as is discussed in the following section.

Table 1. FN-based NPs for the imaging and treatment of tumors.

FN Origin	Purpose	Modifications	Loaded with	In Vivo Tested?	Reference
Human Hc FN	Cancer therapy	BCP1 peptide	DOX	Yes	[31]
Human Hc FN	Cancer therapy	Mutations to enhance the binding of Cu2+	DOX	Yes	[32]
Human Hc FN	Cancer therapy	4 Lysines (C-terminus)	siRNA (EGFR)	Yes	[33]
Human Hc FN	Cancer therapy	PD-L1 binding peptide	DOX	Yes	[34]
Human Hc FN	Cancer therapy	tLyP-1 peptide	PTX	Yes	[35]
Human Hc FN	Cancer therapy	Trastuzumab	DOX	Yes	[36]
Human Hc FN	Cancer therapy	PEGylation (50% subunits)	DOX	Yes	[37]
Human Hc FN	Cancer therapy	PEGylation (75% subunits)	Acriflavine	Yes	[38]
Human Hc FN	Cancer therapy	None	Olaparib	No	[39]
Human Hc FN	Cancer therapy	None	Everolimus	No	[40]
Human Hc FN	Cancer therapy	None	Curcumin	No	[41]
Human Hc FN	Cancer therapy	Anti FAP antibody	Navitoclax	Yes	[42]
Human Hc FN	Cancer therapy	None	DOX	Yes	[43]
Human Hc FN	Cancer therapy	α2β1 targeting peptide	DOX	Yes	[44]
Human Hc FN	Cancer therapy	None	PTX	Yes	[45]
Human Hc FN	Cancer therapy	Trastuzumab or Cetuximab	Empty	No	[46]
Human Hc FN	Cancer therapy	Pout peptide (C terminus)	EPI, Camptothecin	Yes	[47]
Pyrococcus furiosus FN	Cancer therapy	SP94 peptide	DOX	Yes	[48]

Table 1. *Cont.*

FN Origin	Purpose	Modifications	Loaded with	In Vivo Tested?	Reference
Horse spleen FN	Cancer therapy	None	Mertansine	No	[49]
Horse spleen FN	Cancer therapy	None	Arsenoplatin-1	No	[50]
Horse spleen FN	Cancer therapy	Emulsified FN (size 78nm)	Rapamycin and Erastin	Yes	[51]
Horse spleen FN	Cancer therapy	GKRK peptide	Vincristine	Yes	[52]
Unspecified	Cancer therapy	PEG–Panitumumab	Oxaliplatin	Yes	[53]
Unspecified	Cancer therapy	RGD peptide	Resveratrol	Yes	[54]
Unspecified	Cancer therapy	None	Au(III) thiosemicarbazone	Yes	[55]
Pyrococcus furiosus FN	Cancer nanovaccine	SpyCatcher	SpyTagged peptides	Yes	[56]
Human Hc FN	Cancer Immunotherapy	M2pep peptide (N-terminus), cationic peptide (C-terminus)	CpG	Yes	[57]
Human Hc FN	Cancer Theranostic	None	Iron Oxide (core) and IRdye800 or DOX	Yes	[58]
Human Hc FN	Cancer Theranostic	Coated with RBC (functionalized with FA)	Iron Oxide, Cy5.5	Yes	[59]
Horse spleen FN	Cancer Theranostic	2-amino-2-deoxy-glucose	Gold NP	No	[60]
Horse spleen FN	Cancer Theranostic	None	Endogenous Iron	Yes	[61]
Unspecified	Cancer Theranostic	PEG–FA	Perfluoropentane	Yes (imaging only)	[62]
Human Hc FN	Tumor Imaging	None	ICG	Yes	[26,63]
Human Hc FN	Tumor Imaging	SDSSD peptide or hydroxyapatite binding peptide	Cy5	Yes	[64]
Human Hc FN	Tumor Imaging	None	Iron Oxide or Cy5.5	Yes	[65]

3. Production and Modifications of FN

FNs utilized in preclinical studies are usually produced as recombinant protein in *E. coli* strains engineered to express only the human Hc subunit. This procedure involves the transformation of bacteria with a plasmid containing the Hc sequence of interest, which is then purified by anion-exchanger columns after treatment at 70 °C. The resulting FNs are composed, in this case, of 100% Hc subunits [66]. Otherwise, FNs can be purified from the horse spleen, where the ratio between Hc and Lc subunits was found to be ~1/10 [67]. To ensure that purified FNs are not contaminated by endotoxins that could impact both in vitro and in vivo experiments, additional procedures to remove endotoxins might be required [68].

The genetic manipulation of the Hc-FN DNA sequence led to the development of more than one hundred variants to introduce novel functionalities that are able to improve the drug loading, biodistribution and targeting properties of FNs [69]. For example, the self-assembly properties of FNs can be altered to produce novel nanostructures comprising 8 or 48 subunits instead of 24 [70,71]. In addition, ferritin can be modified to produce nanocages that can disassemble at a pH of 4 or 6, instead of 2 [72–74]. Intriguingly, Gu and colleagues developed His-modified ferritins that do not self-assemble at neutral pH. However, metal ions or a pH of 10 induce self-assembly with consequent increases of the drug-loading efficiency, as compared to the standard pH methodology discussed previously [75]. Unfortunately, the stability in serum of His-modified ferritins was not evaluated, and it is unclear if these nanoconstructs are suitable for in vivo studies.

Different strategies have been recently employed to enhance the half-life of FNs in the circulation to provide higher tumor accumulation and reduce clearance by RES. For example, Wang and colleagues developed a novel FN that includes an albumin-binding domain that is able to increase FNs' half-life by 17 times, as compared to the standard FN [76]. In another report, an amino acid sequence rich in proline (P), serine (S) and alanine (A) residues (PAS polypeptide) was inserted by genetic manipulation into FN to increase blood half-life and DOX encapsulation efficiency [77,78]. Interestingly, the insertion of two glutamate residues in the PAS sequence (PASE) further improved FNs' accumulation to the tumor site [79]. In another report, Jin and colleagues introduced in the ferritin construct a blood circulation prolonging (BCP) peptide derived from the phage M13. The generated FN (BCP1–FN) showed improved circulation time compared to standard FN (20 h compared to 2 h). In addition, when loaded with DOX, BCP1–FN–DOX showed superior therapeutic efficacy in a mouse model of melanoma compared to FN–DOX and free DOX. Intriguingly, the authors suggested that the RGD portion of the BCP1 peptide could be responsible for the binding of BCP1–FN to peripheral blood cells, particularly platelets, which are able to protect the nanocages from RES recognition [31]. Nonetheless, blood cells' "hitchhiking" has recently emerged as one of the strategies that can be employed to enhance the delivery of NP to the tumor site [80,81].

Since FNs are composed of different subunits, novel FN-based nanostructures have been developed with combinations of different ferritins resulting in hybrid nanocages with interesting physicochemical properties. Ahn and colleague developed a hybrid FN composed of modified subunits (F160) and standard Hc in a ratio 1:1. F160 was devised to provide large pores to FN and was produced by removing the C-terminal channel forming E-helix from the Hc sequence. The resulting hybrid FN (nicked–FN) allows for the encapsulation of DOX by simple incubation, improving the loading of DOX and the recovery of the nicked–FN–DOX in comparison to encapsulation in unmodified FN or with the pH-mediated disassembly and reassembly methodology [82]. Another strategy to enhance DOX loading into FN was developed by producing a mutant FN that displays an enhanced affinity for copper ions [32]. In another report, FNs were modified with the addition of biotin accepted peptide, which resulted in biotinylated FN that can be more easily modified by the addition of streptavidin-tagged molecules [83]. These modified FN could be used in a variety of immunoassays based on streptavidin-tagged antibodies to increase the sensitivity. It is unsure if they can be employed for in vivo studies.

The delivery of nucleic acids by NP-based delivery systems has always been a primary goal of the research effort in the field of nanotechnology. Interestingly, modified FNs with the addition of a cationic polypeptide were developed to facilitate the incorporation of siRNAs in FNs' nanostructure [33,84]. However, it has also recently been shown that unmodified FNs could incorporate siRNAs by pH-mediated disassembly and reassembly methodology [85].

FNs can also be modified to include immunogenic peptides that are able to induce immune responses against specific antigens. As proof of principle, Kanekiyo and colleagues developed a nanovaccine against the H1N1 virus based of FNs that were modified to include the viral hemagglutinin sequence. Preclinical testing of the developed nanoformulation showed induced protection of animal models to H1N1 infection [86]. More recently, FN-based anticancer nanovaccines have been developed and were tested successfully in preclinical models [56,87,88].

Interestingly, many FN variants have been developed to include novel targeting ligands. For example, Jiang and colleagues introduced, by genetic manipulation, a hepatocellular carcinoma (HCC)-targeting peptide to the FN's structure that was then loaded with DOX. This novel formulation showed superior activity compared to free DOX in reducing HCC tumor growth and metastases in preclinical models [48,89]. In another report, the PD-L1 binding peptide 1 (PD-L1pep1, CLQKTPKQC) was introduced into ferritin's sequence to generate an FN targeted to PD-L1 [34]. Another well-studied tumor-targeting ligand is the tLyp-1 peptide, which binds the receptor Neuropilin 1 expressed in the stroma of many types of tumors [90]. Modified FNs were developed to include the tLyp-1 peptide in the external structure of the nanocage and were subsequently loaded with Paclitaxel (PTX). The resulting FNs (tLyp–FN–PTXs) showed enhanced uptake by tumor cells and were able to control tumor growth in vivo compared to free-PTX or FN, where the sequence of tLyp was mutated (m-tLyp–FN–PTX) [35].

Beyond genetic manipulation, FNs have available primary amines on their surface that can be exploited for chemical conjugation purposes. N-hydroxysuccinimide (NHS) ester or maleimide groups, in combination with 1-ethyl-3-(-3-dimethylaminopropyl) carbodiimide (EDC), are often used to couple peptides, PEG, fluorophores or antibodies to FN in a buffered solution, without the use of organic solvents [42,91]. This coupling methodology is often employed to develop fluorescent versions of FNs that can be utilized in a variety of in vitro and in vivo assays, including flow cytometry, fluorescence microscopy and live imaging, which are critical techniques for NP characterization and for the evaluation of biodistribution and targeting capacity of novel formulations of FNs. In a recent report, FNs were modified with the addition of positively charged polyamine dendrimers (PAMAM) to allow efficient loading of nucleic acids. MiRNA-loaded FNs were successfully used to target leukemia cells and showed promising in vitro results [92].

Overall, both genetic and chemical manipulations can enhance multiple aspects of the intrinsic properties of FNs, such as targeting, loading and half-life. However, it is unknown if these modifications could induce the production of specific anti-ferritin antibodies when administered in humans. Interestingly, it has been shown that modifications such as PEGylation could result in the generation of anti-PEG antibodies [93]. Therefore, it is plausible that some of the developed modifications of the native human Hc subunit, by both genetic and chemical manipulation, could potentially reduce the effectiveness of FN after multiple administrations, limit their targeting capabilities and induce undesirable immunogenic reactions [93]. Furthermore, the FN's origin could be an important factor contributing to immunological side effect. These potential complications should be carefully taken in consideration to ensure the success of modified FNs in the prospect of clinical translation [94].

4. FN-Based NPs for Cancer Treatment in Preclinical Models

One of the main issues in the delivery of chemotherapeutics for cancer treatment is the onset of off-site side effects, which can cause a wide spectrum of complications, such as infections, neuropathies, cytopenias, nephrotoxicity, cardiotoxicity and hepatotoxicity [95–98]. NP-based delivery systems are utilized in oncology primarily to reduce the severity of these side effects, improving drug accumulation at the tumor site. Examples of nanotherapeutics currently used in the clinical practice are Doxil™, Abraxane™, Marqibo™ and DaunoXome™, which are NP-based platforms for the delivery of DOX, PTX, Vincristine and DAU, respectively [1].

Interestingly, FNs have been extensively studied as nanocarriers for DOX, since this hydrophilic drug can be encapsulated efficiently into FNs and can be delivered to tumor cells by the TfR1-mediated intrinsic targeting capabilities of FNs. Our group and others have shown that not only are FN–DOX formulations superior to free DOX or Doxil™ in controlling tumor burden in preclinical models of cancer, but they also dramatically reduced drug cardiotoxic effects, as compared to the free DOX [48,99–101]. In another report, Huang and colleagues developed a hybrid FN–DOX formulation for the treatment of lung cancer. It is composed of PEGylated Hc subunits to provide stealth capabilities and non-PEGylated Hc subunits to allow the binding of the nanocage to TfR1. In vivo results showed that, after intratracheal administration of hybrid FN–DOX, the tumor burden in a orthotopic murine model of lung cancer (3LL) was dramatically reduced when compared to free DOX [37]. Apart from DOX, platinum-based chemotherapeutics (cisplatin, oxaliplatin, Pt(II) terpyridine and carboplatin) have been successfully encapsulated in FNs and have shown encouraging antitumor activity in preclinical models of cancer [53,102,103]. Recently, Ferraro and colleagues developed a novel FN loaded with Arsenoplatin-1 ($Pt(\mu\text{-}NHC(CH_3)O)_2ClAs(OH)_2$), which combines the cytotoxic effects of both cisplatin and arsenic trioxide. Preliminary in vitro results showed that this novel formulation provides selectivity toward cancer cells, but, unfortunately, it was not tested in vivo [50].

The development of drug resistance often occurs after treatment with standard chemotherapeutics, and it can be mediated by the activity of the transporter multidrug resistance protein 1 (MDR1), which is upregulated in the hypoxic areas of tumors and facilitates the excretion of chemotherapeutics outside the tumor cell membrane [104]. Interestingly, hypoxia in the TME can induce the expression of TfR1 mediated by hypoxia-inducible factor-1α (HIF-1α) in tumor cells [105]. Hence, FN-based NPs could be employed to specifically target hypoxic areas in tumors. For this purpose, Huang and colleagues developed a hybrid FN (composed of 75% PEGylated subunits) (Figure 4) for the delivery of the HIF-1α inhibitor Acriflavine(AF). This nanoformulation was particularly effective when used in combination with cisplatin, since the delivery of AF to the TME was able to reduce the expression of MDR1 on tumor cells, thus reducing the development of resistance to cisplatin, which was not effective as standalone treatment in the 3LL lung cancer xenograft model [38].

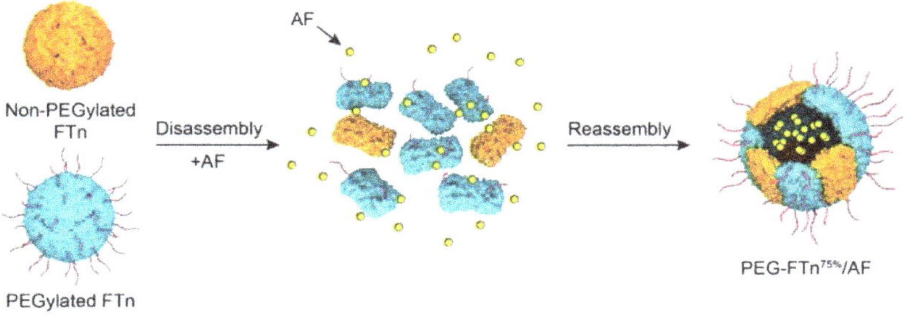

Figure 4. Structure and loading procedure for AF-loaded hybrid PEGylated FN. Hybrid FNs can be developed by utilizing the pH disassembly/reassembly methodology, starting from two different FNs (in this case, a PEGylated FN and non-PEGylated FN). In addition, prior to the reassembly phase, anticancer drugs can be added, resulting in their inclusion inside the FN nanostructure after reassembly. Adapted from [39], American Chemical Society, 2019.

Ferritin was also recently employed to develop a novel nanoformulation containing both Rapamycin, an mTOR inhibitor, and Erastin, a ferroptosis inducer. NPs were produced by the emulsification technique, which was shown to be superior compared to the standard pH disassembly–reassembly methodology in regards to drug loading. Interestingly, the size of the NP formed was 7-fold larger than standard FN (78 compared to 12 nm). Nonetheless, this novel formulation achieved impressive results in controlling the tumor growth in a murine model of breast cancer which recapitulates tumor relapse and metastases formation. Briefly, the primary tumor was allowed to grow, and it was excised to simulate surgery. Subsequently, NPs or the free drugs were included in a thermo-responsive F-127 hydrogel and injected into the tumor resection cavity to test the ability of the nanoformulation to prevent tumor recurrence [51]. Unfortunately, the authors did not evaluate the differences in uptake between standard a FN and the modified version developed. Furthermore, since NPs were not administered intravenously, their biodistribution was not evaluated.

FN has also been explored as a nanocarrier for a large variety of drugs for cancer therapy, such as Olaparib [39], Everolimus [40], Curcumin [41], Oxaliplatin + Panitumumab [53], Mertansine [49], Resveratrol [54] and Navitoclax [42].

4.1. FN-Based NPs for Immunomodulation and Immunotherapy

Another therapeutic strategy which has recently emerged in cancer therapy is the immunomodulation of the TME, in particular, the reprogramming of tumor associated macrophages (TAMs). In solid tumors, TAMs constitute up to 50% of the tumor mass and have been shown to support local immunosuppression and metastases formation [106]. They are recruited from the bloodstream and surrounding tissues by growth factors and chemokines, including colony-stimulating factor 1, C-C motif ligand 2 and vascular endothelial growth factor [107,108]. Interestingly, TAMs are conventionally categorized as anti-inflammatory M2-like macrophages and express high levels of TfR1 compared to pro-inflammatory M1-like macrophages [109,110]. For this reason, TAMs could be effectively targeted by FN-based therapeutics. Of note, macrophages, in general, can be considered as "gate-keepers" of iron metabolism due to their involvement in the recycling of iron from dying erythrocytes [111]. In addition, TAMs-derived FNs have been shown to function as growth factors on malignant mammary epithelium in a process independent of iron [112]. In order to re-educate TAMs and promote a phenotype switch from M2-like to antitumoral M1-like, FNs have been developed to deliver the toll-like receptor 9 agonist CpG, a nucleic acid with M1-polarizing properties [57]. FNs were functionalized with the TAMs-targeting peptide M2pep (YEQDPWGVKWWY), which was combined with a cationic peptide to allow the attachment of the negatively charged CpG. Interestingly, this novel formulation was able to achieve reduction in tumor growth in a murine model of breast cancer. However, a similar effect was seen even when CpG was absent. The authors hypothesized that the antitumoral effect mediated by the M2pep-modifed FN could be mediated by the intrinsic M1-polarizing activity of the cationic peptide included in the modified FN, since the unmodified FN showed only minor antitumor activity [57].

The discovery of the molecular mechanisms underpinning the immunosuppressive state in the TME led to the FDA approval of immune checkpoint inhibitors (ICIs) for cancer therapy, giving rise to novel immunotherapeutic options that are able to induce a strong infiltration of active immune cells in the TME, with consequent control of tumor growth [113]. ICIs currently used in the clinical setting are monoclonal antibodies (mAbs) that are able to block the activity of the programmed cell death protein 1 (PD-1)/PD-L1 interaction or cytotoxic T-lymphocyte antigen-4 expressed by T cells. Recently, a DOX-loaded engineered FN displaying the PD-L1 binding peptide (PpNF) was developed by Seon and colleagues [34]. Interestingly, this novel nanoformulation was able to achieve enhanced tumor-growth reduction in the colon carcinoma CT26 xenograft model, as compared to anti-PD-L1 mAb and free DOX (Figure 5). In addition, the engineered FNs without DOX were shown to be superior to anti-PD-L1 mAb in enhancing the activity of T cells in vitro.

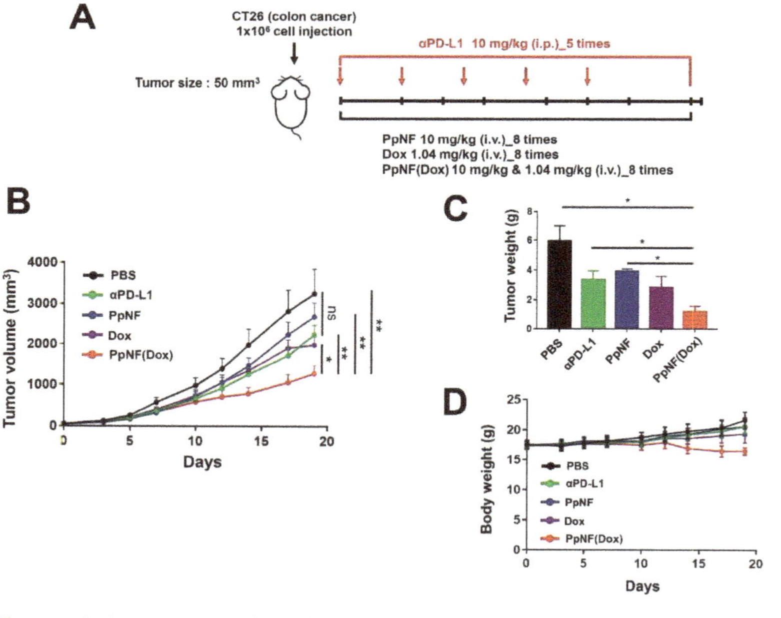

Figure 5. Antitumor activity of FNs displaying the PD-L1 binding peptide (PpNF). (**A**) Experimental schemes for antitumor treatments. Mice bearing s.c. CT26 syngeneic colon tumors were treated with DOX-loaded PpNF (PpNF(Dox)) or PpNF or Dox administered by i.v. injection three times per week. Anti-PD-L1 antibody was administered by i.p. injection twice per week. (**B**) Tumor volumes after treatment. (**C**) The weights of excised tumors from each group at the 19-day post-injection. (**D**) Body weights. The data represent means ± SEM (* $p < 0.05$ and ** $p < 0.01$; *t*-test). Reproduced with permission from [35], Elsevier, 2021.

We hypothesize that, in future years, novel FN-based delivery system will be employed to modulate the activity of immune cells, since a new paradigm for NP-based anticancer therapeutics is emerging [114]. In fact, the expanding arsenal of nanomedicines able to modulate the activity of TME-infiltrating immune cells could be utilized to support standard chemotherapeutics or immunotherapies in order to reactivate the antitumor immunity [115].

4.2. FN-Based NPs for the Treatment of Brain Tumors

The blood–brain barrier (BBB) is a diffusion barrier that impedes the influx of most compounds from the bloodstream to the brain parenchyma and represents a protective interface between the central nervous system and peripheral blood circulation [116]. Interestingly, brain cells require iron for metabolic processes; thus, transferrin and ferritin have to bypass the BBB in a process mediated by ligand–receptor recognition. Within the brain, TfR1 was shown to be expressed by capillary endothelial cells, choroid plexus epithelial cells and neurons, which increase the expression of TfR1 in condition of iron deficiency [117]. In addition, TfR1 has been shown to be overexpressed in brain tumors, particularly in glioblastomas, and its overexpression is associated with worse prognosis [118]. Taking into consideration these experimental evidences, we see that FNs are promising candidates for effective brain tumor therapy, due to their intrinsic targeting capability toward TfR1.

Fan and colleagues demonstrated that DOX-loaded FNs are able to bypass the BBB and deliver DOX to brain tumors in mice, dramatically increasing their survival compared to mice treated with control treatments (free DOX and Doxil™) [43]. Interestingly, FNs maintain their intact structure after crossing the BBB by transcytosis. This process is mediated by endothelial cells and allows the accumulation of FNs in the brain parenchyma in healthy mice. However, FNs co-localize with lysosomes after internalization in glioma cells. These results corroborate the idea that FNs traverse the BBB and effectively deliver therapeutics to brain tumors without affecting the surrounding tissues. The authors speculate that the different fate of FNs between endothelial and tumor cells could be due to the differences in expression of TfR1 [43]. More recently, $\alpha 2\beta 1$-targeted Dox-loaded FN ($\alpha\beta$-FN-DOX) was shown to have enhanced activity compared to FN–DOX in controlling the tumor growth of glioblastoma in an orthotopic model of brain tumor (U-87MG) [44]. Interestingly, $\alpha\beta$-FN–DOX had a higher drug-loading capacity compared to FN–DOX (60 vs. 15%, respectively). The authors speculate that this could be due to the modified integrin $\alpha 2\beta 1$ targeting sequence, which possesses multiple carboxyl groups that could have an impact in ionic interactions between DOX and $\alpha\beta$-FN.

In another report, PTX-loaded FNs were successfully developed by the disassembly and reassembly methodology and were used to treat C6 glioma bearing mice. The results showed enhanced activity of FN–PTX compared to the free drug in controlling tumor growth. Furthermore, treatment with FN–PTX showed no apparent signs of toxicity in the heart, liver, spleen, lung and kidneys of treated animals [45].

Other anticancer compounds, such as Au(III) thiosemicarbazone [55] and vincristine [52], have been successfully loaded into FNs and used to treat brain tumors in various murine models of cancer, achieving impressive results in controlling tumor growth. Interestingly, FNs can also facilitate the delivery of therapeutic mAb through the BBB. Our group has developed FNs coupled with Trastuzumab or Cetuximab, two FDA-approved mABs that are able to target the human epidermal growth factor receptor 2 and the epidermal growth factor receptor, respectively [46]. In addition, our group has developed a methodology to specifically study the translocation of FNs (or potentially other nanocarriers) through the BBB. This ex vivo model is based on layers of primary rat brain microvascular endothelial cells and astrocytes, which are used as a surrogate of the BBB [119].

We speculate that the development of novel FN therapeutics for the treatment of brain tumors will be particularly prominent in the coming years, since FNs have shown to effectively bypass the BBB without disassembly, leading to the release of FN-loaded therapeutics directly to tumor cells and avoiding off-site side effects.

5. Preclinical Exploitation of FN-Based NPs for Tumor Imaging

There is an ever-growing need for novel imaging agents that are able to effectively identify the presence of very small tumors in the early stage of the pathology, when they can be successfully treated by surgery or anticancer therapies. In addition, after successful surgery, patients undergo routine diagnostic tests to reveal the insurgence of metastatic events that could occur even several years after the original diagnosis [120]. Unfortunately, metastatic tumors are often incurable, since they usually become resistant to standard therapies and account for 90% of total cancer death worldwide [121]. Hence, it is critical to identify the presence of metastases with the current imaging modalities (computerized tomography (CT), magnetic resonance imaging (MRI) and positron emission tomography (PET)) which are often utilized together with contrast agents that are able to accumulate specifically in the TME.

In regards to MRI, gadolinium-based contrast agents are widely used, since they are able to identify highly vascularized tissues, such as tumors [122]. However, these types of agents are not cancer-specific and can result in a high rate of false positives. In addition, standard MRI does not have the sufficient spatial resolution to detect micro-metastases, thus leading to possible misdiagnoses of oligometastatic disease [123]. On the other hand, the glucose analog 18F-Fluorodeoxyglucose is a contrast agent utilized in PET/CT imaging to detect tumors, allowing for the visualization of areas with high metabolic activity [124]. Unfortunately, other areas characterized by active metabolic activity (benign tumors, inflammation sites and areas of ongoing infections) can give rise to false-positive results. Lastly, PET scans are quite expensive and require the use of radioactive contrast agents based on glucose that cannot be utilized in pregnant women or diabetic patients, due to off-target side effects [125].

NP-based delivery systems have been employed to enhance the specificity of contrast agents for tumor cells [126–128]. FNs have also been explored as a delivery platform for imaging probes, particularly for metal-based contrast agents. For example, multimodal FNs loaded with superparamagnetic iron oxide (Magnetoferritins) and a near-infrared fluorescence dyes were developed to efficiently detect tumors by multiple imaging modalities [58,59,65]. Interestingly, Magnetoferritins can efficiently identify very small tumors (~1 mm) by MRI in murine models of cancer, thus dramatically increasing the limit of detection of current contrast agents for MRI. Iron-loaded FNs derived from equine spleen (HoS–FN, composed of 85% of Lc and 15% of Hc subunits) were also utilized for MRI visualization of tumors [61]. Interestingly, HoS–FN showed enhanced uptake by SCAR5-positive cells, due to the specific targeting of this receptor mediated by Lc subunits. Furthermore, HoS–FNs were also able to reduce tumor growth, as compared to apoferritin HoS–FN in a murine model of breast cancer.

Imaging agents are also used in fluorescence-guided oncological surgery to assist the surgeon in the identification of metastatic foci, particularly in lymph nodes [129]. Indocyanine-green (ICG) is one of the most used FDA-approved fluorescent dyes for this purpose, since it can be visualized avoiding background autofluorescence (mainly due to hemoglobin) and has a low risk of adverse events [130,131]. Our group has developed ICG-loaded FNs with improved fluorescence accumulation in tumors in comparison to free ICG in a murine model of breast cancer [26,63]. Since specific accumulation of FN–ICG in tumors can be detected up to 24 h after intravenous injection in mice, we speculate that FN–ICG could be administered prior to surgery, and it could be visualized during surgery by fluorescence-guided endoscopy. This methodology could potentially reduce surgery time and improve the detection of small metastases, particularly in lymph nodes.

FNs were also developed to specifically visualize bone metastases by genetic manipulation of ferritin to include osteoblast and hydroxyapatite-binding peptides [64]. In another report, folic acid–functionalized FNs were developed to target tumor cells and deliver perfluoropentane, a compound used for low-intensity focused ultrasound imaging and therapy [62]. Lastly, gold NPs were efficiently encapsulated into 2-amino-2-deoxy-glucose-functionalized FNs to develop a tumor-targeted FN for CT imaging [60].

Overall, FN-based nanostructures can be utilized for the tumor-specific delivery of numerous contrast agents to improve their pharmacokinetic characteristics and enhance tumor accumulation.

6. Drawbacks and Future Perspective of FN

FNs are a versatile drug delivery system for chemotherapeutics and imaging agents. However, one of the major limitations of FNs in regards to drug loading is the low encapsulation capacity for hydrophobic compounds. This is primarily due to the leakage of the loaded hydrophobic drug from FN soon after encapsulation. Nonetheless, hydrophobic drugs can still be loaded inside FN by utilizing methodologies such as the pre-complexation with copper ions or the modification of native ferritin with hydrophobic amino acid sequences, that are able to enhance the affinity of hydrophobic compounds for ferritin. For this purpose, Wang and colleagues designed a novel FN construct (Am-PNCage) by linking the sequence of the Pout peptide (GRGDSKKHHHHHHAFAFAFAFVVVAA) to the C terminus of Hc ferritin through a flexible amino acid sequence GGSG, which replaced the E helix amino acids of Hc. This novel FN was employed to achieve the co-loading of the hydrophilic anthracycline EPI and the hydrophobic topoisomerase inhibitor Camptothecin and showed impressive antitumor activity in different murine models of cancer [47]. This novel FN construct could pave the way for the development of sophisticated FN-based nanostructures that are able to integrate multiple drugs with different mechanisms of action. This combinatorial nanotherapy could synergistically strike solid tumors by taking advantage of specific chemosensitivities to limit the insurgence of resistance to single chemotherapeutics.

An important area that is currently understudied is the relevance of the various modifications of FNs in regards to uptake and toxicity, particularly toward immune cells and erythrocytes in the bloodstream. For example, RGD-modified NP (nano-emulsions and liposomes) have been shown to be taken up by phagocytes in the bloodstream that are then able to transport NPs to specific sites in the body where there is ongoing inflammation and/or angiogenesis, such as the TME [132,133]. In fact, a majority of research efforts have been focused on showing that FN-loaded drugs can induce fewer side effects as compared to the free drug. This has been shown extensively for DOX, since FN–DOXs have an encouraging minimal effect on cardiomyocytes when compared to DOX [99,100]. However, it remains unclear if FN modifications can impact their uptake on different cell types present in the bloodstream. This area of study could be particularly relevant to pursue since it has been recently speculated that the enhanced accumulation of nanotherapeutics in the TME could be mediated not only by the EPR effect but also by the phenomenon of NP hitchhiking [81,114,133,134].

Indeed, FNs have favorable and interesting characteristics as an NP-based delivery system. Their efficient loading capacity for different drugs used in oncology, intrinsic targeting toward TfR1 and biocompatibility make them an ideal nano-platform for the treatment and imaging of tumors. Unfortunately, to date, FNs have not yet reached the clinical stage. In fact, the current high cost of production somewhat limits their translational potential. However, we speculate that the recent advancement concerning drug-loading efficiency and customization capabilities could facilitate the interest of pharmaceutical industries in developing novel production protocols for FNs that are aimed at enhancing purity, while, at the same time, reducing the costs of production. Collectively, the number of experimental evidences in support of the use of FNs as nano-delivery systems are ever-increasing, making their translation from bench to bedside a reasonable possibility. Lastly, the opportunity of co-encapsulating different drugs into FNs allows for the development of novel FN-based theranostic agents that are able to combine both imaging and therapeutic functionality in a fully biocompatible nanosystem. For these reasons, we believe that, in the near future, the clinical application of FNs could play a pivotal role in the diagnosis and treatment of solid tumors.

Author Contributions: F.C., F.M. and S.M. conceived of and designed the work; F.M., A.B., M.S., L.S., M.T., F.C. and S.M. wrote the manuscript; M.T., S.M., F.M. and F.C. revised the paper. All authors have read and agreed to the published version of the manuscript.

Funding: The research leading to these results has received funding from AIRC under IG 2017—ID. 20172 project—P.I. Corsi Fabio.

Institutional Review Board Statement: Not applicable.

Informed Consent Statement: Not applicable.

Acknowledgments: The authors acknowledge AIRC for F.M. postdoctoral fellowship and University of Milan for A.B., M.S. and L.S. doctoral and postdoctoral fellowship, respectively. Figures 1 and 2 were created with BioRender.com (accessed on 2 November 2021).

Conflicts of Interest: The authors declare no conflict of interest.

References

1. Anselmo, A.C.; Mitragotri, S. Nanoparticles in the clinic: An update. *Bioeng. Transl. Med.* **2019**, *4*, e10143. [CrossRef]
2. Wang, J.; Li, Y.; Nie, G. Multifunctional biomolecule nanostructures for cancer therapy. *Natl. Rev. Mater.* **2021**, *6*, 766–783. [CrossRef] [PubMed]
3. Wei, G.; Wang, Y.; Yang, G.; Wang, Y.; Ju, R. Recent progress in nanomedicine for enhanced cancer chemotherapy. *Theranostics* **2021**, *11*, 6370–6392. [CrossRef] [PubMed]
4. Gustafson, H.H.; Holt-Casper, D.; Grainger, D.W.; Ghandehari, H. Nanoparticle uptake: The phagocyte problem. *Nano Today* **2015**, *10*, 487–510. [CrossRef] [PubMed]
5. Nooraei, S.; Bahrulolum, H.; Hoseini, Z.S.; Katalani, C.; Hajizade, A.; Easton, A.J.; Ahmadian, G. Virus-Like particles: Preparation, immunogenicity and their roles as nanovaccines and drug nanocarriers. *J. Nanobiotechnol.* **2021**, *19*, 59. [CrossRef]
6. Bhaskar, S.; Lim, S. Engineering protein nanocages as carriers for biomedical applications. *NPG Asia Mater.* **2017**, *9*, e371. [CrossRef]
7. Truffi, M.; Fiandra, L.; Sorrentino, L.; Monieri, M.; Corsi, F.; Mazzucchelli, S. Ferritin nanocages: A biological platform for drug delivery, imaging and theranostics in cancer. *Pharmacol. Res.* **2016**, *107*, 57–65. [CrossRef] [PubMed]
8. Arosio, P.; Ingrassia, R.; Cavadini, P. Ferritins: A family of molecules for iron storage, antioxidation and more. *Biochim. Biophys. Acta Gen. Subj.* **2009**, *1790*, 589–599. [CrossRef]
9. Harrison, P.M.; Arosio, P. The ferritins: Molecular properties, iron storage function and cellular regulation. *Biochim. Biophys. Acta Bioenerg.* **1996**, *1275*, 161–203. [CrossRef]
10. Chasteen, N.D.; Harrison, P.M. Mineralization in ferritin: An efficient means of iron storage. *J. Struct. Biol.* **1999**, *126*, 182–194. [CrossRef]
11. Cohen, L.A.; Gutierrez, L.; Weiss, A.; Leichtmann-Bardoogo, Y.; Zhang, D.L.; Crooks, D.R.; Sougrat, R.; Morgenstern, A.; Galy, B.; Hentze, M.W.; et al. Serum ferritin is derived primarily from macrophages through a nonclassical secretory pathway. *Blood* **2010**, *116*, 1574–1584. [CrossRef]
12. Moreira, A.C.; Mesquita, G.; Gomes, M.S. Ferritin: An inflammatory player keeping iron at the core of pathogen-host interactions. *Microorganisms* **2020**, *8*, 589. [CrossRef]
13. Cullis, J.O.; Fitzsimons, E.J.; Griffiths, W.J.H.; Tsochatzis, E.; Thomas, D.W. Investigation and management of a raised serum ferritin. *Br. J. Haematol.* **2018**, *181*, 331–340. [CrossRef]
14. Camaschella, C.; Girelli, D. The changing landscape of iron deficiency. *Mol. Asp. Med.* **2020**, *75*, 100861. [CrossRef]
15. Asano, T.; Komatsu, M.; Yamaguchi-Iwai, Y.; Ishikawa, F.; Mizushima, N.; Iwai, K. Distinct Mechanisms of Ferritin Delivery to Lysosomes in Iron-Depleted and Iron-Replete Cells. *Mol. Cell. Biol.* **2011**, *31*, 2040–2052. [CrossRef]
16. Cai, C.; Ching, A.; Lagace, C.; Linsenmayer, T. Nuclear ferritin-mediated protection of corneal epithelial cells from oxidative damage to DNA. *Dev. Dyn.* **2008**, *237*, 2676–2683. [CrossRef]
17. Ahmad, S.; Moriconi, F.; Naz, N.; Sultan, S.; Sheikh, N.; Ramadori, G.; Malik, I.A. Ferritin L and ferritin H are differentially located within hepatic and extra hepatic organs under physiological and acute phase conditions. *Int. J. Clin. Exp. Pathol.* **2013**, *6*, 622–629.
18. Darshan, D.; Vanoaica, L.; Richman, L.; Beermann, F.; Kühn, L.C. Conditional deletion of ferritin H in mice induces loss of iron storage and liver damage. *Hepatology* **2009**, *50*, 852–860. [CrossRef]
19. Ryu, M.S.; Zhang, D.; Protchenko, O.; Shakoury-Elizeh, M.; Philpott, C.C. PCBP1 and NCOA4 regulate erythroid iron storage and heme biosynthesis. *J. Clin. Investig.* **2017**, *127*, 1786–1797. [CrossRef]
20. Surguladze, N.; Patton, S.; Cozzi, A.; Fried, M.G.; Connor, J.R. Characterization of nuclear ferritin and mechanism of translocation. *Biochem. J.* **2005**, *388*, 731–740. [CrossRef]

21. Zhang, L.; Li, L.; Di Penta, A.; Carmona, U.; Yang, F.; Schöps, R.; Brandsch, M.; Zugaza, J.L.; Knez, M. H-Chain Ferritin: A Natural Nuclei Targeting and Bioactive Delivery Nanovector. *Adv. Healthc. Mater.* **2015**, *4*, 1305–1310. [CrossRef]
22. Bellini, M.; Mazzucchelli, S.; Galbiati, E.; Sommaruga, S.; Fiandra, L.; Truffi, M.; Rizzuto, M.A.; Colombo, M.; Tortora, P.; Corsi, F.; et al. Protein nanocages for self-triggered nuclear delivery of DNA-targeted chemotherapeutics in Cancer Cells. *J. Control. Release* **2014**, *196*, 184–196. [CrossRef]
23. Kurzątkowska, K.; Pazos, M.A., II; Herschkowitz, J.I.; Hepel, M. Cancer-Targeted Controlled Delivery of Chemotherapeutic Anthracycline Derivatives Using Apoferritin Nanocage Carriers. *Int. J. Mol. Sci.* **2021**, *22*, 1362. [CrossRef]
24. Santambrogio, P.; Levi, S.; Arosio, P.; Palagi, L.; Vecchio, G.; Lawson, D.M.; Yewdall, S.J.; Artymiuk, P.J.; Harrison, P.M.; Jappelli, R.; et al. Evidence that a salt bridge in the light chain contributes to the physical stability difference between heavy and light human ferritins. *J. Biol. Chem.* **1992**, *267*, 14077–14083. [CrossRef]
25. Yang, R.; Liu, Y.; Meng, D.; Wang, D.; Blanchard, C.L.; Zhou, Z. Effect of atmospheric cold plasma on structure, activity, and reversible assembly of the phytoferritin. *Food Chem.* **2018**, *264*, 41–48. [CrossRef]
26. Sitia, L.; Sevieri, M.; Bonizzi, A.; Allevi, R.; Morasso, C.; Foschi, D.; Corsi, F.; Mazzucchelli, S. Development of Tumor-Targeted Indocyanine Green-Loaded Ferritin Nanoparticles for Intraoperative Detection of Cancers. *ACS Omega* **2020**, *5*, 12035–12045. [CrossRef] [PubMed]
27. Jiang, B.; Chen, X.; Sun, G.; Chen, X.; Yin, Y.; Jin, Y.; Mi, Q.; Ma, L.; Yang, Y.; Yan, X.; et al. A natural drug entry channel in the ferritin nanocage. *Nano Today* **2020**, *35*, 100948. [CrossRef]
28. Zhen, Z.; Tang, W.; Chen, H.; Lin, X.; Todd, T.; Wang, G.; Cowger, T.; Chen, X.; Xie, J. RGD-Modified Apoferritin Nanoparticles for Efficient Drug Delivery to Tumors. *ACS Nano* **2013**, *7*, 4830–4837. [CrossRef]
29. He, J.; Fan, K.; Yan, X. Ferritin drug carrier (FDC) for tumor targeting therapy. *J. Control. Release* **2019**, *311–312*, 288–300. [CrossRef] [PubMed]
30. Zhang, J.; Cheng, D.; He, J.; Hong, J.; Yuan, C.; Liang, M. Cargo loading within ferritin nanocages in preparation for tumor-targeted delivery. *Nat. Protoc.* **2021**, *16*, 4878–4896. [CrossRef]
31. Jin, P.; Sha, R.; Zhang, Y.; Liu, L.; Bian, Y.; Qian, J.; Qian, J.; Lin, J.; Ishimwe, N.; Hu, Y.; et al. Blood Circulation-Prolonging Peptides for Engineered Nanoparticles Identified via Phage Display. *Nano Lett.* **2019**, *19*, 1467–1478. [CrossRef]
32. Wang, Z.; Dai, Y.; Wang, Z.; Jacobson, O.; Zhang, F.; Yung, B.C.; Zhang, P.; Gao, H.; Niu, G.; Liu, G.; et al. Metal ion assisted interface re-engineering of a ferritin nanocage for enhanced biofunctions and cancer therapy. *Nanoscale* **2018**, *10*, 1135–1144. [CrossRef]
33. Huang, H.; Yuan, S.; Ma, Z.; Ji, P.; Ma, X.; Wu, Z.; Qi, X. Genetic recombination of poly(l-lysine) functionalized apoferritin nanocages that resemble viral capsid nanometer-sized platforms for gene therapy. *Biomater. Sci.* **2020**, *8*, 1759–1770. [CrossRef]
34. Jeon, I.S.; Yoo, J.D.; Gurung, S.; Kim, M.; Lee, C.; Park, E.J.; Park, R.W.; Lee, B.; Kim, S. Anticancer nanocage platforms for combined immunotherapy designed to harness immune checkpoints and deliver anticancer drugs. *Biomaterials* **2021**, *270*, 120685. [CrossRef]
35. Ma, Y.; Li, R.; Dong, Y.; You, C.; Huang, S.; Li, X.; Wang, F.; Zhang, Y. tLyP-1 peptide functionalized human H chain ferritin for targeted delivery of paclitaxel. *Int. J. Nanomed.* **2021**, *16*, 789–802. [CrossRef]
36. Andreata, F.; Bonizzi, A.; Sevieri, M.; Truffi, M.; Monieri, M.; Sitia, L.; Silva, F.; Sorrentino, L.; Allevi, R.; Zerbi, P.; et al. Co-Administration of H-ferritin-doxorubicin and Trastuzumab in neoadjuvant setting improves efficacy and prevents cardiotoxicity in HER2 + murine breast cancer model. *Sci. Rep.* **2020**, *10*, 11425. [CrossRef]
37. Huang, X.; Chisholm, J.; Zhuang, J.; Xiao, Y.; Duncan, G.; Chen, X.; Suk, J.S.; Hanes, J. Protein nanocages that penetrate airway mucus and tumor tissue. *Proc. Natl. Acad. Sci. USA* **2017**, *114*, E6595–E6602. [CrossRef]
38. Huang, X.; Zhuang, J.; Chung, S.W.; Huang, B.; Halpert, G.; Negron, K.; Sun, X.; Yang, J.; Oh, Y.; Hwang, P.M.; et al. Hypoxia-tropic Protein Nanocages for Modulation of Tumor- and Chemotherapy-Associated Hypoxia. *ACS Nano* **2019**, *13*, 236–247. [CrossRef]
39. Mazzucchelli, S.; Truffi, M.; Baccarini, F.; Beretta, M.; Sorrentino, L.; Bellini, M.; Rizzuto, M.A.; Ottria, R.; Ravelli, A.; Ciuffreda, P.; et al. H-Ferritin-nanocaged olaparib: A promising choice for both BRCA-mutated and sporadic triple negative breast cancer. *Sci. Rep.* **2017**, *7*, 7505. [CrossRef]
40. Bonizzi, A.; Truffi, M.; Sevieri, M.; Allevi, R.; Sitia, L.; Ottria, R.; Sorrentino, L.; Sottani, C.; Negri, S.; Grignani, E.; et al. Everolimus nanoformulation in biological nanoparticles increases drug responsiveness in resistant and low-responsive breast cancer cell lines. *Pharmaceutics* **2019**, *11*, 384. [CrossRef]
41. Pandolfi, L.; Bellini, M.; Vanna, R.; Morasso, C.; Zago, A.; Carcano, S.; Avvakumova, S.; Bertolini, J.; Rizzuto, M.; Colombo, M.; et al. H-Ferritin Enriches the Curcumin Uptake and Improves the Therapeutic Efficacy in Triple Negative Breast Cancer Cells. *Biomacromolecules* **2017**, *18*, 3318–3330. [CrossRef]
42. Sitia, L.; Bonizzi, A.; Mazzucchelli, S.; Negri, S.; Sottani, C.; Grignani, E.; Rizzuto, M.A.; Prosperi, D.; Sorrentino, L.; Morasso, C.; et al. Selective Targeting of Cancer-Associated Fibroblasts by Engineered H-Ferritin Nanocages Loaded with Navitoclax. *Cells* **2021**, *10*, 328. [CrossRef]
43. Fan, K.; Jia, X.; Zhou, M.; Wang, K.; Conde, J.; He, J.; Tian, J.; Yan, X. Ferritin Nanocarrier Traverses the Blood Brain Barrier and Kills Glioma. *ACS Nano* **2018**, *12*, 4105–4115. [CrossRef] [PubMed]
44. Huang, C.-W.; Chuang, C.-P.; Chen, Y.-J.; Wang, H.-Y.; Lin, J.-J.; Huang, C.-Y.; Wei, K.-C.; Huang, F.-T. Integrin $\alpha 2\beta 1$-targeting ferritin nanocarrier traverses the blood–brain barrier for effective glioma chemotherapy. *J. Nanobiotechnol.* **2021**, *19*, 180. [CrossRef]
45. Liu, W.; Lin, Q.; Fu, Y.; Huang, S.; Guo, C.; Li, L.; Wang, L.; Zhang, Z.; Zhang, L. Target delivering paclitaxel by ferritin heavy chain nanocages for glioma treatment. *J. Control. Release* **2020**, *323*, 191–202. [CrossRef]

46. Rizzuto, M.A.; Magro, R.D.; Barbieri, L.; Pandolfi, L.; Sguazzini-Viscontini, A.; Truffi, M.; Salvioni, L.; Corsi, F.; Colombo, M.; Re, F.; et al. H-Ferritin nanoparticle-mediated delivery of antibodies across a BBB in vitro model for treatment of brain malignancies. *Biomater. Sci.* **2021**, *9*, 2032–2042. [CrossRef]
47. Wang, Z.; Zhang, S.; Zhang, R.; Chen, X.; Sun, G.; Zhou, M.; Han, Q.; Zhang, B.; Zhao, Y.; Jiang, B.; et al. Bioengineered Dual-Targeting Protein Nanocage for Stereoscopical Loading of Synergistic Hydrophilic/Hydrophobic Drugs to Enhance Anticancer Efficacy. *Adv. Funct. Mater.* **2021**, *31*, 2102004. [CrossRef]
48. Jiang, B.; Zhang, R.; Zhang, J.; Hou, Y.; Chen, X.; Zhou, M.; Tian, X.; Hao, C.; Fan, K.; Yan, X. GRP78-targeted ferritin nanocaged ultra-high dose of doxorubicin for hepatocellular carcinoma therapy. *Theranostics* **2019**, *9*, 2167–2182. [CrossRef] [PubMed]
49. Tan, T.; Wang, Y.; Wang, H.; Cao, H.; Wang, Z.; Wang, J.; Li, J.; Li, Y.; Zhang, Z.; Wang, S. Apoferritin nanocages loading mertansine enable effective eradiation of cancer stem-like cells in vitro. *Int. J. Pharm.* **2018**, *553*, 201–209. [CrossRef]
50. Ferraro, G.; Pratesi, A.; Cirri, D.; Imbimbo, P.; Monti, D.M.; Messori, L.; Merlino, A. Arsenoplatin-Ferritin nanocage: Structure and cytotoxicity. *Int. J. Mol. Sci.* **2021**, *22*, 1874. [CrossRef] [PubMed]
51. Li, Y.; Wang, X.; Yan, J.; Liu, Y.; Yang, R.; Pan, D.; Wang, L.; Xu, Y.; Li, X.; Yang, M. Nanoparticle ferritin-bound erastin and rapamycin: A nanodrug combining autophagy and ferroptosis for anticancer therapy. *Biomater. Sci.* **2019**, *7*, 3779–3787. [CrossRef]
52. Zhai, M.; Wang, Y.; Zhang, L.; Liang, M.; Fu, S.; Cui, L.; Yang, M.; Gong, W.; Li, Z.; Yu, L.; et al. Glioma targeting peptide modified apoferritin nanocage. *Drug Deliv.* **2018**, *25*, 1013–1024. [CrossRef] [PubMed]
53. Lin, C.Y.; Yang, S.J.; Peng, C.L.; Shieh, M.J. Panitumumab-Conjugated and Platinum-Cored pH-Sensitive Apoferritin Nanocages for Colorectal Cancer-Targeted Therapy. *ACS Appl. Mater. Interfaces* **2018**, *10*, 6096–6106. [CrossRef] [PubMed]
54. Zheng, Q.; Cheng, W.; Zhang, X.; Shao, R.; Li, Z. A pH-Induced Reversible Assembly System with Resveratrol-Controllable Loading and Release for Enhanced Tumor-Targeting Chemotherapy. *Nanoscale Res. Lett.* **2019**, *14*, 305. [CrossRef]
55. Zhang, J.; Zhang, Z.; Jiang, M.; Li, S.; Yuan, H.; Sun, H.; Yang, F.; Liang, H. Developing a Novel Gold(III) Agent to Treat Glioma Based on the Unique Properties of Apoferritin Nanoparticles: Inducing Lethal Autophagy and Apoptosis. *J. Med. Chem.* **2020**, *63*, 13695–13708. [CrossRef]
56. Wang, W.; Liu, Z.; Zhou, X.; Guo, Z.; Zhang, J.; Zhu, P.; Yao, S.; Zhu, M. Ferritin nanoparticle-based SpyTag/SpyCatcher-enabled click vaccine for tumor immunotherapy. *Nanomed. Nanotechnol. Biol. Med.* **2019**, *16*, 69–78. [CrossRef]
57. Shan, H.; Dou, W.; Zhang, Y.; Qi, M. Targeted ferritin nanoparticle encapsulating CpG oligodeoxynucleotides induces tumor-associated macrophage M2 phenotype polarization into M1 phenotype and inhibits tumor growth. *Nanoscale* **2020**, *12*, 22268–22280. [CrossRef]
58. Jiang, B.; Jia, X.; Ji, T.; Zhou, M.; He, J.; Wang, K.; Tian, J.; Yan, X.; Fan, K. Ferritin nanocages for early theranostics of tumors via inflammation-enhanced active targeting. *Sci. China Life Sci.* **2021**. online ahead of print. [CrossRef]
59. Song, R.; Ruan, M.; Dai, J.; Xue, W. Biomimetic magnetofluorescent ferritin nanoclusters for magnetic resonance and fluorescence-dual modal imaging and targeted tumor therapy. *J. Mater. Chem. B* **2021**, *9*, 2494–2504. [CrossRef]
60. Aslan, T.N.; Aşık, E.; Güray, N.T.; Volkan, M. The potential application of gold-apoferritin nanocages conjugated with 2-amino-2-deoxy-glucose for imaging of breast cancer cells. *JBIC J. Biol. Inorg. Chem.* **2020**, *25*, 1139–1152. [CrossRef]
61. Bitonto, V.; Alberti, D.; Ruiu, R.; Aime, S.; Geninatti Crich, S.; Cutrin, J.C. L-ferritin: A theranostic agent of natural origin for MRI visualization and treatment of breast cancer. *J. Control. Release* **2020**, *319*, 300–310. [CrossRef]
62. Li, J.; Ji, H.; Jing, Y.; Wang, S. pH- and acoustic-responsive platforms based on perfluoropentane-loaded protein nanoparticles for ovarian tumor-targeted ultrasound imaging and therapy. *Nanoscale Res. Lett.* **2020**, *15*, 31. [CrossRef]
63. Sevieri, M.; Sitia, L.; Bonizzi, A.; Truffi, M.; Mazzucchelli, S.; Corsi, F. Tumor Accumulation and Off-Target Biodistribution of an Indocyanine-Green Fluorescent Nanotracer: An Ex Vivo Study on an Orthotopic Murine Model of Breast Cancer. *Int. J. Mol. Sci.* **2021**, *22*, 1601. [CrossRef]
64. Kim, J.-W.; Lee, K.-K.; Park, K.-W.; Kim, M.; Lee, C.-S. Genetically Modified Ferritin Nanoparticles with Bone-Targeting Peptides for Bone Imaging. *Int. J. Mol. Sci.* **2021**, *22*, 4854. [CrossRef]
65. Cao, C.; Wang, X.; Cai, Y.; Sun, L.; Tian, L.; Wu, H.; He, X.; Lei, H.; Liu, W.; Chen, G.; et al. Targeted In Vivo Imaging of Microscopic Tumors with Ferritin-based Nanoprobes Across Biological Barriers. *Adv. Mater.* **2014**, *26*, 2566–2571. [CrossRef]
66. Zhang, C.; Zhang, X.; Zhao, G. Ferritin nanocage: A versatile nanocarrier utilized in the field of food, nutrition, and medicine. *Nanomaterials* **2020**, *10*, 1894. [CrossRef]
67. Plath, L.D.; Ozdemir, A.; Aksenov, A.A.; Bier, M.E. Determination of Iron Content and Dispersity of Intact Ferritin by Superconducting Tunnel Junction Cryodetection Mass Spectrometry. *Anal. Chem.* **2015**, *87*, 8985–8993. [CrossRef]
68. Silva, F.; Sitia, L.; Allevi, R.; Bonizzi, A.; Sevieri, M.; Morasso, C.; Truffi, M.; Corsi, F.; Mazzucchelli, S. Combined method to remove endotoxins from protein nanocages for drug delivery applications: The case of human ferritin. *Pharmaceutics* **2021**, *13*, 229. [CrossRef]
69. Jin, Y.; He, J.; Fan, K.; Yan, X. Ferritin variants: Inspirations for rationally designing protein nanocarriers. *Nanoscale* **2019**, *11*, 12449–12459. [CrossRef]
70. Zhang, S.; Zang, J.; Zhang, X.; Chen, H.; Mikami, B.; Zhao, G. "Silent" Amino Acid Residues at Key Subunit Interfaces Regulate the Geometry of Protein Nanocages. *ACS Nano* **2016**, *10*, 10382–10388. [CrossRef]
71. Wang, W.; Wang, L.; Chen, H.; Zang, J.; Zhao, X.; Zhao, G.; Wang, H. Selective Elimination of the Key Subunit Interfaces Facilitates Conversion of Native 24-mer Protein Nanocage into 8-mer Nanorings. *J. Am. Chem. Soc.* **2018**, *140*, 14078–14081. [CrossRef]

72. Choi, S.H.; Choi, K.; Chan Kwon, I.; Ahn, H.J. The incorporation of GALA peptide into a protein cage for an acid-inducible molecular switch. *Biomaterials* **2010**, *31*, 5191–5198. [CrossRef]
73. Wang, W.; Wang, L.; Li, G.; Zhao, G.; Zhao, X.; Wang, H. AB loop engineered ferritin nanocages for drug loading under benign experimental conditions. *Chem. Commun.* **2019**, *55*, 12344–12347. [CrossRef]
74. Zang, J.; Chen, H.; Zhang, X.; Zhang, C.; Guo, J.; Du, M.; Zhao, G. Disulfide-Mediated conversion of 8-mer bowl-like protein architecture into three different nanocages. *Nat. Commun.* **2019**, *10*, 778. [CrossRef] [PubMed]
75. Gu, C.; Zhang, T.; Lv, C.; Liu, Y.; Wang, Y.; Zhao, G. His-Mediated Reversible Self-Assembly of Ferritin Nanocages through Two Different Switches for Encapsulation of Cargo Molecules. *ACS Nano* **2020**, *14*, 17080–17090. [CrossRef]
76. Wang, C.; Zhang, C.; Li, Z.; Yin, S.; Wang, Q.; Guo, F.; Zhang, Y.; Yu, R.; Liu, Y.; Su, Z. Extending Half Life of H-Ferritin Nanoparticle by Fusing Albumin Binding Domain for Doxorubicin Encapsulation. *Biomacromolecules* **2018**, *19*, 773–781. [CrossRef]
77. Falvo, E.; Tremante, E.; Arcovito, A.; Papi, M.; Elad, N.; Boffi, A.; Morea, V.; Conti, G.; Toffoli, G.; Fracasso, G.; et al. Improved Doxorubicin Encapsulation and Pharmacokinetics of Ferritin-Fusion Protein Nanocarriers Bearing Proline, Serine, and Alanine Elements. *Biomacromolecules* **2016**, *17*, 514–522. [CrossRef]
78. Yin, S.; Wang, Y.; Zhang, B.; Qu, Y.; Liu, Y.; Dai, S.; Zhang, Y.; Wang, Y.; Bi, J. Engineered human heavy-chain ferritin with half-life extension and tumor targeting by PAS and RGDK peptide functionalization. *Pharmaceutics* **2021**, *13*, 521. [CrossRef]
79. Falvo, E.; Malagrinò, F.; Arcovito, A.; Fazi, F.; Colotti, G.; Tremante, E.; Di Micco, P.; Braca, A.; Opri, R.; Giuffrè, A.; et al. The presence of glutamate residues on the PAS sequence of the stimuli-sensitive nano-ferritin improves in vivo biodistribution and mitoxantrone encapsulation homogeneity. *J. Control. Release* **2018**, *275*, 177–185. [CrossRef]
80. Zheng, L.; Hu, X.; Wu, H.; Mo, L.; Xie, S.; Li, J.; Peng, C.; Xu, S.; Qiu, L.; Tan, W. In Vivo Monocyte/Macrophage-Hitchhiked Intratumoral Accumulation of Nanomedicines for Enhanced Tumor Therapy. *J. Am. Chem. Soc.* **2020**, *142*, 382–391. [CrossRef]
81. Zelepukin, I.V.; Yaremenko, A.V.; Shipunova, V.O.; Babenyshev, A.V.; Balalaeva, I.V.; Nikitin, P.I.; Deyev, S.M.; Nikitin, M.P. Nanoparticle-Based drug delivery via RBC-hitchhiking for the inhibition of lung metastases growth. *Nanoscale* **2019**, *11*, 1636–1646. [CrossRef] [PubMed]
82. Ahn, B.; Lee, S.G.; Yoon, H.R.; Lee, J.M.; Oh, H.J.; Kim, H.M.; Jung, Y. Four-Fold Channel-Nicked Human Ferritin Nanocages for Active Drug Loading and pH-Responsive Drug Release. *Angew. Chem. Int. Ed.* **2018**, *57*, 2909–2913. [CrossRef]
83. Men, D.; Zhang, T.-T.; Hou, L.-W.; Zhou, J.; Zhang, Z.-P.; Shi, Y.-Y.; Zhang, J.-L.; Cui, Z.-Q.; Deng, J.-Y.; Wang, D.-B.; et al. Self-Assembly of Ferritin Nanoparticles into an Enzyme Nanocomposite with Tunable Size for Ultrasensitive Immunoassay. *ACS Nano* **2015**, *9*, 10852–10860. [CrossRef] [PubMed]
84. Lee, E.J.; Lee, S.J.; Kang, Y.S.; Ryu, J.H.; Kwon, K.C.; Jo, E.; Yhee, J.Y.; Kwon, I.C.; Kim, K.; Lee, J. Engineered proteinicles for targeted delivery of siRNA to cancer cells. *Adv. Funct. Mater.* **2015**, *25*, 1279–1286. [CrossRef]
85. Li, L.; Muñoz-Culla, M.; Carmona, U.; Lopez, M.P.; Yang, F.; Trigueros, C.; Otaegui, D.; Zhang, L.; Knez, M. Ferritin-Mediated siRNA delivery and gene silencing in human tumor and primary cells. *Biomaterials* **2016**, *98*, 143–151. [CrossRef]
86. Kanekiyo, M.; Wei, C.J.; Yassine, H.M.; McTamney, P.M.; Boyington, J.C.; Whittle, J.R.R.; Rao, S.S.; Kong, W.P.; Wang, L.; Nabel, G.J. Self-Assembling influenza nanoparticle vaccines elicit broadly neutralizing H1N1 antibodies. *Nature* **2013**, *499*, 102–106. [CrossRef]
87. Lee, B.-R.; Ko, H.K.; Ryu, J.H.; Ahn, K.Y.; Lee, Y.-H.; Oh, S.J.; Na, J.H.; Kim, T.W.; Byun, Y.; Kwon, I.C.; et al. Engineered Human Ferritin Nanoparticles for Direct Delivery of Tumor Antigens to Lymph Node and Cancer Immunotherapy. *Sci. Rep.* **2016**, *6*, 35182. [CrossRef] [PubMed]
88. Han, J.A.; Kang, Y.J.; Shin, C.; Ra, J.S.; Shin, H.H.; Hong, S.Y.; Do, Y.; Kang, S. Ferritin protein cage nanoparticles as versatile antigen delivery nanoplatforms for dendritic cell (DC)-based vaccine development. *Nanomed. Nanotechnol. Biol. Med.* **2014**, *10*, 561–569. [CrossRef]
89. Jiang, B.; Yan, L.; Zhang, J.; Zhou, M.; Shi, G.; Tian, X.; Fan, K.; Hao, C.; Yan, X. Biomineralization Synthesis of the Cobalt Nanozyme in SP94-Ferritin Nanocages for Prognostic Diagnosis of Hepatocellular Carcinoma. *ACS Appl. Mater. Interfaces* **2019**, *11*, 9747–9755. [CrossRef]
90. Chuckran, C.A.; Liu, C.; Bruno, T.C.; Workman, C.J.; Vignali, D.A. Neuropilin-1: A checkpoint target with unique implications for cancer immunology and immunotherapy. *J. Immunother. Cancer* **2020**, *8*, e000967. [CrossRef]
91. Khoshnejad, M.; Greineder, C.F.; Pulsipher, K.W.; Villa, C.H.; Altun, B.; Pan, D.C.; Tsourkas, A.; Dmochowski, I.J.; Muzykantov, V.R. Ferritin Nanocages with Biologically Orthogonal Conjugation for Vascular Targeting and Imaging. *Bioconjug. Chem.* **2018**, *29*, 1209–1218. [CrossRef]
92. Palombarini, F.; Masciarelli, S.; Incocciati, A.; Liccardo, F.; Di Fabio, E.; Iazzetti, A.; Fabrizi, G.; Fazi, F.; Macone, A.; Bonamore, A.; et al. Self-Assembling ferritin-dendrimer nanoparticles for targeted delivery of nucleic acids to myeloid leukemia cells. *J. Nanobiotechnol.* **2021**, *19*, 172. [CrossRef]
93. Suk, J.S.; Xu, Q.; Kim, N.; Hanes, J.; Ensign, L.M. PEGylation as a strategy for improving nanoparticle-based drug and gene delivery. *Adv. Drug Deliv. Rev.* **2016**, *99*, 28–51. [CrossRef] [PubMed]
94. Tosi, G.; Belletti, D.; Pederzoli, F.; Ruozi, B. Apoferritin nanocage as drug reservoir: Is it a reliable drug delivery system? *Expert Opin. Drug Deliv.* **2016**, *13*, 1341–1343. [CrossRef] [PubMed]
95. Ramadori, G.; Cameron, S. Effects of systemic chemotherapy on the liver. *Ann. Hepatol.* **2010**, *9*, 133–143. [CrossRef]
96. Oun, R.; Moussa, Y.E.; Wheate, N.J. The side effects of platinum-based chemotherapy drugs: A review for chemists. *Dalt. Trans.* **2018**, *47*, 6645–6653. [CrossRef]

97. Branca, J.J.V.; Carrino, D.; Gulisano, M.; Ghelardini, C.; Di Cesare Mannelli, L.; Pacini, A. Oxaliplatin-Induced Neuropathy: Genetic and Epigenetic Profile to Better Understand How to Ameliorate This Side Effect. *Front. Mol. Biosci.* **2021**, *8*, 643824. [CrossRef]
98. Jagieła, J.; Bartnicki, P.; Rysz, J. Nephrotoxicity as a Complication of Chemotherapy and Immunotherapy in the Treatment of Colorectal Cancer, Melanoma and Non-Small Cell Lung Cancer. *Int. J. Mol. Sci.* **2021**, *22*, 4618. [CrossRef]
99. Mazzucchelli, S.; Bellini, M.; Fiandra, L.; Truffi, M.; Rizzuto, M.A.; Sorrentino, L.; Longhi, E.; Nebuloni, M.; Prosperi, D.; Corsi, F.; et al. Nanometronomic treatment of 4T1 breast cancer with nanocaged doxorubicin prevents drug resistance and circumvents cardiotoxicity. *Oncotarget* **2016**, *8*, 8383–8396. [CrossRef]
100. Liang, M.; Fan, K.; Zhou, M.; Duan, D.; Zheng, J.; Yang, D.; Feng, J.; Yan, X. H-ferritin-nanocaged doxorubicin nanoparticles specifically target and kill tumors with a single-dose injection. *Proc. Natl. Acad. Sci. USA* **2014**, *111*, 14900–14905. [CrossRef]
101. Yang, Z.; Wang, X.; Diao, H.; Zhang, J.; Li, H.; Sun, H.; Guo, Z. Encapsulation of platinum anticancer drugs by apoferritin. *Chem. Commun.* **2007**, *33*, 3453–3455. [CrossRef]
102. Ferraro, G.; Pica, A.; Petruk, G.; Pane, F.; Amoresano, A.; Cilibrizzi, A.; Vilar, R.; Monti, D.M.; Merlino, A. Preparation, structure, cytotoxicity and mechanism of action of ferritin-Pt(II) terpyridine compound nanocomposites. *Nanomedicine* **2018**, *13*, 2995–3007. [CrossRef]
103. Comerford, K.M.; Wallace, T.J.; Karhausen, J.; Louis, N.A.; Montalto, M.C.; Colgan, S.P. Hypoxia-Inducible factor-1-dependent regulation of the multidrug resistance (MDR1) gene. *Cancer Res.* **2002**, *62*, 3387–3394.
104. Tacchini, L.; Bianchi, L.; Bernelli-Zazzera, A.; Cairo, G. Transferrin Receptor Induction by Hypoxia: HIF-1-mediated transcriptional activation and cell-specific post-transcriptional regulation. *J. Biol. Chem.* **1999**, *274*, 24142–24146. [CrossRef]
105. Vitale, I.; Manic, G.; Coussens, L.M.; Kroemer, G.; Galluzzi, L. Macrophages and Metabolism in the Tumor Microenvironment. *Cell Metab.* **2019**, *30*, 36–50. [CrossRef]
106. Allavena, P.; Anfray, C.; Ummarino, A.; Andón, F.T. Therapeutic manipulation of tumor-associated macrophages: Facts and hopes from a clinical and translational perspective. *Clin. Cancer Res.* **2021**, *27*, 3291–3297. [CrossRef]
107. Malfitano, A.M.; Pisanti, S.; Napolitano, F.; Di Somma, S.; Martinelli, R.; Portella, G. Tumor-Associated macrophage status in cancer treatment. *Cancers* **2020**, *12*, 1987. [CrossRef]
108. Corna, G.; Campana, L.; Pignatti, E.; Castiglioni, A.; Tagliafico, E.; Bosurgi, L.; Campanella, A.; Brunelli, S.; Manfredi, A.A.; Apostoli, P.; et al. Polarization dictates iron handling by inflammatory and alternatively activated macrophages. *Haematologica* **2010**, *95*, 1814. [CrossRef]
109. Marques, O.; Porto, G.; Rêma, A.; Faria, F.; Paula, A.C.; Gomez-Lazaro, M.; Silva, P.; da Silva, B.M.; Lopes, C. Local iron homeostasis in the breast ductal carcinoma microenvironment. *BMC Cancer* **2016**, *16*, 187. [CrossRef]
110. Cairo, G.; Recalcati, S.; Mantovani, A.; Locati, M. Iron trafficking and metabolism in macrophages: Contribution to the polarized phenotype. *Trends Immunol.* **2011**, *32*, 241–247. [CrossRef]
111. Alkhateeb, A.A.; Han, B.; Connor, J.R. Ferritin stimulates breast cancer cells through an iron-independent mechanism and is localized within tumor-associated macrophages. *Breast Cancer Res. Treat.* **2013**, *137*, 733–744. [CrossRef]
112. Wei, S.C.; Duffy, C.R.; Allison, J.P. Fundamental mechanisms of immune checkpoint blockade therapy. *Cancer Discov.* **2018**, *8*, 1069–1086. [CrossRef]
113. Sofias, A.M.; Combes, F.; Koschmieder, S.; Storm, G.; Lammers, T. A paradigm shift in cancer nanomedicine: From traditional tumor targeting to leveraging the immune system. *Drug Discov. Today* **2021**, *26*, 1482–1489. [CrossRef]
114. Mainini, F.; De Santis, F.; Fucà, G.; Di Nicola, M.; Rivoltini, L.; Eccles, M. Nanobiotechnology and Immunotherapy: Two Powerful and Cooperative Allies against Cancer. *Cancers* **2021**, *13*, 3765. [CrossRef]
115. Sweeney, M.D.; Zhao, Z.; Montagne, A.; Nelson, A.R.; Zlokovic, B.V. Blood-Brain Barrier: From Physiology to Disease and Back. *Physiol. Rev.* **2019**, *99*, 21–78. [CrossRef]
116. Moos, T.; Morgan, E.H. Transferrin and Transferrin Receptor Function in Brain Barrier Systems. *Cell. Mol. Neurobiol.* **2000**, *20*, 77–95. [CrossRef]
117. Rosager, A.M.; Sørensen, M.D.; Dahlrot, R.H.; Hansen, S.; Schonberg, D.L.; Rich, J.N.; Lathia, J.D.; Kristensen, B.W. Transferrin receptor-1 and ferritin heavy and light chains in astrocytic brain tumors: Expression and prognostic value. *PLoS ONE* **2017**, *12*, e0182954. [CrossRef]
118. Fiandra, L.; Mazzucchelli, S.; Truffi, M.; Bellini, M.; Sorrentino, L.; Corsi, F. In Vitro Permeation of FITC-loaded Ferritins Across a Rat Blood-brain Barrier: A Model to Study the Delivery of Nanoformulated Molecules. *J. Vis. Exp.* **2016**, *2016*, 54279. [CrossRef]
119. Phan, T.G.; Croucher, P.I. The dormant cancer cell life cycle. *Nat. Rev. Cancer* **2020**, *20*, 398–411. [CrossRef]
120. Pienta, K.J.; Hammarlund, E.U.; Axelrod, R.; Amend, S.R.; Brown, J.S. Convergent Evolution, Evolving Evolvability, and the Origins of Lethal Cancer. *Mol. Cancer Res.* **2020**, *18*, 801–810. [CrossRef]
121. Jeong, Y.; Hwang, H.S.; Na, K. Theranostics and contrast agents for magnetic resonance imaging. *Biomater. Res.* **2018**, *22*, 20. [CrossRef]
122. Millet, I.; Pages, E.; Hoa, D.; Merigeaud, S.; Doyon, F.C.; Prat, X.; Taourel, P. Pearls and pitfalls in breast MRI. *Br. J. Radiol.* **2012**, *85*, 197–207. [CrossRef]
123. Farwell, M.D.; Pryma, D.A.; Mankoff, D.A. PET/CT imaging in cancer: Current applications and future directions. *Cancer* **2014**, *120*, 3433–3445. [CrossRef] [PubMed]

124. Kikano, E.G.; Avril, S.; Marshall, H.; Jones, R.S.; Montero, A.J.; Avril, N. PET/CT Variants and Pitfalls in Breast Cancers. *Semin. Nucl. Med.* **2021**, *51*, 474–484. [CrossRef] [PubMed]
125. Ma, T.; Zhang, P.; Hou, Y.; Ning, H.; Wang, Z.; Huang, J.; Gao, M. "Smart" Nanoprobes for Visualization of Tumor Microenvironments. *Adv. Healthc. Mater.* **2018**, *7*, e1800391. [CrossRef] [PubMed]
126. Li, Y.; Zhou, Y.; Yue, X.; Dai, Z. Cyanine Conjugate-Based Biomedical Imaging Probes. *Adv. Healthc. Mater.* **2020**, *9*, e2001327. [CrossRef] [PubMed]
127. Morato, Y.L.; Paredes, K.O.; Chamizo, L.L.; Marciello, M.; Filice, M. Recent Advances in Multimodal Molecular Imaging of Cancer Mediated by Hybrid Magnetic Nanoparticles. *Polymers* **2021**, *13*, 2989. [CrossRef] [PubMed]
128. Thammineedi, S.R.; Saksena, A.R.; Nusrath, S.; Iyer, R.R.; Shukla, S.; Patnaik, S.C.; Reddy, R.P.; Boleneni, N.; Sharma, R.M.; Smith, L.; et al. Fluorescence-Guided cancer surgery—A new paradigm. *J. Surg. Oncol.* **2021**, *123*, 1679–1698. [CrossRef] [PubMed]
129. Son, G.M.; Ahn, H.-M.; Lee, I.Y.; Ha, G.W. Multifunctional Indocyanine Green Applications for Fluorescence-Guided Laparoscopic Colorectal Surgery. *Ann. Coloproctol.* **2021**, *37*, 133–140. [CrossRef] [PubMed]
130. Egloff-Juras, C.; Bezdetnaya, L.; Dolivet, G.; Lassalle, H.-P. NIR fluorescence-guided tumor surgery: New strategies for the use of indocyanine green. *Int. J. Nanomed.* **2019**, *14*, 7823–7838. [CrossRef]
131. Sofias, A.M.; Bjørkøy, G.; Ochando, J.; Sønstevold, L.; Hegvik, M.; Davies, C.dL.; Haraldseth, O.; Lammers, T.; Mulder, W.J.M.; Hak, S. Cyclic Arginine–Glycine–Aspartate-Decorated Lipid Nanoparticle Targeting toward Inflammatory Lesions Involves Hitchhiking with Phagocytes. *Adv. Sci.* **2021**, *8*, 2100370. [CrossRef]
132. Sofias, A.M.; Toner, Y.C.; Meerwaldt, A.E.; van Leent, M.M.T.; Soultanidis, G.; Elschot, M.; Gonai, H.; Grendstad, K.; Flobak, Å.; Neckmann, U.; et al. Tumor Targeting by α v β 3-Integrin-Specific Lipid Nanoparticles Occurs via Phagocyte Hitchhiking. *ACS Nano* **2020**, *14*, 7832–7846. [CrossRef]
133. Dong, X.; Chu, D.; Wang, Z. Leukocyte-Mediated Delivery of Nanotherapeutics in Inflammatory and Tumor Sites. *Theranostics* **2017**, *7*, 751–763. [CrossRef] [PubMed]
134. Moore, T.L.; Hauser, D.; Gruber, T.; Rothen-Rutishauser, B.; Lattuada, M.; Petri-Fink, A.; Lyck, R. Cellular Shuttles: Monocytes/Macrophages Exhibit Transendothelial Transport of Nanoparticles under Physiological Flow. *ACS Appl. Mater. Interfaces* **2017**, *9*, 18501–18511. [CrossRef]

pharmaceutics

MDPI

Review

Nanoformulations of α-Mangostin for Cancer Drug Delivery System

Lisna Meylina [1,2], Muchtaridi Muchtaridi [3], I Made Joni [4,5], Ahmed Fouad Abdelwahab Mohammed [6] and Nasrul Wathoni [1,*]

1 Department of Pharmaceutics and Pharmaceutical Technology, Faculty of Pharmacy, Universitas Padjadjaran, Sumedang 45363, Indonesia; lisna@farmasi.unmul.ac.id
2 Department of Pharmaceutics and Pharmaceutical Technology, Faculty of Pharmacy, Universitas Mulawarman, Samarinda 75119, Indonesia
3 Department of Pharmaceutical Analysis and Medicinal Chemistry, Faculty of Pharmacy, Universitas Padjadjaran, Sumedang 45363, Indonesia; muchtaridi@unpad.ac.id
4 Department of Physics, Faculty of Mathematics and Natural Sciences, Universitas Padjadjaran, Sumedang 45363, Indonesia; imadejoni@phys.unpad.ac.id
5 Functional Nano Powder University Center of Excellence, Universitas Padjadjaran, Sumedang 45363, Indonesia
6 Department of Pharmaceutics, Faculty of Pharmacy, Minia University, Minia 61519, Egypt; ahmed.mohamed1@minia.edu.eg
* Correspondence: nasrul@unpad.ac.id; Tel.: +62-2842-888-888 (ext. 3510)

Abstract: Natural compounds are emerging as effective agents for the treatment of malignant diseases. The active constituent of α-mangostin from the pericarp of *Garcinia mangostana* L. has earned significant interest as a plant base compound with anticancer properties. Despite α-mangostin's superior properties as an anticancer agent, its applications are limited due to its poor solubility and physicochemical stability, rapid systemic clearance, and low cellular uptake. Our review aimed to summarize and discuss the nanoparticle formulations of α-mangostin for cancer drug delivery systems from published papers recorded in Scopus, PubMed, and Google Scholar. We investigated various types of α-mangostin nanoformulations to improve its anticancer efficacy by improving bioavailability, cellular uptake, and localization to specific areas These nanoformulations include nanofibers, lipid carrier nanostructures, solid lipid nanoparticles, polymeric nanoparticles, nanomicelles, liposomes, and gold nanoparticles. Notably, polymeric nanoparticles and nanomicelles can increase the accumulation of α-mangostin into tumors and inhibit tumor growth in vivo. In addition, polymeric nanoparticles with the addition of target ligands can increase the cellular uptake of α-mangostin. In conclusion, nanoformulations of α-mangostin are a promising tool to enhance the cellular uptake, accumulation in cancer cells, and the efficacy of α-mangostin as a candidate for anticancer drugs.

Keywords: *Garcinia mangostana* L.; nanotechnology; drug delivery; cancer therapy

Citation: Meylina, L.; Muchtaridi, M.; Joni, I.M.; Mohammed, A.F.A.; Wathoni, N. Nanoformulations of α-Mangostin for Cancer Drug Delivery System. *Pharmaceutics* **2021**, *13*, 1993. https://doi.org/10.3390/pharmaceutics13121993

Academic Editor: Mazzucchelli Serena

Received: 30 September 2021
Accepted: 8 November 2021
Published: 24 November 2021

1. Introduction

Cancer is the second leading cause of death globally and was responsible for an estimated 9.6 million worldwide deaths in 2018 [1]. The incidence of cancer cases is gradually increasing because of population growth and life expectancy with a projected increase in cases of up to 75% and is predicted to be the main cause of death by 2030 [2–4]. Currently, cancer treatments are surgery, radiotherapy, and anticancer drugs (chemotherapy and immunotherapy). Surgery and radiotherapy are most effective for the treatment of local and non-metastatic cancers, but less efficient for metastatic cancers [5,6]. Furthermore, anticancer drugs are the current choice for the treatment of metastatic cancer, because they can reach every organ in the body through the bloodstream [7]. However, anticancer drugs such as chemotherapy have many limitations, including the high incidence of side effects, limited effectiveness, multidrug resistance, and is highly toxic to growing healthy cells due to their non-specific targeting of cancer cells [8,9].

The development of natural product-based compounds is a promising strategy for cancer treatment. Various plant-based molecules such as α-mangostin, curcumin, and taxanes have chemopreventive or anticancer properties based on in vitro and in vivo studies [10–21]. α-mangostin is a xanthone derivate extracted from the skin of the fruit of *Garcinia mangostana* Linn known as the queen of fruits. It has revealed several anticancer properties [22–26] in various kinds of cancers such as colon [27,28], lung [29], pancreas [30], breast [22,31], skin [32], and blood [33]. However, α-mangostin has some physicochemical properties drawback, especially limited water solubility, resulting in poor absorption and low bioavailability on intravenous and oral administration, which can affect its effectiveness as an anticancer therapeutic agent [34–36].

Thus, to overcome those limitations, nanoparticles or nano-structured materials were introduced as a drug delivery system as an alternative to conventional drug delivery systems to achieve high water solubility and specific biodistribution [37,38]. Various nanotechnology-based drug delivery systems such as polymeric nanoparticles, solid lipid nanoparticles, liposomes, and nanomicelles have been designed to improve the low-water solubility of α-mangostin [39]. Many researchers reported that these formulations can enhance the therapeutic efficacy achieved from enhanced targeted localization and improved cellular internalization of α-mangostin nanoformulation in cancer cells. Despite many developments in nanoformulation of drug delivery systems for α-mangostin, the engineering on morphology and chemical structure related to their targeted or selective recognition of cancer cells and control released drug delivery system remain challenges. This review highlights various nanotechnological approaches used for α-mangostin delivery focused on cancer therapy.

2. Methodology

This review is based on the literature obtained from Google Scholar, PubMed, and Scopus using the keywords "nanoformulation of α-mangostin for cancer drug delivery system", "nanoparticle formulation of α-mangostin for cancer drug delivery system", and "α-mangostin nanoparticle for cancer drug delivery system" published in the last 10 years. Opinions, assessments, and unrelated subjects such as pharmacological characteristics and bioactivities were utilized as exclusion criteria. The flowchart of the methodology is shown in Figure 1. The distribution of articles based on the year of publication can be seen in Figure 2.

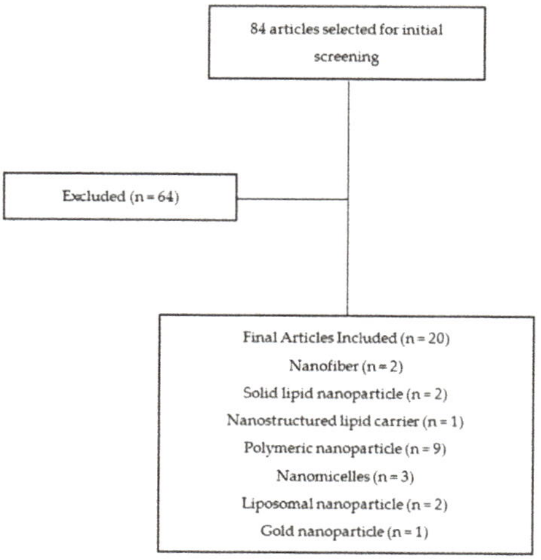

Figure 1. Flowchart of methodology.

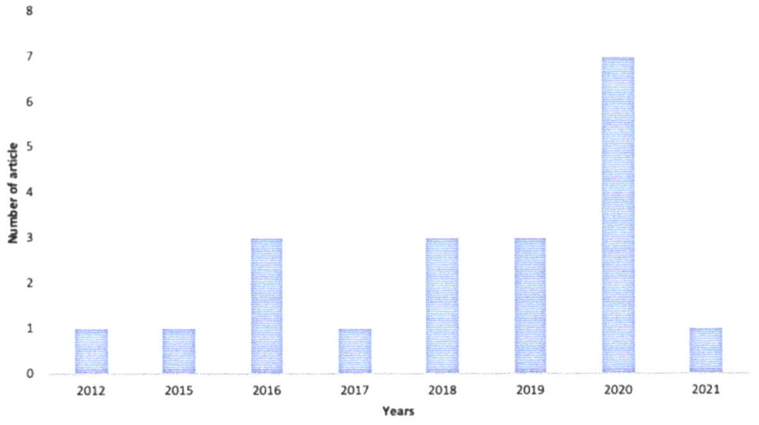

Figure 2. Distribution of articles based on the year of publication nanoformulation of α-mangostin for cancer.

3. α-Mangostin

α-mangostin (Figure 3) is a metabolite of 1,3,6,7-tetrahydroxy-2,8-di(3-methyl-2-butenyl) xanthones isolated from mangosteen pericarps [23,40–42]. α-mangostin is soluble in methanol and has a water solubility of 2.03×10^{-4} mg/L at 25 °C (Table 1) [43,44]. The low solubility of α-mangostin in water causes low bioavailability. Pharmacokinetic studies performed in mice after a single oral dose (20 mg/kg) showed low bioavailability (F = 2.29%) which was thought to be due to low gastrointestinal absorption and rapid metabolism of α-mangostin in the liver and small intestine [45].

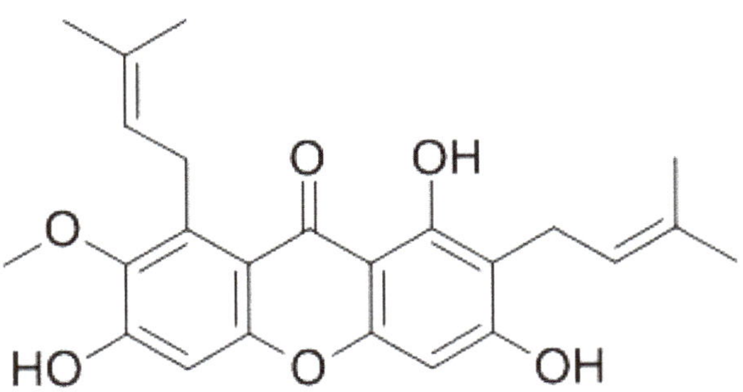

Figure 3. Chemical structure of α-mangostin.

α-mangostin has been found to possess a wide range of biological activities such as anticancer [22,27,46–49], antioxidant [50], antibacterial [51,52], hepatoprotective [53], cardioprotective [54], antimalarial [55], anti-obesity activities [56], and neuroprotective properties in Alzheimer's disease [57]. Previous studies have shown that α-mangostin acts against cancer cells through several mechanism pathways such as suppressing fatty acid synthase [31], downregulating the PI3K/Akt pathway [46], and β-catenin gene regulation [28]. Despite having the above-mentioned anticancer properties, its poor oral bioavailability remains the main drawback, limiting its clinical potential. As a result, these issues have led to the development of α-mangostin nanoformulations in the quest of improving α-mangostin delivery for better therapeutic outcomes.

Table 1. The physicochemical properties of α-mangostin [39,58].

Property	Description
Molecular formula	$C_{24}H_{26}O_6$
IUPAC name	1,3,6-Trihydroxy-7-methoxy-2,8-bis(3-methylbut-2-en-1-yl)-9H-xanthen-9-one
Molecular weight	410.5
Color/Form	Faint yellow to yellow powder
Melting point	180–181 °C
Solubility	Soluble in methanol, in water (2.03×10^{-4} mg/L at 25 °C)
LogP	log Kow = 7.71
Stability	Stable under normal temperatures and pressures
Dissociation constants	pKa1 = 3.68 (primary carbonyl); pKa2 = 7.69 (secondary carbonyl); pKa3 = 9.06 (tertiary carbonyl)

4. α-Mangostin Nanoformulation

The application of nanotechnology to medicine has resulted in the development of nanoparticle therapeutic carriers [59]. The recommended particle size for cancer treatment is 10–200 nm, which allows it to easily infiltrate tumor blood vessels that leak and collect in tumor tissue, reducing side effects [60].

Several types of nanoparticles are in different stages of the development process as cancer drug delivery systems, including nano polymers, micelles, liposomes, lipid-based carriers (lipid emulsions and lipid–drug conjugates), and several ligand-targeted products (such as an antibody, folate, and transferrin conjugated molecules) [59,61–63]. Nanoparticles usually consist of two or more components, and at least one of them is an active pharmaceutical ingredient. The materials and methods selection for nanoparticle assembly is mainly based on the size and shape of the nanoparticles, physicochemical properties of the active ingredient, the target delivery, and their stability and safety. Components and methods selected are very crucial to ensure the intended administration method and drug targeting abilities [64].

Nano-based cancer treatment, which employs a mixture of nanomaterials and chemotherapeutic drugs, plays a significant role in the therapeutic effects against cancer [60,65,66]. Nanoparticles preferentially accumulate in tumors due to their enhanced permeability and retention effect (EPR) (Figure 4), resulting in differences in the biodistribution between conventional chemotherapeutics and nanoparticle drug carriers, where nanoparticle-based chemotherapy can achieve higher drug concentrations in the intratumor environment, and eventually better therapeutic efficiency and lower toxicity [67–69]. Various nanoformulations are being explored to enhance the delivery of α-mangostin to tumor sites. α-mangostin nanoformulations for cancer should enhance anticancer activity and selectivity compared to free α-mangostin, and at the same time be non-toxic to normal cells. α-mangostin nanoformulations for cancer that have been reported in the literature include nanofibers, nanostructures lipid carrier, solid lipid nanoparticles, polymeric nanoparticles, nanomicelles, liposomes, and gold nanoparticles. These nanoformulations are described in Figure 5 and summarized in Table 2.

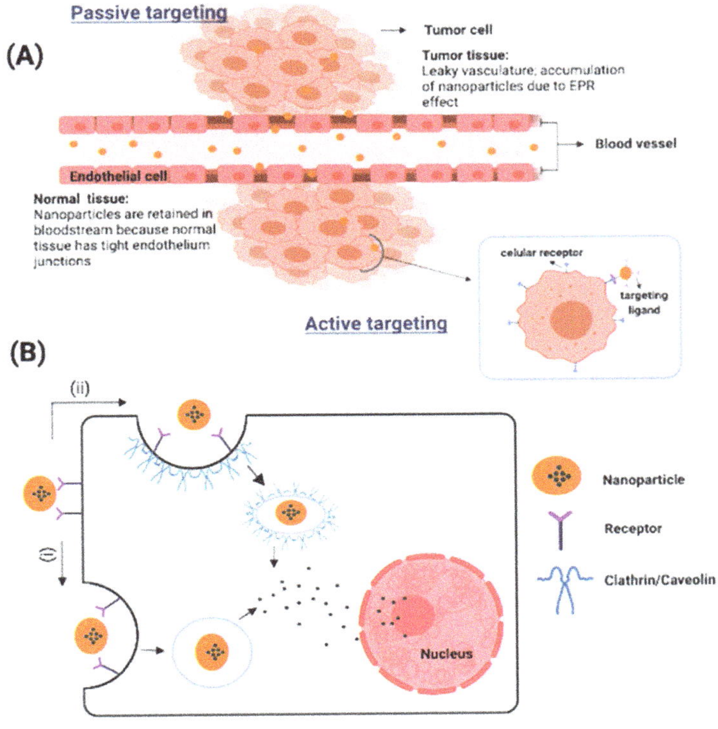

Figure 4. (**A**) Nanoparticles are intended to exploit the enhanced permeability and retention effect to exit the blood vessels through leaky vasculature, accumulate within tumor tissues, and enter the cells via endocytosis before releasing their 'drug'. Conversely, owing to the tight endothelium junctions in normal tissues, the nanoparticles would remain in the bloodstream. (**B**) Small nanoparticles could be internalized through many pathways such as (i) clathrin or caveolin-mediated endocytosis and (ii) clathrin/caveolin-independent endocytosis.

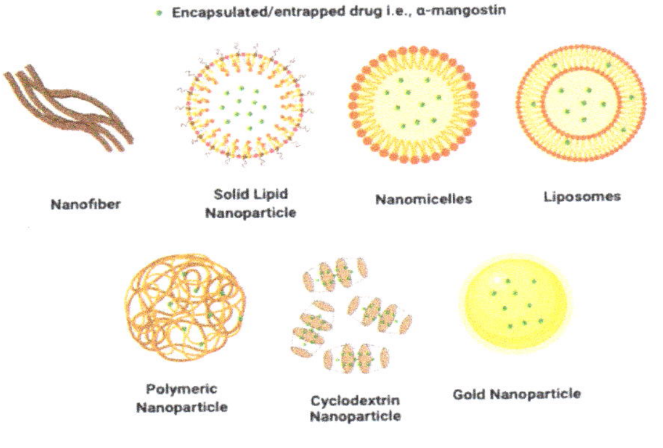

Figure 5. Nanoformulation of α-mangostin for cancer drug delivery.

Table 2. Summary of α-mangostin nanoformulations.

Carrier	Cell Line	Outcome	Ref.
Silk fibroin N-(3-Dimethylaminopropyl)-N′-ethylcarbodiimide hydrochloride (EDC) and polyethylenimine (crosslinker)	Caco-2 MCF-7	• Increased solubility • Sustained release • Reduced hematotoxicity up to 90% • Stable for up to 24 h when dispersed in IV diluent and 6 months when preserved as lyophilized powder at 4 °C • Improved cytotoxicity and apoptosis in vitro	[3]
Acetobacter xylinum	B16F10; MCF-7; hGF; HaCaT	• Significantly lower toxicity to normal cells than to cancer cells • Slightly toxic to HaCaT cell (normal cells)	[70]
PLGA, soybean lecithin, DSPE–PEG2000–COOH Thioaptamer (ligand)	MCF-7	• Particle size 150–300 nm • Internalization of nanoparticles • Strong disaggregation of MCF-7 multicellular tumor spheroids	[71]
Miglyol 812, cetyl paomitate, montanov 82, and oleoyl chitosan (coating agent)	Caco-2 Hela	• Particle size < 200 nm • High physical stability • Excellent encapsulation efficiency (EE > 90%) • High level cellular internalization • Improved cytotoxicity • Downregulation of cyclin D1-(CCND1) and anti-apoptotic gene BCL2	[72]
PLGA	Pancreatic cancer cell line (AsPC-1, PANC-1, and Mia-Paca-2) Cancer stem cells (CSCs)	• Particle size < 200 nm • Can easily enter into the cells • Downregulation of pluripotency maintaining factors components of Shh pathway, Gli targets, EMT markers, transcription factors, and upregulation of E-cadherin • Inhibit proliferation, colony formation; cell motility, migration, and invasion • Induced apoptosis, inhibits the growth, inhibits development and metastasis of pancreatic cancer	[35]
PLGA	HCT116 and HT29 Normal epithelial cells (CRL-1831)	• Internalization of nanoparticles • Suppressing the expression of Notch receptors and their ligands, γ-secretase complex protein and downstream target • Induced cancer cells and did not induce apoptosis in normal cells	[73]
α, β dan γ cyclodextrin (CD) Epichlorohydrid (ECH) as linker	CT26WT	• The highest level of solubility and complexation efficiency of α-mangostin was shown in complexation with βCD • CD nanoparticle α-mangostin complexes showed larger loading ratio than CDs themselves • Rapid release and slow release • Decreased cytotoxicity	[74,75]
β-cyclodextrin	A549	• Particles size < 50 nm • Nanoparticles taken up into the cancer cells and affected nuclear morphology • Improved cytotoxicity	[76]
Chitosan and alginate Genipin as crosslinker	HT-29	• Particle size around 400–500 nm • Genipin as a crosslinker significantly increases the loading efficiency and loading capacity of the nanoparticles • Improved cytotoxicity	[77]

Table 2. *Cont.*

Carrier	Cell Line	Outcome	Ref.
Poly-(ethylene glycol)–poly(l-lactide) (PEG–PLA) CREKA peptide (ligand)	PANC-1 NIH3T3, PANC-1-Luc2	• Particle size 100 nm • Controlled and continuous release • Increased intracellular delivery • Suppressed NIH3T3 activation, decreased fibronectin expression (in vivo), promotes tumor vascular normalization and enhances blood perfusion (in vivo) • Blocked TGF-β signaling pathways by inhibiting phosphorylated Smad2 and Smad3 protein synthesis (in vitro and in vivo); cancer-associated fibroblast (CAF) suppression and collagen downregulation effect	[78]
Chitosan and Kappa Carrageenan	MCF-7	• Particle size 200–400 nm • Excellent encapsulation efficiency (EE > 97%) • Increased solubility • Initial burst release • Improved cytotoxicity	[79]
Polyvinylpyrrolidone (PVP)	HCT 116	• Significantly increases the solubility of alpha mangostin (1000-fold) • Particles size < 130 nm • Nanomicelles can enter into cancer cell • The half maximal inhibitory concentrations for α-mangostin nanomicelles was higher than raw α-mangostin	[43]
Monomethoxy poly (ethylene glycol)-polycaprolactones (MPEG-PCL)	A375 and B16 non-tumor cell lines (LO2, Vero, and HEK293T cells)	• Mean particle size = 30 nm as a monodisperse system • Drug loading up to 99.1% • Sustained drug-release profile • Excellent cellular uptake • Induced apoptosis via the mitochondrial-mediated intrinsic pathway and the exogenous apoptosis pathway • Pharmacokinetic study shows a slow excretion behavior from blood vessels	[80]
Methoxy poly(ethylene glycol)-poly(lactide) (MPEG-PLA)	U87	• Mean particle size 32 nm and EE = 99.5% • Sustained release • Improved the pharmacokinetics of α-mangostin • Suppression of protein bcl-2 decreased, pro-apoptotic protein Bax and cleaved-caspase-3, 8 and 9	[81]
Phosphatidylcholine and cholesterol	Calu-3; HT-29; MCF-7; Caco-2; HaCaT; HDF	• Particle size around 100 nm • Lower toxicity in normal cells than α-mangostin in aqueous solution • Decreased cytotoxicity, induced apoptosis	[82]
Dioleoylphosphatidylcholine and cholesterol	Hep-G2	• Particle size around 100 nm, high entrapment efficiency, slow and sustained release • Improved cytotoxicity	[83]
Gold citrate	PC-3 DU145	• Induced DNA fragmentation • Improved cytotoxicity	[84]

4.1. *Nanofibers*

Nanofibers are one-dimensional nanomaterial with a size of less than 500 nm [85]. Currently, nanofibrous polymer materials are being thoroughly investigated for a variety of medical uses. The use of nanofiber system drug carriers in anticancer therapy offers numerous benefits, including the ability to form fibers with varying diameters ranging from nanometers to sub-microns, surface modification, having a large outer surface ratio and unique porosity to maximize the drug encapsulation process, controlled and sustained drug release in the desired workplace to improve efficacy [86–91].

In 2019, D.T. Pham et al., tried to develop a fibroin nanoformulation for α-mangostin using a natural polymer (silk fibroin) as a carrier, EDC or PEI as a crosslinker, and utilize the desolvation method. Fibroin is a natural biocompatible and biodegradable protein extracted from *Bombyx mori* silk. This material has been used as a carrier for anticancer drugs [92,93]. The observed mean size was around 300 nm with a narrow size distribution with PI < 0.3 and zeta potentials from −15 to +30 mV proportionally to the increase of EDC content. The formula that uses a crosslinker had a higher entrapment efficiency and loading capacity (70%) and (7%) than the formula without a crosslinker (<50% and <5%). Compared to the free α-mangostin, nanoparticles increased the drug's solubility up to threefold. In vitro cytotoxicity of α-mangostin and α-mangostin loaded nanoparticles was conducted in Caco-2 and MCF-7 cells. The results showed that the nanoparticle maintained the apoptotic effect of α-mangostin and exhibited improved cytotoxicity (i.e., lower IC_{50}) than the free α-mangostin. It is related to differences in the cellular uptake mechanisms. Free α-mangostin in solution penetrates the cells via a passive transport pathway, limited by cell lipid bilayers and the efflux pumps. Whereas, silk fibroin nanoparticles loaded with α-mangostin effectively enter the cell via surface adsorption and endocytosis pathways, where it is possible that two amino acids located near the N-terminal of the fibroin heavy chain can bind to surface receptor integrin. In particular, the cell surface receptor integrin was overexpressed in many cancer cells, including colon and breast cancer. This binding will induce the endocytosis process. Therefore, the formulation of α-mangostin using fibroin nanoparticles can enhance IC_{50} so that it can boost the effectiveness of α-mangostin as a cancer medicine [3].

Taokaew et al. formulated bacterial cellulosic nanofiber film synthesized from *Acetobacter xylinum* (Gram-negative aerobic bacteria). This bacterial-based nanofiber has several advantages, such as its similarity to the natural extracellular matrix, high loading capacity, non-toxicity, and resistance to heating during the sterilization process. Bacterial cellulosic cytotoxicity testing was performed on cancer cells (B16F10 melanoma and MCF-7 breast cancer cells) and normal cells (HaCat fibroblast and hGF keratinocyte cells). The loading capacity of α-mangostin was up to 7.33% and the loading efficiency ~67.3% with a concentration proportion of 0.01–1% (2, 24, and 250 mg/cm^3). The characteristics of bacterial cellulosic nanofiber are cellulose fibers with a diameter of 50–100 nm with an empty cavity that stretches between the fibers to absorb the active ingredients. The cytotoxicity of nanofibers on cancer and normal cells was tested at three α-mangostin loading concentrations, namely 2, 24, and 250 mg/cm^3. The findings revealed that a concentration of 2 mg/cm^3 had no cytotoxic effect on B16F10 melanoma, but a higher concentration of α-mangostin inhibited growth and caused changes in cell morphology that suggested membrane damage in B16F10 melanoma and MCF-7 breast cancer cells. Furthermore, testing on normal cells showed a decrease in viability, namely ~40% in hGF cells and 38% in HaCaT cells during 24 h of observation. After incubation for 48 h, HaCaT cells experienced an increase the viability by 111% while hGF cells by 53%. It suggests that the α-mangostin nanofiber is slightly more toxic to HaCaT cells than hGF cells. It was concluded that bacterial cellulosic nanofiber α-mangostin has lower cytotoxicity against normal cells compared to cancer cells [70].

4.2. Solid Lipid Nanoparticle

Solid lipid nanoparticles (SLN) are colloidal particles constructed from biodegradable physiological lipids which are solid at room and body temperature. It has sizes ranging from 50 to 1000 nm according to the manufacturing method and type of lipid used. SLN is widely used as a carrier for cancer medicines because it has many benefits, such as solvent-free preparation, being composed of biocompatible and biodegradable materials, site-specific targeting, physical stability, and controlled release for both hydrophilic and lipophilic drugs [94–96].

F. Bonafè et al. attempted to develop solid lipid nanoparticles of α-mangostin conjugated with CD44 thioaptamer as a ligand targeting multicellular tumor spheroids (MCTSs) by MCF-7 cell. Using a nanoprecipitation technique, lipid nanoparticles were synthesized from PLGA, soybean lecithin, and DSPE–PEG2000–COOH, and the thioaptamer was conjugated to nanoparticles using the two catalysts, 1-ethyl-3-(3-dimethylaminopropyl) carbodiimide, and N-hydroxysuccinimide [71]. PLGA deployed as the core can carry hydrophobic drugs with high loading capacity. Then, hydrophilic coating by soybean lecithin and DSPE-PEG2000-COOH provides steric protection capable of reducing systemic clearance rates, prolonged circulation half-life in vivo, and functional groups for surface modification attachment of ligands [97]. The mean diameters of the nanoparticles were 227.0 ± 88 and 174.0 ± 29 nm (after filtered by 200 nm cut-off). α-mangostin nanoparticles (0.1 μg/mL) induced significant dissociation of MCTSs at a dose 10 times lower than the α-mangostin (1 μg/mL). These results suggested that nanocarrier is a suitable vehicle for α-mangostin to suppress tumors at low concentrations. Moreover, the nanoparticles conjugated to the CD44 thioaptamer could reduce the size of the spheroids. It suggests that several cells died or slowed down their process of duplication [71]. Another solid lipid nanoparticle was prepared by V. Kumar and tested against diethylnitrosamine-induced hepatocellular cancer. This method generated particles with a size of 182.3 nm and a polydispersity index of 0.203, and it reduced hepatic nodules by 84.5% and 93.4%, respectively. The nanoparticle α-mangostin controlled the PI3K and Akt pathways, which are involved in the inhibition of hepatic cancer growth and proliferation, as well as its chemoprotective action [98].

4.3. Nanostructured Lipid Carriers

Nanostructured lipid carriers are novel therapeutic formulations consisting of physiological and biocompatible lipids, surfactants, and co-surfactants. The use of NCL as a carrier for chemotherapy agents is very promising because it is a bio-compatible and bio-degradable lipid-based nanoparticle that is able to improve its physical and chemical stability, and significantly increase the therapeutic capacity with low pharmacokinetic properties [99–101].

The mucoadhesive NLC was developed for the possible oral delivery of α-mangostin. Nanostructure lipid carrier coated with oleoyl-quaternized-chitosan (NLC-CS) with high-pressure homogenization process was used for the nanoparticles prepared in a range of nanoparticle sizes from 200 to 400 nm, low polydispersity, zeta potentials 40.9 mV, with excellent encapsulation efficiency (>90%). Results demonstrated that the NLC-CS has a higher toxicity than the NLC against Hela and Caco-2 cells. The CS-NLC particles resulted in better cellular uptake than the NLC particles due to the mucoadhesive properties of the CS-NLC particles. Positively charged nanoparticles have a higher internalization rate than neutral or negatively charged species. This behavior is due to the strong electrostatic adhesion of positively charged chitosan to negatively charged mucosal surfaces. Furthermore, CS-NLC outperformed NLC due to improved mucoadhesion of the particles with the cells, allowing for internalization. These findings support the use of surface-modified CS-NLC nanoparticles as mucoadhesive carriers for drugs to cancer cells [72].

4.4. Polymeric Nanoparticle

Polymeric nanoparticles are particles in the size range from 1 to 1000 nm and can be loaded with active compounds trapped in or surface adsorbed onto the polymer core [102]. These systems are typically structured through spontaneous assembly in which the therapeutic compound is trapped within the core of the nanoparticle structure [103]. Generally, polymeric nanoparticles are composed of natural polymers (chitosan and alginate) and synthetic polymers (PLGA and cyclodextrins) [104]. There are many advantages to using polymeric nanoparticles for the delivery of anticancer drugs in cancer treatment, among others, they can be used to deliver various types of drugs such as hydrophilic and hydrophobic drugs, peptides, and biological macromolecules via several routes of administration [105], improve the drug solubility [106], provide controlled release, increase bioavailability, therapeutic index, can transport active ingredients to targeted tissues or organs at a specified concentration [102,107], surface modification with ligand linking for stealth and targeted drug delivery, biocompatibility, biodegradability, and low toxicity [108].

4.4.1. PLGA Nanoparticle

Verma et al. developed α-mangostin encapsulated PLGA nanoparticles to improve the bioactivities of α-mangostin for inhibiting pancreatic cancer stem cells (Pan CSCs) in human and KC mice (PdxCre; LSL-KrasG12D) (Pan CSC). This formula produces a particle size of 186.3 ± 6.42 nm, a zeta potential of 0.03 ± 0.005 mV, and a drug encapsulation of 51.16 ± 2.61%. The results indicated that α-mangostin nanoparticles inhibited cell proliferation of Pan CSCs and pancreatic cancer cell lines more effectively than α-mangostin and had no significant effect on human pancreatic normal ductal epithelial [35]. In another study, V. Chandra Boinpelly et al. reported that α-mangostin-encapsulated PLGA nanoparticles inhibit human colorectal cancer cells and colony deformation in colorectal cancer HCT116 and HT29 cells in a dose-dependent manner at a dose of 0–10 mol/L [73]. These findings suggest that α-mangostin encapsulated PLGA nanoparticles are appropriate for use as carriers to advance the effectiveness and therapeutic effects of α-mangostin for pancreatic cancer and colorectal cancer.

4.4.2. PEG-PLA Nanoparticles

In 2020, J. Feng et al. developed cancer-associated fibroblasts (CAFs) targeting polymer nanoparticle (methoxy poly (ethylene glycol)3000-poly (lactic acid)34000 (Me-PEG-PLA) and maleimide-poly (ethylene glycol)3000-poly (lactic acid)34000 (Male-PEG-PLA) coated with CREKA peptide and loaded with α-mangostin. The concept of using CAFs-targeting nanoformulation α-mangostin is an effective method of modifying tumor environment to enhance pancreatic ductal adenocarcinoma chemotherapy. Nanoparticles with spherical shape particles and average particle size diameter of α-mangostin nanoparticles [NP(α-M)] is 103.76 ± 7.45 nm; α-mangostin nanoparticles coated with CREKA peptide [CRE-NP(α-M)] is 106.93 ± 3.69 nm with a zeta potential of −31.77 ± 2.12 mV, respectively. The negative charge generated by the nanoparticles renders an optimal possibility to develop the EPR effect when examined in vivo. The nanoparticles effectively reduced the production of secreted mass extracellular matrix both in vitro (NIH3T3 cells) (5.8-fold) and in vivo by blocking the TGF-β signaling pathway and significantly inhibiting the tumor growth (<70%), inducing apoptosis and necrosis on an orthotopic mice model. Biodegradable nanocarriers (PEG-PLA) demonstrate controlled and continuous release behavior. Additionally, the use of PEG can avoid the elimination of nanoparticles by the reticuloendothelial system, which further increases the circulation time in the body. Moreover, CREKA peptide as a target ligand can bind to the fibroin–fibronectin complex that is overexpressed in the tumor mass extracellular matrix, thereby increasing the accumulation in the tumor region. From the results achieved, we can infer that this formula could be amplified as an efficient method to enhance the therapeutic effects [78].

4.4.3. Chitosan-Alginate Nanoparticles

Samprasit et al. formulated α-mangostin-loaded chitosan/alginate (CS/ALG) nanoparticles cross-linked with genipin (GP), used the ionotropic gelation method, and evaluated antitumor activity against the colorectal cancer cells (HT-29) [77]. Chitosan is composed of amino groups with a polycationic charge that interacts with negatively charged polymers such as alginate, forming colloid nanoparticles. Chitosan and alginate complexes have attracted attention due to their easy processing [109–111]. Genipin acts as a cross-linking agent that reacts with the primary amine group of chitosan to form a rigid nanoparticle structure with a slow degradation rate, making it capable of controlling swelling, degradation, and drug release [111]. The mean particle sizes of GP nanoparticles and non-GP nanoparticles were 477.2 ± 32.2 and 437.6 ± 50.3 nm. The percentage of loading efficiency and capacity of GP nanoparticles is three times higher than that of non-GP nanoparticles. It demonstrates that GP enhanced the α-mangostin loading, wherein during the cross-linking process α-mangostin may be trapped within the GP nanoparticles, resulting in increased loading of α-mangostin. It is also in line with the increase in the particle size of the GP nanoparticles. In vitro release tests simulated the state of the digestive tract at pH 1.2, 6.8, and 7.4. The result is that the release rate of α-mangostin is slower than the non-GP nanoparticles at all pH conditions. Genipin can control the degradation, diffusion, erosion of the nanoparticles, thereby inhibiting α-mangostin release into the medium. From the entire pH range tested, the release of α-mangostin at low pH was significantly lower than at high pH. At high pH, the amine group of chitosan undergoes deprotonation and loses its charge, causing weak electrostatic interaction between chitosan and alginate and unstable nanoparticles, which implies a higher release of α-mangostin. The pattern of slow release of α-mangostin from GP nanoparticles in acid medium and rapid release of α-mangostin at higher pH illustrates that GP nanoparticles can concentrate on the release of α-mangostin in the small intestine and colon [77,111,112].

The viability of the cells treated with varying concentrations of blank GP nanoparticles showed no substantial decrease in cell viability. As a result, blank GP nanoparticles have little effect on cell viability and protected drug carrier. The cytotoxicity of the α-mangostin-loaded GP nanoparticles was dose-dependent. Cell viability was lower at concentrations of 200 and 300 μg/mL of α-mangostin-loaded GP nanoparticles compared to the control. These findings are associated with α-mangostin's cytotoxicity, confirming that α-mangostin-loaded nanoparticles and α-mangostin have antitumor action against colorectal adenocarcinoma cells. It concluded that chitosan/alginate nanoparticles cross-linked with genipin could be promising candidates for a controlled release drug delivery system of α-mangostin to the large intestine [77].

4.4.4. Chitosan-Kappa Carrageenan Nanoparticles

Wathoni et al. developed an enteric-coated nanoparticle of α-mangostin-loaded chitosan-kappa carrageenan. Kappa carrageenan is a natural polymer used as an encapsulator that will protect nanoparticles from gastric acid because of its low sensitivity to pH and ionic strength. This formula yields an average particle size of 200–400 nm, high entrapment efficiency, and increased solubility. In vitro release testing showed that the nanoparticles produced an initial burst release for pH 1.2 and 7.4, respectively, followed by a slow and sustained release. Cytotoxicity in MCF-7 cells showed an increase in nanoparticle activity compared to free α-mangostin. The IC_{50} of α-mangostin, α-mangostin-chitosan nanoparticle, and α-mangostin-chitosan-kappa carrageenan was 8.2, 6.7, and 4.7 g/mL, respectively [79].

4.5. Cyclodextrin Nanoparticles

Cyclodextrins are non-reducing oligosaccharides of starch-modified products with a ring-shaped chemical structure and are formed through the cyclization process by CGTase activity (cyclodextrin glycosyltransferase). Suitable reaction conditions will produce three main groups of cyclodextrin: α-, β-, and γ-cyclodextrins consisting of 6, 7, and 8 units of (1,4) linked D(+)-glucopyranose. As there are numerous hydroxyl groups linked to the top and bottom of the molecule, it is water-soluble, but the inside of the cyclic structure is hydrophobic and may trap hydrophobic molecules; this configuration is known as the "molecular pocket". The number of glucose units determines the cavity size, and the cavity diameters of α-, β-, and γ-cyclodextrin are 5.7, 7.8, and 9.5 Å, respectively [113–115]. The cyclodextrins molecular pocket is used to trap hydrophobic medicines like α-mangostin without interfering with their bioactivity [74].

Cyclodextrin-based nanoparticles (CDNP) encapsulating α-mangostin were developed by a polyaddition reaction using epichlorohydrin and examined cytotoxicity in CT26WT cancer cells. The cyclodextrin-containing hyperbranched polymer was produced by a poly-addition reaction between cyclodextrin and epichlorohydrin (ECH) that capably integrates numerous cyclodextrins into the polymer using a relatively easy synthesis technique. The results showed that βCDNP showed a higher loading ratio of α-mangostin.

The cell viability results after 24 and 48 h of incubation indicated the CDNP themselves are inert, while in the case of CDNP α-mangostin complexes, the cell viability is different depending on the type of CD. The αCDNP/α-mangostin and γCDNP/α-mangostin showed high cytotoxicity for cells after 24 h. However, cell viability decreased significantly for βCDNP/α-mangostin after 48 h. It indicates that the α-mangostin in the αCDNP and γCDNP systems can be easily released from the polymer network structure so that almost no cells survived after 24 h of incubation. In contrast, because βCDNP can strongly resist α-mangostin release, the βCDNP system does not show any toxicity after 24 h. However, due to the presence of other candidates of the guest molecule for βCD, such as cholesterol in the cell membrane, the α-mangostin in βCDNP was probably gradually exchanged by the guest molecules, resulting in releasing α-mangostin and inducing higher toxicity after 48 h [74]. Continuing the previous study, V.T.H.Doan et al. focused on the anticancer activity of CDNP/α-mangostin. They have evaluated in vitro and in vivo anticancer efficacy using a CT26WT cell line. There was an increase in the IC50 value from CDNP compared to α-mangostin in the monolayer culture, in the order of α-mangostin (~14.5 μM), αCDNP/α-mangostin (~27.7 μM), γCDNP/α-mangostin (~43.5 μM), and βCDNP/α-mangostin (~50.4 μM). It occurs due to nanoparticles retention capability for α-mangostin release from the system. In this study, they examined the in vitro cytotoxicity of α-mangostin and CDNP/α-mangostin on CT26WT spheroid cells. The IC50 values increased ~33 times in α-mangostin and ~9 times in CDNPs/α-mangostin. In the spheroid cells, the drug must penetrate the spheroid cellular layer to migrate inside and kill the cells. This is because of the hydrophobicity of α-mangostin that was easily adsorbed on the spheroid cells and causes a decrease in drug concentration in the spheroid. Whereas, CDNP reduces the hydrophobicity of α-mangostin so that it has better cellular penetration than α-mangostin, which has better cytotoxicity and lower IC50 than that observed for α-mangostin. The contrast in IC50 values between βCDNP/α-mangostin and γCDNP/α-mangostin is due to the higher α-mangostin retention capability of βCDNP than γCDNP. α-mangostin may not be released much in the case of βCDNP/α-mangostin, appearing in a higher IC_{50} than γCDNP/α-mangostin. It is relevant to the cavity size of CDs; γCDNP/α-mangostin have a larger cavity than βCDNP/α-mangostin. Impressively, in vivo anticancer efficacy showed changes in the tumor volume and tumor growth ratio after i.v. (10 mg/kg) of α-mangostin and βCDNP/α-mangostin into BALB/c mice bearing CT26WT tumors. α-mangostin showed tumor volume and tumor growth ratio gradually increased over 22 days and reached 1145 ± 194 mm^3 in tumor volume and tumor growth, indicating an increase of ~10-fold. While, the βCDNP/α-mangostin showed appreciable suppression of tumor growth. The tumor volume was 501 ± 372 mm^3 and the tumor growth ratio was ~4 [75]. The increase in tumor size and volume on α-mangostin treatment was due to α-mangostin swiftly clearing from the blood within 3.5 h after intravenous injection with a single dose [36]. Consequently, to achieve significant anticancer efficacy a frequent α-mangostin administration is required. Whereas for βCDNP/α-mangostin, most of the particles remained in the serum after 6 h, showing good circulation in the blood, which may be due to the hydrophilic surface of βCDNP. Biodistribution testing showed βCDNP appeared to accumulate in the tumor [75].

In another study, M.P. Nguyen Thi et al. synthesized ßCDNP/α-mangostin with particles size of <50 nm and zeta potential of −38 mV. Cytotoxicity against lung cancer cells A549 demonstrated that βCDNP/α-mangostin exhibited lower IC_{50} values than α-mangostin (2.34 and 4.86 g/mL, respectively). Furthermore, fluorescence microscopy revealed that βCDNP/α-mangostin is carried into cancer cells and changes the shape of the cell's nucleus. These findings suggest that βCDNP/α-mangostin improves bioavailability and anticancer efficacy [76].

4.6. Nanomicelles

Nanomicelles are colloidal nano-sized structures with a hydrophobic core and a hydrophilic shell [116]. Nanomicelles have been widely used as the delivery system for anticancer drugs, in particular for the delivery of water-insoluble bioactive compounds [117]. Micelles structures formed amphiphilic block copolymers containing hydrophilic blocks and hydrophobic blocks, some hydrophilic and nonionic polymers, such as PEG, poly (N-vinyl pyrrolidone) (PVP), poly (N-isopropyl acrylamide) (PNIPAM), and poly (hydroxypropyl methacrylamide) (PHPMA) is widely used as shell-forming material. Nanomicelles have many advantages, including being structurally stable, having the ability to trap a large number of hydrophobic drugs, their surface can be conjugated with the targeting ligand [118], and they have a suitable size distribution to avoid rapid renal excretion, allowing accumulation into the tumor tissue through the effect of EPR [119].

In a study from A.F.A. Aisha et al., α-mangostin was prepared in PVP by solvent evaporation method and the intracellular delivery through endocytosis that may enhance the antitumor efficacy of α-mangostin was examined. The cellular uptake assay on human colon tumor cell line 116 (HCT16) showed high permeability of α-mangostin through the cytoplasmic membrane around the treated cells whereas nanomicelles showed fluorescence in the form of spherical particles. These results suggest that cellular uptake from nanomicelles can be mediated through endocytosis. The cytotoxic effect of α-mangostin and nanomicelles exhibited a significant cytotoxic effect in a dose-dependent manner with IC_{50} of 7.7 ± 0.1 and 8.9 ± 0.2 μg/mL, respectively. Cytotoxicity results show intracellular release of drug payload, which occurred due to the degradative effect of lysosomal enzymes and also show the interaction of PVP and α-mangostin does not affect α-mangostin's cytotoxicity [43].

Another study of nanomicelles nanoformulation was used to improve the α-mangostin anti-melanoma effect. The α-mangostin/MPEG-PCL has a core structure with PCL (hydrophobic) which absorbs α-mangostin and MPEG (hydrophilic) as shells. The molecular modeling examined the interaction between the αM and MPEG-PCL as carriers which were observed by comparing the interactions in the aquatic environment and the tumor environment. The results showed that the interaction between α-mangostin and MPEG-PCL in an aqueous environment tended to persist with a spherical particle shape, whereas in the tumor environment, MPEG-PCL did not appear to be able to maintain their spherical particles so that it was easier for αM to exit the nanoparticle system. In in vitro cytotoxic testing, α-mangostin nanomicelles possessed a stronger inhibitory effect compared to the α-mangostin on A375 and B16 melanoma cells and exhibited low cytotoxicity for non-tumor cell lines (LO2, Vero, and HEK293T cells). The growth inhibitory effect of the α-mangostin and α-mangostin nanomicelles shows inhibition on melanoma cell colonies. Surprisingly, no colony formation was observed in A375 and B16 cells treated by α-mangostin nanomicelles. However, several colony formations under the same condition were formed in the α-mangostin group. These results suggest that α-mangostin nanomicelles have a more intense restraining effect on the colony formation rate in melanoma cells compared to α-mangostin. Moreover, the improved apoptotic activity of α-mangostin nanomicelles was determined compared with α-mangostin. It is because the nanomicelles had a better cellular uptake behavior than the α-mangostin. Furthermore, in vivo anticancer activity of the α-mangostin nanomicelles was observed using A375 cells injected subcutaneously into the female BALB/c athymic mice. The α-mangostin nanomicelles possessed a better effect on tumor growth inhibition (tumor weight and volume = 0.35 g and 400 mm^3) compared to the free α-mangostin (0.7 g and 910 mm^3). The α-mangostin nanomicelles held smaller tumors than α-mangostin. In conclusion, a strategy was presented in which the α-mangostin delivery system in nanomicelles comprehensively increases the anti-tumor activity of α-mangostin in vivo [80].

In another study, α-mangostin encapsulated with MPEG-PLA nanomicelles was formulated by S. Zheng et al. The nanomicelles show a spherical, monodisperse, and narrow particle size structure with an average size of 32 nm. In vitro antitumor activity of α-mangostin and nanomicelles on U87 cells indicate the inhibition of the proliferation process and promoted apoptosis. The antitumor effect of nanomicelles was higher when compared to α-mangostin, with apoptosis rates, at concentrations of 10 and 20 μg/mL, of

36.6% and 54.9% for nanomicelles and 20.4% and 45.5% for α-mangostin. Furthermore, the antitumor activity test on female C57/BL6 mice showed nanomicelles increased the antitumor effect of α-mangostin (65% reduction of tumor volume) and indicated by a decrease in tumor weight (0.62 g for nanomicelles and 1.14 g for α-mangostin), inhibited tumor growth, decreased proliferation, suppression of angiogenesis, and an increase in the apoptosis index that was two times higher than that of α-mangostin. This increase in activity is due to the particle size of nanomicelles at the nanoscale, being able to pass through the inter-endothelial junctions of a tumor by passive diffusion, and the use of PEG to increase the stability of nanoparticles in blood circulation and high retention in the tumor region [81].

4.7. Liposomal Nanoparticles

Liposomes are membrane vesicles composed of amphiphilic lipids that surround the water nucleus. Liposomes form spontaneously as lipid molecules are dispersed in a liquid medium, resulting in nanometer to micrometer size. Liposomes are formed by a phospholipid sheath composed of one or more lipid bilayers with hydrophilic head groups and hydrophobic tail groups [120,121]. This vesicle bilayer system enables the liposome to trap lipophilic and hydrophilic molecules, enabling these vesicles to bundle different drugs. Via the effect of increased permeability and preservation, encapsulation in the liposome structure can shield compounds from early inactivation, oxidation in the bloodstream, improve the half-life, and increase aggregation in the tumor [120,122,123].

R. Benjakul et al. developed liposomes nanoformulation and evaluated their cytotoxic effect and mechanism inducing cell death in various human carcinomas (human lung epithelial carcinoma (Calu-3), human colon carcinoma (HT-29), human breast carcinoma (MCF-7), and human colon carcinoma (Caco-2) cells). Liposome was prepared with phosphatidylcholine and cholesterol using the reverse-phase evaporation method. The particle size, polydispersity index, and zeta potential were 113.98 ± 2.95 nm, 0.13 ± 0.01, and −25.6 ± 0.07 mV, respectively. The cytotoxic effect trends of α-mangostin liposomes on human carcinoma cell lines revealed that α-mangostin liposomes were less toxic than α-mangostin. In all four measured cells, the IC_{50} was 2–4 times greater than α-mangostin. It is due to the disparity in the cellular absorption process, where α-mangostin can reach cells through passive diffusion, while α-mangostin liposomes can enter cells through adsorption, endocytosis, or the fusion mechanisms, then accompanied by an α-mangostin release from its carriers. Although free α-mangostin showed relatively higher toxicity than α-mangostin liposomes, the use of liposomes as nanocarriers can still be used as a promising α-mangostin delivery system for anticancer therapy to avoid direct toxicity of α-mangostin to normal cells (drug targeting) [82].

In another study from M.P. Nguyen Thi et al., α-mangostin liposomes were formulated with dioleoylphosphatidylcholine and cholesterol, and a cytotoxicity test was conducted on human hepatocellular carcinoma (Hep-G2) cells. There was a significant decrease in the IC_{50} value of 2.4 times between α-mangostin and α-mangostin liposomes (1.9 and 4.6 μM, respectively). Furthermore, the cell viability significantly reduced after 48 and 96 h in the α-mangostin liposome group compared with the α-mangostin. Due to the α-mangostin liposome group, α-mangostin is released slowly and the cells could absorb it gradually so that it kills fewer cells than free α-mangostin within 24 h and length of the period (48 and 96 h). It is consistent with the release of in vitro liposome that exhibits a sustained drug release profile. As for α-mangostin, cell viability varied with the concentration during the first 24 h. The results of this study indicate that liposomes exhibited a more effective cytotoxic effect against Hep-G2 cells as compared with free α-mangostin [83].

4.8. Gold Nanoparticles

Gold nanoparticles are excellent carrier molecules for cancer drug delivery. They can be synthesized in a variety of sizes and surface characteristics, which make them promising candidates as drug delivery vehicles. Multifunctional gold nanoparticles are now widely used in cancer therapy because of their inertness and biocompatibility. The surface of the gold nanoparticles can be easily modified to provide a controlled release strategy using internal or external stimuli [124–126]. S. Qiu et al. developed α-mangostin gold nanoparticles using gold citrate and coated with PEI and sulfated β-cyclodextrin to complex α-mangostin. This formula produces an average particle size of around 100 nm with a particle size distribution that is monodisperse. The zeta potential of the nanoparticles is 30 ± 3 mV, meaning that they are stable. In vitro cell viability assays of gold α-mangostin nanoparticles against prostate cancer cell lines (PC-3 and DU145) revealed a 15% and 50% improvement in activity relative to α-mangostin, respectively. Furthermore, α-mangostin nanoparticles are known to trigger an apoptotic death pathway that enables tumor cell damage and elimination through phagocytosis. According to the findings of this research, this drug delivery mechanism (gold nanoparticles) could increase the cytotoxicity of α-mangostin against prostate cancer cell lines [84].

5. The Perspective of the Authors

Recently, nanoformulations have been taken into consideration as a drug carrier as they have improved the pharmacokinetic properties of the drug to increase its efficiency and reduce side effects. In cancer treatment, targeted treatment only targeted the cancer cells to be killed and less harm to the normal cells is increasingly desirable. Nanotechnology has improved cancer therapy in many ways, such as selective recognition of cancer cells, targeted drug delivery, and overcoming the limitations of α-mangostin.

It is generally known that nanoparticles with smaller sizes and modified surface properties can reach or penetrate certain areas that normally cannot be passed through. Surface properties, particularly the surface charge of the particles, can also influence the distribution of the nanoparticles and possibly lead to higher toxicity in both cancer and non-cancer cells. Therefore, future research can perform surface functionalization of α-mangostin nanoformulations with specific antibodies, proteins, nucleic acids, and small molecules that can actively attach to cancer cells. Currently, CREKA peptide and thioaptamer are ligands that use α-mangostin nanoformulations. The use of these two ligands showed to increase the intracellular delivery and the efficacy of α-mangostin. There are many types of target ligands (such as antibodies and small molecules) that can develop in nanoformulations of α-mangostin to increase the anticancer effectiveness of α-mangostin.

Polymeric nanoformulation (PEG-PLA-CREKA), β-cyclodextrin nanoparticles, liposomes, gold nanoparticles, and nanomicelles (MPEG-PCL and MPEG-PLA) have particle size characteristics >150 nm, which are suitable for entry into cancer cells and increased cellular uptake of α-mangostin nanoparticles. Cytotoxicity testing in vitro also showed increased activity in the tested cancer cell lines, except for in β-cyclodextrin nanoparticles, there was a decrease in cytotoxicity in the CT26WT cell line and liposomes (phosphatidylcholine and cholesterol) in Calu-3, HT-29, MCF-7, and Caco-2 cell lines. It is due to differences in the cellular mechanism for uptake of free α-mangostin, which can pass into cells by passive transport (diffusion) mechanism, the β-cyclodextrin and liposome nanoformulations can be taken up into cells by adsorption and endocytosis and related to the α-mangostin release from the carrier. MPEG-PLA and MPEG-PCL nanomicelles were proved to improve the bioavailability of α-mangostin in mice which showed significantly reduced α-mangostin elimination. In addition, the in vivo anticancer activity showed better inhibition of tumor growth than α-mangosin.

Polymeric, fibroin, solid lipid nanoparticles are very promising to be developed due to biodegradable components, high-loading capacity (up to 17%), and easy preparation, but these formulas produce large particle sizes up to 400 nm. The development of nanoformulations for α-mangostin such as enteric coating nanoparticles (chitosan-kappa carrageenan) and mucoadhesive (NLC-CS) is an alternative for oral delivery of α-mangostin, by modifying the release stimulated by pH and the surface charge interactions of the nanoparticles that can target the release at specific areas of the gastrointestinal tract. Even though in vitro studies have been shown to increase cytotoxicity activity through the activity of inhibiting proliferation and increasing apoptosis against cancer cells, further studies regarding anticancer activity in vivo are needed.

Although recent reports on α-mangostin nanoformulation to date appear promising, most of the published data come from in vitro and in vivo (Table 3) trials and not clinical trials. Therefore, there are concerns regarding the toxicity of nanoparticles, as little is known about the behavior of nanoparticles in humans. While the unique properties of nanoparticles brought about by their small size provide enormous opportunity for medicinal uses, safety concerns have surfaced since their physicochemical properties might lead to altered pharmacokinetics, with the ability to overcome biological barriers. Furthermore, the inherent toxicity of several of the compounds, as well as their capacity to collect and stay in the human body, has hampered their translation. The use of biological capping materials like such as chitosan further reduce toxicity while their biocompatibility and biodegradative capacity making them an intuitive choice for a nanocarrier. The stability, circulation time, access and bioavailability to disease locations, and safety profile of these nanoparticles are critical for successful clinical translation. Thus, clinical trials are ultimately needed to understand the in vivo behavior of the α-mangostin nanoformulation, which can ultimately lead to the design of a suitable formulation with superior therapeutic efficacy.

Table 3. The volume reduction of tumor treatment with different α-mangostin nanoformulations.

Formulation	Type of Tumor	Reduction of Tumor Volume	Ref.
PLGA	Pancreatic	More than 60% of tumor reduction with 20 mg/kg dosage	[35]
Cyclodextrin nanoparticle	Colon	Approximately 56% of tumor reduction with 10 mg/kg dosage	[75]
PEG-PLA nanomicelles coated with CREKA peptide	Pancreatic	More than 70% of tumor reduction with 20 mg/kg dosage	[78]
MPEG-PCL nanomicelles	Melanoma	Almost 50% of tumor growth reduction with 50 mg/kg dosage	[80]
MPEG-PLA nanomicelles	Glioma	Approximately 65% tumor reduction with 50 mg/kg dosage	[81]

6. Conclusions

In recent years, various nanoformulations of α-mangostin have been developed to improve bioavailability and effectiveness in cancer treatment. α-mangostin exhibits excellent anticancer properties but its poor solubility, rapid elimination, and poor pharmacokinetic properties hinder its usability as a potent drug against cancer. Many techniques have been used to address this issue, one of which is the development of nanosized delivery vehicles. For α-mangostin formulated as nanoparticles, this development is very welcome because they are not only able to improve the dispersion of α-mangostin in aqueous solution, but also provide other advantages not obtained from conventional delivery techniques. These advantages include changeable particle size as well as modifiable surface characteristics. For example, the conjugation of α-mangostin nanoparticles with targeting ligands such as peptides and aptamers can provide specific targeting to cancer cells. With the nanoformulation, the researchers can improve the bioavailability and therapeutic properties of α-mangostin nanoparticles to achieve high therapeutic efficacy. Moreover, the α-mangostin nanoformulation appears to have good affinity and selectivity against cancer cells while imparting negligible toxicity to normal cells.

Author Contributions: Conceptualization, L.M. and N.W.; methodology, L.M.; software, L.M.; validation, N.W., M.M. and I.M.J.; formal analysis, L.M.; investigation, L.M.; resources, L.M.; data curation, L.M.; writing—original draft preparation, L.M.; writing—review and editing, N.W., M.M., I.M.J. and A.F.A.M.; visualization, L.M.; supervision, N.W.; project administration, L.M.; funding acquisition, N.W. All authors have read and agreed to the published version of the manuscript.

Funding: This research was funded by the College Superior Applied Research (Penelitian Terapan Unggulan Perguruan Tinggi), Ministry of Culture and Higher Education, Republic of Indonesia, grant number 1207/UN6.3.1/PT.00/2021.

Institutional Review Board Statement: Not applicable.

Informed Consent Statement: Not applicable.

Data Availability Statement: Not applicable.

Conflicts of Interest: The authors declare no conflict of interest.

References

1. Bray, F.; Ferlay, J.; Soerjomataram, I.; Siegel, R.L.; Torre, L.A.; Jemal, A. Global cancer statistics 2018: GLOBOCAN estimates of incidence and mortality worldwide for 36 cancers in 185 countries. *CA Cancer J. Clin.* **2018**, *68*, 394–424. [CrossRef] [PubMed]
2. Majolo, F.; de Oliveira Becker Delwing, L.K.; Marmitt, D.J.; Bustamante-Filho, I.C.; Goettert, M.I. Medicinal plants and bioactive natural compounds for cancer treatment: Important advances for drug discovery. *Phytochem. Lett.* **2019**, *31*, 196–207. [CrossRef]
3. Pham, D.T.; Saelim, N.; Tiyaboonchai, W. Alpha mangostin loaded crosslinked silk fibroin-based nanoparticles for cancer chemotherapy. *Colloids Surfaces B Biointerfaces* **2019**, *181*, 705–713. [CrossRef]
4. Sung, H.; Ferlay, J.; Siegel, R.L.; Laversanne, M.; Soerjomataram, I.; Jemal, A.; Bray, F. Global Cancer Statistics 2020: GLOBOCAN Estimates of Incidence and Mortality Worldwide for 36 Cancers in 185 Countries. *CA Cancer J. Clin.* **2021**, *71*, 209–249. [CrossRef] [PubMed]
5. Baskar, R.; Lee, K.A.; Yeo, R.; Yeoh, K.W. Cancer and radiation therapy: Current advances and future directions. *Int. J. Med. Sci.* **2012**, *9*, 193–199. [CrossRef]
6. Tohme, S.; Simmons, R.L.; Tsung, A. Surgery for Cancer: A Trigger for Metastases. *Cancer Res.* **2018**, *77*, 1548–1552. [CrossRef]
7. Swain, S.M. Chemotherapy: Updates and New Perspectives. *Oncologist* **2011**, *16*, 30–39. [CrossRef] [PubMed]
8. Pérez-Herrero, E.; Fernández-Medarde, A. Advanced targeted therapies in cancer: Drug nanocarriers, the future of chemotherapy. *Eur. J. Pharm. Biopharm.* **2015**, *93*, 52–79. [CrossRef]
9. Huang, C.Y.; Ju, D.T.; Chang, C.F.; Muralidhar Reddy, P.; Velmurugan, B.K. A review on the effects of current chemotherapy drugs and natural agents in treating non-small cell lung cancer. *Biomedicine* **2017**, *7*, 12–23. [CrossRef] [PubMed]
10. Lin, S.R.; Chang, C.H.; Hsu, C.F.; Tsai, M.J.; Cheng, H.; Leong, M.K.; Sung, P.J.; Chen, J.C.; Weng, C.F. Natural compounds as potential adjuvants to cancer therapy: Preclinical evidence. *Br. J. Pharmacol.* **2020**, *177*, 1409–1423. [CrossRef]
11. Nirmala, M.J.; Samundeeswari, A.; Sankar, P.D. Natural plant resources in anti-cancer therapy—A review. *Res. Plant Biol.* **2011**, *1*, 1–14.
12. Khan, T.; Gurav, P. PhytoNanotechnology: Enhancing delivery of plant based anti-cancer drugs. *Front. Pharmacol.* **2018**, *8*, 1002. [CrossRef]
13. Subramaniam, S.; Selvaduray, K.R.; Radhakrishnan, A.K. Bioactive compounds: Natural defense against cancer? *Biomolecules* **2019**, *9*, 758. [CrossRef]
14. Rejhová, A.; Opattová, A.; Čumová, A.; Slíva, D.; Vodička, P. Natural compounds and combination therapy in colorectal cancer treatment. *Eur. J. Med. Chem.* **2018**, *144*, 582–594. [CrossRef] [PubMed]
15. Bishayee, A.; Sethi, G. Bioactive natural products in cancer prevention and therapy: Progress and promise. *Semin. Cancer Biol.* **2016**, *40–41*, 1–3. [CrossRef] [PubMed]
16. Davatgaran-Taghipour, Y.; Masoomzadeh, S.; Farzaei, M.H.; Bahramsoltani, R.; Karimi-Soureh, Z.; Rahimi, R.; Abdollahi, M. Polyphenol nanoformulations for cancer therapy: Experimental evidence and clinical perspective. *Int. J. Nanomed.* **2017**, *12*, 2689–2702. [CrossRef] [PubMed]
17. Mary Lazer, L.; Sadhasivam, B.; Palaniyandi, K.; Muthuswamy, T.; Ramachandran, I.; Balakrishnan, A.; Pathak, S.; Narayan, S.; Ramalingam, S. Chitosan-based nano-formulation enhances the anticancer efficacy of hesperetin. *Int. J. Biol. Macromol.* **2018**, *107*, 1988–1998. [CrossRef]
18. Cosco, D.; Mare, R.; Paolino, D.; Salvatici, M.C.; Cilurzo, F.; Fresta, M. Sclareol-loaded hyaluronan-coated PLGA nanoparticles: Physico-chemical properties and in vitro anticancer features. *Int. J. Biol. Macromol.* **2019**, *132*, 550–557. [CrossRef]
19. Hamishehkar, H.; Bahadori, M.B.; Vandghanooni, S.; Eskandani, M.; Nakhlband, A.; Eskandani, M. Preparation, characterization and anti-proliferative effects of sclareol-loaded solid lipid nanoparticles on A549 human lung epithelial cancer cells. *J. Drug Deliv. Sci. Technol.* **2018**, *45*, 272–280. [CrossRef]

20. Chowdhury, P.; Nagesh, P.K.B.; Hatami, E.; Wagh, S.; Dan, N.; Tripathi, M.K.; Khan, S.; Hafeez, B.B.; Meibohm, B.; Chauhan, S.C.; et al. Tannic acid-inspired paclitaxel nanoparticles for enhanced anticancer effects in breast cancer cells. *J. Colloid Interface Sci.* **2019**, *535*, 133–148. [CrossRef]
21. Gagliardi, A.; Bonacci, S.; Paolino, D.; Celia, C.; Procopio, A.; Fresta, M.; Cosco, D. Paclitaxel-loaded sodium deoxycholate-stabilized zein nanoparticles: Characterization and in vitro cytotoxicity. *Heliyon* **2019**, *5*, e02422. [CrossRef]
22. Kritsanawong, S.; Innajak, S.; Imoto, M.; Watanapokasin, R. Antiproliferative and apoptosis induction of α-mangostin in T47D breast cancer cells. *Int. J. Oncol.* **2016**, *48*, 2155–2165. [CrossRef] [PubMed]
23. Ibrahim, M.Y.; Hashim, N.M.; Mariod, A.A.; Mohan, S.; Abdulla, M.A.; Abdelwahab, S.I.; Arbab, I.A. α-Mangostin from *Garcinia mangostana* Linn: An updated review of its pharmacological properties. *Arab. J. Chem.* **2016**, *9*, 317–329. [CrossRef]
24. Wang, M.H.; Zhang, K.J.; Gu, Q.L.; Bi, X.L.; Wang, J.X. Pharmacology of mangostins and their derivatives: A comprehensive review. *Chin. J. Nat. Med.* **2017**, *15*, 81–93. [CrossRef]
25. Ovalle-Magallanes, B.; Eugenio-Pérez, D.; Pedraza-Chaverri, J. Medicinal properties of mangosteen (*Garcinia mangostana* L.): A comprehensive update. *Food Chem. Toxicol.* **2017**, *109*, 102–122. [CrossRef] [PubMed]
26. Shan, T.; Ma, Q.; Guo, K.; Liu, J.; Li, W.; Wang, F.; Wu, E. Xanthones from Mangosteen Extracts as Natural Chemopreventive Agents: Potential Anticancer Drugs. *Curr. Mol. Med.* **2011**, *11*, 666–677. [CrossRef] [PubMed]
27. Watanapokasin, R.; Jarinthanan, F.; Nakamura, Y.; Sawasjirakij, N.; Jaratrungtawee, A.; Suksamrarn, S. Effects of α-mangostin on apoptosis induction of human colon cancer. *World J. Gastroenterol.* **2011**, *17*, 2086–2095. [CrossRef] [PubMed]
28. Yoo, J.H.; Kang, K.; Jho, E.H.; Chin, Y.W.; Kim, J.; Nho, C.W. α- and γ-Mangostin inhibit the proliferation of colon cancer cells via β-catenin gene regulation in Wnt/cGMP signalling. *Food Chem.* **2011**, *129*, 1559–1566. [CrossRef]
29. Zhang, C.; Yu, G.; Shen, Y. The naturally occurring xanthone α-mangostin induces ROS-mediated cytotoxicity in non-small scale lung cancer cells. *Saudi J. Biol. Sci.* **2018**, *25*, 1090–1095. [CrossRef]
30. Ma, Y.; Yu, W.; Shrivastava, A.; Srivastava, R.K.; Shankar, S. Inhibition of pancreatic cancer stem cell characteristics by α-Mangostin: Molecular mechanisms involving Sonic hedgehog and Nanog. *J. Cell. Mol. Med.* **2019**, *23*, 2719–2730. [CrossRef]
31. Li, P.; Tian, W.; Ma, X. Alpha-mangostin inhibits intracellular fatty acid synthase and induces apoptosis in breast cancer cells. *Mol. Cancer* **2014**, *13*, 138. [CrossRef] [PubMed]
32. Wang, J.J.; Sanderson, B.J.S.; Zhang, W. Significant anti-invasive activities of α-mangostin from the mangosteen pericarp on two human skin cancer cell lines. *Anticancer Res.* **2012**, *32*, 3805–3816. [PubMed]
33. Chen, J.J.; Long, Z.J.; Xu, D.F.; Xiao, R.Z.; Liu, L.L.; Xu, Z.F.; Qiu, S.X.; Lin, D.J.; Liu, Q. Inhibition of autophagy augments the anticancer activity of α-mangostin in chronic myeloid leukemia cells. *Leuk. Lymphoma* **2014**, *55*, 628–638. [CrossRef] [PubMed]
34. Sodalee, K.; Sapsuphan, P.; Wongsirikul, R.; Puttipipatkhachorn, S. Preparation and evaluation of alpha-mangostin solid self-emulsifying drug delivery system. *Asian J. Pharm. Sci.* **2016**, *11*, 225–226. [CrossRef]
35. Verma, R.K.; Yu, W.; Shrivastava, A.; Shankar, S.; Srivastava, R.K. α-Mangostin-encapsulated PLGA nanoparticles inhibit pancreatic carcinogenesis by targeting cancer stem cells in human, and transgenic (KrasG12D, and KrasG12D/tp53R270H) mice. *Sci. Rep.* **2016**, *6*, 32743. [CrossRef]
36. Li, L.; Brunner, I.; Han, A.R.; Hamburger, M.; Kinghorn, A.D.; Frye, R.; Butterweck, V. Pharmacokinetics of α-mangostin in rats after intravenous and oral application. *Mol. Nutr. Food Res.* **2011**, *55*, 67–74. [CrossRef]
37. Xu, S.; Olenyuk, B.Z.; Okamoto, C.T.; Hamm-Alvarez, S.F. Targeting receptor-mediated endocytotic pathways with nanoparticles: Rationale and advances. *Adv. Drug* **2013**, *65*, 121–138. [CrossRef]
38. Cerqueira, B.B.S.; Lasham, A.; Shelling, A.N.; Al-Kassas, R. Nanoparticle therapeutics: Technologies and methods for overcoming cancer. *Eur. J. Pharm. Biopharm.* **2015**, *97*, 140–151. [CrossRef]
39. Wathoni, N.; Rusdin, A.; Motoyama, K.; nJoni, I.M.; Lesmana, R.; Muchtaridi, M. Nanoparticle drug delivery systems for α-mangostin. *Nanotechnol. Sci. Appl.* **2020**, *13*, 23–36. [CrossRef]
40. Muchtaridi, M.; Wijaya, C.A. Anticancer potential of α-mangostin. *Asian J. Pharm. Clin. Res.* **2017**, *10*, 440–445. [CrossRef]
41. Ahmad, M.; Yamin, B.M.; Mat Lazim, A. A study on dispersion and characterisation of α-mangostin loaded pH sensitive microgel systems. *Chem. Cent. J.* **2013**, *7*, 2–7. [CrossRef] [PubMed]
42. Guo, M.; Wang, X.; Lu, X.; Wang, H.; Brodelius, P.E. α-Mangostin extraction from the native mangosteen (*Garcinia mangostana* L) and the binding mechanisms of α-mangostin to HSAorTRF. *PLoS ONE* **2016**, *11*, e0161566. [CrossRef] [PubMed]
43. Aisha, A.F.A.; Ismail, Z.; Abu-Salah, K.M.; Majid, A.M.S.A. Solid Dispersions of α-Mangostin Improve Its Aqueous Solubility Through Self-Assembly of Nanomicelles. *J. Pharm. Sci.* **2012**, *101*, 815–825. [CrossRef] [PubMed]
44. Bumrung, J.; Chanchao, C.; Intasanta, V.; Palaga, T.; Wanichwecharungruang, S. Water-dispersible unadulterated α -mangostin particles for biomedical applications. *R. Soc. Open Sci.* **2020**, *7*, 200543. [CrossRef] [PubMed]
45. Choi, Y.H.; Han, S.Y.; Kim, Y.J.; Kim, Y.M.; Chin, Y.W. Absorption, tissue distribution, tissue metabolism and safety of α-mangostin in mangosteen extract using mouse models. *Food Chem. Toxicol.* **2014**, *66*, 140–146. [CrossRef] [PubMed]
46. Xu, Q.; Ma, J.; Lei, J.; Duan, W.; Sheng, L.; Chen, X.; Hu, A.; Wang, Z.; Wu, Z.; Wu, E.; et al. α-mangostin suppresses the viability and epithelial-mesenchymal transition of pancreatic cancer cells by downregulating the PI3K/Akt pathway. *Biomed. Res. Int.* **2014**, *2014*, 546353. [CrossRef]
47. Johnson, J.J.; Petiwala, S.M.; Syed, D.N.; Rasmussen, J.T.; Adhami, V.M.; Siddiqui, I.A.; Kohl, A.M.; Mukhtar, H. A-Mangostin, a Xanthone From Mangosteen Fruit, Promotes Cell Cycle Arrest in Prostate Cancer and Decreases Xenograft Tumor Growth. *Carcinogenesis* **2012**, *33*, 413–419. [CrossRef]

48. Lee, H.N.; Jang, H.Y.; Kim, H.J.; Shin, S.A.; Choo, G.S.; Park, Y.S.; Kim, S.K.; Jung, J.Y. Antitumor and apoptosis-inducing effects of α-mangostin extracted from the pericarp of the mangosteen fruit (*Garcinia mangostana* L.) in YD-15 tongue mucoepidermoid carcinoma cells. *Int. J. Mol. Med.* **2016**, *37*, 939–948. [CrossRef]
49. Lee, C.H.; Ying, T.H.; Chiou, H.L.; Hsieh, S.C.; Wen, S.H.; Chou, R.H.; Hsieh, Y.H. Alpha-mangostin induces apoptosis through activation of reactive oxygen species and ASK1/p38 signaling pathway in cervical cancer cells. *Oncotarget* **2017**, *8*, 47425–47439. [CrossRef]
50. Herrera-Aco, D.R.; Medina-Campos, O.N.; Pedraza-Chaverri, J.; Sciutto-Conde, E.; Rosas-Salgado, G.; Fragoso-González, G. Alpha-mangostin: Anti-inflammatory and antioxidant effects on established collagen-induced arthritis in DBA/1J mice. *Food Chem. Toxicol.* **2019**, *124*, 300–315. [CrossRef]
51. Sakagami, Y.; Thevanesam, V.; Programme, N.P.; Lanka, S.; Lanka, S.; Sakagami, Y. Natural Product Research: Formerly Natural Product Letters Antibacterial activity of xanthones from *Garcinia mangostana* (L.) and their structure—Activity relationship studies. *Nat. Prod. Res.* **2013**, *27*, 938–941.
52. Sivaranjani, M.; Leskinen, K.; Aravindraja, C.; Saavalainen, P.; Pandian, S.K.; Skurnik, M.; Ravi, A.V. Deciphering the antibacterial mode of action of alpha-mangostin on Staphylococcus epidermidis RP62A through an integrated transcriptomic and proteomic approach. *Front. Microbiol.* **2019**, *10*, 1–16. [CrossRef]
53. Fu, T.; Li, H.; Zhao, Y.; Cai, E.; Zhu, H.; Li, P.; Liu, J. Hepatoprotective effect of α-mangostin against lipopolysaccharide/D-galactosamine-induced acute liver failure in mice. *Biomed. Pharmacother.* **2018**, *106*, 896–901. [CrossRef] [PubMed]
54. Buelna-Chontal, M.; Correa, F.; Hernández-Reséndiz, S.; Zazueta, C.; Pedraza-Chaverri, J. Protective effect of α-mangostin on cardiac reperfusion damage by attenuation of oxidative stress. *J. Med. Food* **2011**, *14*, 1370–1374. [CrossRef]
55. Upegui, Y.; Robledo, S.M.; Gil Romero, J.F.; Quiñones, W.; Archbold, R.; Torres, F.; Escobar, G.; Nariño, B.; Echeverri, F. In vivo Antimalarial Activity of α-Mangostin and the New Xanthone δ-Mangostin. *Phyther. Res.* **2015**, *29*, 1195–1201. [CrossRef]
56. Taher, M.; Mohamed Amiroudine, M.Z.A.; Tengku Zakaria, T.M.F.S.; Susanti, D.; Ichwan, S.J.A.; Kaderi, M.A.; Ahmed, Q.U.; Zakaria, Z.A. α -mangostin improves glucose uptake and inhibits adipocytes differentiation in 3T3-L1 cells via PPAR γ, GLUT4, and leptin expressions. *Evid-Based Complement. Altern. Med.* **2015**, *2015*, 740238. [CrossRef] [PubMed]
57. Wang, Y.; Xia, Z.; Xu, J.R.; Wang, Y.X.; Hou, L.N.; Qiu, Y.; Chen, H.Z. α-Mangostin, a polyphenolic xanthone derivative from mangosteen, attenuates β-amyloid oligomers-induced neurotoxicity by inhibiting amyloid aggregation. *Neuropharmacology* **2012**, *62*, 871–881. [CrossRef] [PubMed]
58. National Center for Biotechnology Information. PubChem Compound Summary for CID 5281650, Alpha-Mangostin. 2021. Available online: https://pubchem.ncbi.nlm.nih.gov/compound/alpha-Mangostin (accessed on 10 November 2021).
59. Ravindran, R. Nano Technology in Cancer Diagnosis and Treatment: An Overview. *Oral Maxillofac. Pathol. J.* **2011**, *2*, 101–106.
60. Karra, N.; Benita, S. The Ligand Nanoparticle Conjugation Approach for Targeted Cancer Therapy. *Curr. Drug Metab.* **2011**, *13*, 22–41. [CrossRef]
61. Jeevanandam, J.; Chan, Y.S.; Danquah, M.K. Nano-formulations of drugs: Recent developments, impact and challenges. *Biochimie* **2016**, *128*, 99–112. [CrossRef]
62. Siddiqui, I.A.; Sanna, V.; Ahmad, N.; Sechi, M.; Mukhtar, H. Resveratrol nanoformulation for cancer prevention and therapy. *Ann. N. Y. Acad. Sci.* **2015**, *1348*, 20–31. [CrossRef]
63. Aghebati-Maleki, A.; Dolati, S.; Ahmadi, M.; Baghbanzhadeh, A.; Asadi, M.; Fotouhi, A.; Yousefi, M.; Aghebati-Maleki, L. Nanoparticles and cancer therapy: Perspectives for application of nanoparticles in the treatment of cancers. *J. Cell. Physiol.* **2020**, *235*, 1962–1972. [CrossRef]
64. Lopalco, A.; Denora, N. Nanoformulations for Drug Delivery: Safety, Toxicity, and Efficacy. In *Computational Toxicology: Methods and Protocols*; Nicolotti, O., Ed.; Springer: New York, NY, USA, 2018; pp. 347–365. ISBN 978-1-4939-7899-1.
65. Gurunathan, S.; Kang, M.H.; Qasim, M.; Kim, J.H. Nanoparticle-mediated combination therapy: Two-in-one approach for cancer. *Int. J. Mol. Sci.* **2018**, *19*, 3264. [CrossRef] [PubMed]
66. Tiwari, G.; Tiwari, R.; Bannerjee, S.; Bhati, L.; Pandey, S.; Pandey, P.; Sriwastawa, B. Drug delivery systems: An updated review. *Int. J. Pharm. Investig.* **2012**, *2*, 2–11. [CrossRef] [PubMed]
67. Russell, L.M.; Dawidczyk, C.M.; Searson, P.C. Quantitative Evaluation of the Enhanced Permeability and Retention (EPR) Effect. *Cancer Nanotechnol.* **2017**, *1530*, 41–195. [CrossRef]
68. Acharya, S.; Sahoo, S.K. PLGA nanoparticles containing various anticancer agents and tumour delivery by EPR effect. *Adv. Drug Deliv. Rev.* **2011**, *63*, 170–183. [CrossRef]
69. Khobragade, D.S.; Trambak Patil, A.; Shamrao Khobragade, D.; Annaji Chafle, S.; Prasadrao Ujjainkar, A.; Niranjanrao Umathe, S.; Laxminarayan Lakhotia, C. Development and evaluation of a hot-melt coating technique for enteric coating. *Brazilian J. Pharm. Sci.* **2012**, *48*, 69–77.
70. Taokaew, S.; Chiaoprakobkij, N.; Siripong, P.; Sanchavanakit, N.; Pavasant, P.; Phisalaphong, M. Multifunctional cellulosic nanofiber film with enhanced antimicrobial and anticancer properties by incorporation of ethanolic extract of *Garcinia mangostana* peel. *Mater. Sci. Eng. C* **2021**, *120*, 111783. [CrossRef]
71. Bonafè, F.; Pazzini, C.; Marchionni, S.; Guarnieri, C.; Muscari, C. Complete disaggregation of MCF-7-derived breast tumour spheroids with very low concentrations of α-mangostin loaded in CD44 thioaptamer-tagged nanoparticles. *Int. J. Med. Sci.* **2019**, *16*, 33–42. [CrossRef]

72. Yostawonkul, J.; Surassmo, S.; Iempridee, T.; Pimtong, W.; Suktham, K.; Sajomsang, W.; Gonil, P.; Ruktanonchai, U.R. Surface modification of nanostructure lipid carrier (NLC) by oleoyl-quaternized-chitosan as a mucoadhesive nanocarrier. *Colloids Surfaces B Biointerfaces* **2017**, *149*, 301–311. [CrossRef]
73. Chandra Boinpelly, V.; Verma, R.K.; Srivastav, S.; Srivastava, R.K.; Shankar, S. α-Mangostin-encapsulated PLGA nanoparticles inhibit colorectal cancer growth by inhibiting Notch pathway. *J. Cell. Mol. Med.* **2020**, *24*, 11343–11354. [CrossRef] [PubMed]
74. Doan, V.T.H.; Lee, J.H.; Takahashi, R.; Nguyen, P.T.M.; Nguyen, V.A.T.; Pham, H.T.T.; Fujii, S.; Sakurai, K. Cyclodextrin-based nanoparticles encapsulating α-mangostin and their drug release behavior: Potential carriers of α-mangostin for cancer therapy. *Polym. J.* **2020**, *52*, 457–466. [CrossRef]
75. Doan, V.T.H.; Takano, S.; Doan, N.A.T.; Nguyen, P.T.M.; Nguyen, V.A.T.; Pham, H.T.T.; Nakazawa, K.; Fujii, S.; Sakurai, K. Anticancer efficacy of cyclodextrin-based hyperbranched polymer nanoparticles containing alpha-mangostin. *Polym. J.* **2020**, *53*, 481–492. [CrossRef]
76. Nguyen Thi, M.P.; Tran Dai, L.; Ta Thu, M.; Nguyen Trung, H. Cytotoxicity of α-mangostin encapsulated polymeric nanoparticles against lung cancer cells. *Acad. J. Biol.* **2018**, *40*, 108–114. [CrossRef]
77. Samprasit, W.; Akkaramongkolporn, P.; Jaewjira, S.; Opanasopit, P. Design of alpha mangostin-loaded chitosan/alginate controlled-release nanoparticles using genipin as crosslinker. *J. Drug Deliv. Sci. Technol.* **2018**, *46*, 312–321. [CrossRef]
78. Feng, J.; Xu, M.; Wang, J.; Zhou, S.; Liu, Y.; Liu, S.; Huang, Y.; Chen, Y.; Chen, L.; Song, Q.; et al. Sequential delivery of nanoformulated α-mangostin and triptolide overcomes permeation obstacles and improves therapeutic effects in pancreatic cancer. *Biomaterials* **2020**, *241*, 119907. [CrossRef]
79. Wathoni, N.; Meylina, L.; Rusdin, A.; Abdelwahab Mohammed, A.F.; Tirtamie, D.; Herdiana, Y.; Motoyama, K.; Panatarani, C.; Joni, I.M.; Lesmana, R.; et al. The potential cytotoxic activity enhancement of α-mangostin in chitosan-kappa carrageenan-loaded nanoparticle against mcf-7 cell line. *Polymers* **2021**, *13*, 1681. [CrossRef]
80. Yang, S.; Gao, X.; He, Y.; Hu, Y.; Xu, B.; Cheng, Z.; Xiang, M.; Xie, Y. Applying an innovative biodegradable self-assembly nanomicelles to deliver α-mangostin for improving anti-melanoma activity. *Cell Death Dis.* **2019**, *10*, 146. [CrossRef]
81. Zheng, S.; Liu, J.; Faried, A.; Richard, S.A.; Gao, X. Novel chemically synthesized, alpha-mangostin-loaded nano-particles, enhanced cell death through multiple pathways against malignant glioma. *J. Biomed. Nanotechnol.* **2018**, *14*, 1866–1882. [CrossRef]
82. Benjakul, R.; Kongkaneramit, L.; Sarisuta, N.; Moongkarndi, P.; Müller-Goymann, C.C. Cytotoxic effect and mechanism inducing cell death of α-mangostin liposomes in various human carcinoma and normal cells. *Anticancer. Drugs* **2015**, *26*, 824–834. [CrossRef]
83. Trang Phan, T.K.; Tran, T.Q.; Nguyen Pham, D.T.; Nguyen, D.T. Characterization, Release Pattern, and Cytotoxicity of Liposomes Loaded With α-Mangostin Isolated from Pericarp of Mangosteen (*Garcinia mangostana* L.). *Nat. Prod. Commun.* **2020**, *15*, 1934578X20974559. [CrossRef]
84. Qiu, S.; Granet, R.; Mbakidi, J.P.; Brégier, F.; Pouget, C.; Micallef, L.; Sothea-Ouk, T.; Leger, D.Y.; Liagre, B.; Chaleix, V.; et al. Delivery of tanshinone IIA and α-mangostin from gold/PEI/cyclodextrin nanoparticle platform designed for prostate cancer chemotherapy. *Bioorg. Med. Chem. Lett.* **2016**, *26*, 2503–2506. [CrossRef] [PubMed]
85. Adam Khan, P.; Sasikanth, K.; Nama, S.; Suresh, P.; Brahmaiah, B. Nanofibers—A New Trend in Nano Drug Delivery Systems. *Int. J. Pharm. Res. Anal.* **2013**, *3*, 47–55.
86. Poláková, L.; Širc, J.; Hobzová, R.; Cocârță, A.I.; Heřmánková, E. Electrospun nanofibers for local anticancer therapy: Review of in vivo activity. *Int. J. Pharm.* **2019**, *558*, 268–283. [CrossRef] [PubMed]
87. Cavo, M.; Serio, F.; Kale, N.R.; D'Amone, E.; Gigli, G.; Del Mercato, L.L. Electrospun nanofibers in cancer research: From engineering of: From vitro 3D cancer models to therapy. *Biomater. Sci.* **2020**, *8*, 4887–4905. [CrossRef]
88. Sasikala, A.R.K.; Unnithan, A.R.; Park, C.H.; Kim, C.S. *Nanofiber-Based Anticancer Drug Delivery Platform*; Elsevier Inc.: Amsterdam, The Netherlands, 2019; ISBN 9780128149447.
89. Balaji, A.; Vellayappan, M.V.; John. A.A.; Subramanian, A.P.; Jaganathan, S.K.; Supriyanto, E.; Razak, S.I.A. An insight on electrospun-nanofibers-inspired modern drug delivery system in the treatment of deadly cancers. *RSC Adv.* **2015**, *5*, 57984–58004. [CrossRef]
90. Hu, X.; Liu, S.; Zhou, G.; Huang, Y.; Xie, Z.; Jing, X. Electrospinning of polymeric nanofibers for drug delivery applications. *J. Control. Release* **2014**, *185*, 12–21. [CrossRef]
91. Prabhu, P. *Nanofibers for Medical Diagnosis and Therapy*; Springer: Berlin, Germany, 2019; ISBN 9783319536552.
92. Jastrzebska, K.; Kucharczyk, K.; Florczak, A.; Dondajewska, E.; Mackiewicz, A.; Dams-Kozlowska, H. Silk as an innovative biomaterial for cancer therapy. *Reports Pract. Oncol. Radiother.* **2015**, *20*, 87–98. [CrossRef]
93. Mottaghitalab, F.; Farokhi, M.; Shokrgozar, M.A.; Atyabi, F.; Hosseinkhani, H. Silk fibroin nanoparticle as a novel drug delivery system. *J. Control. Release* **2015**, *206*, 161–176. [CrossRef]
94. García-Pinel, B.; Porras-Alcalá, C.; Ortega-Rodríguez, A.; Sarabia, F.; Prados, J.; Melguizo, C.; López-Romero, J.M. Lipid-based nanoparticles: Application and recent advances in cancer treatment. *Nanomaterials* **2019**, *9*, 638. [CrossRef]
95. Nasirizadeh, S.; Malaekeh-Nikouei, B. Solid lipid nanoparticles and nanostructured lipid carriers in oral cancer drug delivery. *J. Drug Deliv. Sci. Technol.* **2020**, *55*, 101458. [CrossRef]
96. Bayón-Cordero, L.; Alkorta, I.; Arana, L. Application of solid lipid nanoparticles to improve the efficiency of anticancer drugs. *Nanomaterials* **2019**, *9*, 474. [CrossRef] [PubMed]
97. Chan, J.M.; Zhang, L.; Yuet, K.P.; Liao, G.; Rhee, J.W.; Langer, R.; Farokhzad, O.C. PLGA-lecithin-PEG core-shell nanoparticles for controlled drug delivery. *Biomaterials* **2009**, *30*, 1627–1634. [CrossRef]

98. Kumar, V. Solid Lipid Nanoparticle of Alpha-Mangostin Exerts Diethylnitrosamine-Induced Hepatocellular Carcinoma via Alteration of PI3K/Akt Pathway. *Gut Liver.* **2019**, *13*, 209.
99. Haider, M.; Abdin, S.M.; Kamal, L.; Orive, G. Nanostructured lipid carriers for delivery of chemotherapeutics: A review. *Pharmaceutics* **2020**, *12*, 288. [CrossRef] [PubMed]
100. Salvi, V.R.; Pawar, P. Nanostructured lipid carriers (NLC) system: A novel drug targeting carrier. *J. Drug Deliv. Sci. Technol.* **2019**, *51*, 255–267. [CrossRef]
101. Rizwanullah, M.; Ahmad, J.; Amin, S. Nanostructured Lipid Carriers: A Novel Platform for Chemotherapeutics. *Curr. Drug Deliv.* **2016**, *13*, 4–26. [CrossRef]
102. Zieli'nska, A.; Carreiró, F.; Oliveira, A.M.; Neves, A.; Pires, B.; Venkates, S.N.; Durazzo, A.; Lucarini, M.; Ede, P.; Silva, A.M.; et al. Polymeric Nanoparticles: Production, Characterization, Toxicology and Ecotoxicology. *Molecules* **2020**, *25*, 3731. [CrossRef]
103. Madkour, L.H. Polymer nanoparticle drug-nucleic acid combinations. In *Nucleic Acids as Gene Anticancer Drug Delivery Therapy*; Academic Press: London, UK, 2019; pp. 241–255. [CrossRef]
104. Kumar, S.; Dilbaghi, N.; Saharan, R.; Bhanjana, G. Nanotechnology as Emerging Tool for Enhancing Solubility of Poorly Water-Soluble Drugs. *Bionanoscience* **2012**, *2*, 227–250. [CrossRef]
105. Peltonen, L.; Singhal, M.; Hirvonen, J. *Principles of Nanosized Drug Delivery Systems*; Elsevier Ltd.: Amsterdam, The Netherlands, 2020; ISBN 9780081029855.
106. Li, B.; Li, Q.; Mo, J.; Dai, H. Drug-loaded polymeric nanoparticles for cancer stem cell targeting. *Front. Pharmacol.* **2017**, *8*, 1–12. [CrossRef]
107. Kamaly, N.; Xiao, Z.; Valencia, P.M.; Radovic-Moreno, A.F.; Farokhzad, O.C. Targeted polymeric therapeutic nanoparticles: Design, development and clinical translation. *Chem. Soc. Rev.* **2012**, *41*, 2971–3010. [CrossRef]
108. Suriya Prabha, A.; Dorothy, R.; Jancirani, S.; Rajendran, S.; Singh, G.; Senthil Kumaran, S. Recent Advances in the Study of Toxicity of Polymer-Based Nanomaterials. *Nanotoxicity* **2020**, *2020*, 143–165.
109. Azevedo, M.A.; Bourbon, A.I.; Vicente, A.A.; Cerqueira, M.A. Alginate/chitosan nanoparticles for encapsulation and controlled release of vitamin B2. *Int. J. Biol. Macromol.* **2014**, *71*, 141–146. [CrossRef]
110. Katuwavila, N.P.; Perera, A.D.L.C.; Samarakoon, S.R.; Soysa, P.; Karunaratne, V.; Amaratunga, G.A.J.; Karunaratne, D.N. Chitosan-Alginate Nanoparticle System Efficiently Delivers Doxorubicin to MCF-7 Cells. *J. Nanomater.* **2016**, *2016*, 3178904. [CrossRef]
111. Samprasit, W.; Opanasopit, P. Chitosan-Based Nanoparticles for Controlled-Release Delivery of α–Mangostin. *Int. J. Pharma Med. Biol. Sci.* **2020**, *9*, 1–5. [CrossRef]
112. Anirudhan, T.S.; Anila, M.M.; Franklin, S. Synthesis characterization and biological evaluation of alginate nanoparticle for the targeted delivery of curcumin Synthesis characterization and biological evaluation of alginate nanoparticle for the targeted delivery of curcumin. *Mater. Sci. Eng. C* **2019**, *78*, 1125–1134. [CrossRef] [PubMed]
113. Duchêne, D. Cyclodextrins and Their Inclusion Complexes. In *Cyclodextrins in Pharmaceutics, Cosmetics, and Biomedicine: Current and Future Industrial Applications*; Bilensoy, E., Ed.; John Wiley & Sons, Inc.: Hoboken, NJ, USA, 2011; pp. 1–18. [CrossRef]
114. Hashidzume, A.; Takashima, Y.; Yamaguchi, H.; Harada, A. Comprehensive Supramolecular Chemistry II: Cyclodextrin. In *Comprehensive Supramolecular Chemistry II*; Elsevier: Amsterdam, The Netherlands, 2017; Volume 9, pp. 269–316, ISBN 9780128031988.
115. Gadade, D.D.; Pekamwar, S.S. Cyclodextrin based nanoparticles for drug delivery and theranostics. *Adv. Pharm. Bull.* **2020**, *10*, 166–183. [CrossRef] [PubMed]
116. Amirmahani, N.; Mahmoodi, N.O.; Mohammadi Galangash, M.; Ghavidast, A. Advances in nanomicelles for sustained drug delivery. *J. Ind. Eng. Chem.* **2017**, *55*, 21–34. [CrossRef]
117. Mohamed, S.; Parayath, N.N.; Taurin, S.; Greish, K. Polymeric nano-micelles: Versatile platform for targeted delivery in cancer. *Ther. Deliv.* **2014**, *5*, 1101–1121. [CrossRef]
118. Miyata, K.; Christie, R.J.; Kataoka, K. Polymeric micelles for nano-scale drug delivery. *React. Funct. Polym.* **2011**, *71*, 227–234. [CrossRef]
119. Trinh, H.M.; Joseph, M.; Cholkar, K.; Mitra, R.; Mitra, A.K. *Nanomicelles in Diagnosis and Drug Delivery*; Elsevier: Amsterdam, The Netherlands, 2017; Volume 1, pp. 45–58, ISBN 9780323429979.
120. Saraf, S.; Jain, A.; Tiwari, A.; Verma, A.; Panda, P.K.; Jain, S.K. Advances in liposomal drug delivery to cancer: An overview. *J. Drug Deliv. Sci. Technol.* **2020**, *56*, 101549. [CrossRef]
121. Pawar, H.R.; Bhosale, S.S.; Derle, N.D. Use of liposomes in cancer therapy: A review. *Int. J. Pharm. Sci. Res.* **2012**, *3*, 3585–3590.
122. Sercombe, L.; Veerati, T.; Moheimani, F.; Wu, S.Y.; Sood, A.K.; Hua, S. Advances and Challenges of Liposome Assisted Drug Delivery. *Front. Pharmacol.* **2015**, *329*, 286. [CrossRef] [PubMed]
123. Li, M.; Du, C.; Guo, N.; Teng, Y.; Meng, X.; Sun, H.; Li, S.; Yu, P.; Galons, H. Composition design and medical application of liposomes. *Eur. J. Med. Chem.* **2019**, *164*, 640–653. [CrossRef] [PubMed]
124. Chandran, P.R.; Thomas, R.T. *Gold Nanoparticles in Cancer Drug Delivery*; Elsevier Inc.: Amsterdam, The Netherlands, 2015; ISBN 9780323353038.
125. Bai, X.; Wang, Y.; Song, Z.; Feng, Y.; Chen, Y.; Zhang, D.; Feng, L. The basic properties of gold nanoparticles and their applications in tumor diagnosis and treatment. *Int. J. Mol. Sci.* **2020**, *21*, 2480. [CrossRef]
126. Dreaden, E.C.; Austin, L.A.; MacKey, M.A.; El-Sayed, M.A. Size matters: Gold nanoparticles in targeted cancer drug delivery. *Ther. Deliv.* **2012**, *3*, 457–478. [CrossRef]

pharmaceutics

Review

Natural Carbon Nanodots: Toxicity Assessment and Theranostic Biological Application

Ming-Hsien Chan [1,†], Bo-Gu Chen [2,†], Loan Thi Ngo [2,3], Wen-Tse Huang [2], Chien-Hsiu Li [1], Ru-Shi Liu [2,*] and Michael Hsiao [1,4,*]

1 Genomics Research Center, Academia Sinica, Taipei 115, Taiwan; ahsien0718@gate.sinica.edu.tw (M.-H.C.); g803020178g@gate.sinica.edu.tw (C.-H.L.)
2 Department of Chemistry, National Taiwan University, Taipei 106, Taiwan; r09223161@ntu.edu.tw (B.-G.C.); d10223121@ntu.edu.tw (L.T.N.); d09223104@ntu.edu.tw (W.-T.H.)
3 Nano Science and Technology Program, Taiwan International Graduate Program, Academia Sinica and National Taiwan University, Taipei 115, Taiwan
4 Department of Biochemistry, College of Medicine, Kaohsiung Medical University, Kaohsiung 807, Taiwan
* Correspondence: rsliu@ntu.edu.tw (R.-S.L.); mhsiao@gate.sinica.edu.tw (M.H.)
† These authors contributed equally to this work.

Citation: Chan, M.-H.; Chen, B.-G.; Ngo, L.T.; Huang, W.-T.; Li, C.-H.; Liu, R.-S.; Hsiao, M. Natural Carbon Nanodots: Toxicity Assessment and Theranostic Biological Application. *Pharmaceutics* **2021**, *13*, 1874. https://doi.org/10.3390/pharmaceutics13111874

Academic Editor: Mazzucchelli Serena

Received: 30 September 2021
Accepted: 29 October 2021
Published: 5 November 2021

Publisher's Note: MDPI stays neutral with regard to jurisdictional claims in published maps and institutional affiliations.

Abstract: This review outlines the methods for preparing carbon dots (CDs) from various natural resources to select the process to produce CDs with the best biological application efficacy. The oxidative activity of CDs mainly involves photo-induced cell damage and the destruction of biofilm matrices through the production of reactive oxygen species (ROS), thereby causing cell auto-apoptosis. Recent research has found that CDs derived from organic carbon sources can treat cancer cells as effectively as conventional drugs without causing damage to normal cells. CDs obtained by heating a natural carbon source inherit properties similar to the carbon source from which they are derived. Importantly, these characteristics can be exploited to perform non-invasive targeted therapy on human cancers, avoiding the harm caused to the human body by conventional treatments. CDs are attractive for large-scale clinical applications. Water, herbs, plants, and probiotics are ideal carbon-containing sources that can be used to synthesize therapeutic and diagnostic CDs that have become the focus of attention due to their excellent light stability, fluorescence, good biocompatibility, and low toxicity. They can be applied as biosensors, bioimaging, diagnosis, and treatment applications. These advantages make CDs attractive for large-scale clinical application, providing new technologies and methods for disease occurrence, diagnosis, and treatment research.

Keywords: natural carbon nanodots; tumor-targeting probes; biosensing; cancer theranostic; toxicity assessment

1. Introduction

The materials studied in the past were at the micrometer scale; however, in the past few decades, nanometer-scale development has changed, which has become an essential direction of scientific and technological development. Nano-sized materials are widely used in various fields, such as medicine, biosensor development, energy research, and catalysis. As fluorescent nanomaterials can emit light, they have the potential for application in biomarking technology, such as semiconductor-based quantum dots containing cadmium sulfide (CdS) or cadmium selenide (CdSe). However, most materials with a high quantum yield (QY, ϕ) on the market include heavy metal components, which pose risks related to their biological toxicity and cytotoxicity. Contamination may occur during the synthesis process, and they are not adequately recycled after use; as such, the associated harm to the environment should not be underestimated. The above shortcomings limit the application range as semiconductor quantum dots.

In recent years, nanotechnology has been widely used in the field of biomedicine [1]. Generally, when the scale of a substance drops to the nanometer level, its physical and chemical properties also undergo tremendous changes, especially in terms of its optical properties, providing an ideal development space for applications in the field of biomarkers and optical imaging of diseases [2]. Since their inception, carbon nanomaterials have attracted attention from researchers in materials science and biomedicine, especially considering their excellent optical properties, facilitating their application in bioimaging, biomarkers, and sensors [3]. Fluorescent carbon dots (CDs) are another new type of carbon nanomaterial, with a typical particle size of about 10 nm. In addition to the advantages related to the particle size and wavelength-dependent luminescence characterizing quantum dots, fluorescent CDs also have high light stability and present no light scintillation phenomena. Their surface is easily functionalized and modified, and the preparation materials are widely available [4–6]. Wang et al. evaluated the toxicity of CDs and showed that they do not cause any abnormalities or damage to tissues and organs [7]. Kang et al. compared CDs, single-walled carbon tubes, and carbon dioxide [8]. After the cytotoxicities of silicon and zinc oxide were determined, CDs showed the lowest toxicity compared with the materials mentioned above. CDs' unique optical properties and excellent biological properties confer outstanding advantages and good development prospects in biomedical optical imaging and tumor diagnosis. In addition, the ability to introduce multiple functional groups onto the surface of CDs also provides possibilities for their modification.

Green manufacturing is a modern manufacturing model that comprehensively considers the environmental impact and resource benefits. Due to the increasing awareness of environmental protection, green manufacturing is becoming an increasingly important process in all walks of life. The chemical synthesis of CDs can be similar to that produced from various natural resources in the nanoscale process. For instance, Hsu et al. used coffee grounds to synthesize CDs by calcining in 2012. The average particle size of the synthesized CDs was about 5 nm, and they were successfully applied to biological imaging. Unlike the previous need for complex processing procedures, a rapid and straightforward synthesis method was developed. Later, Crista et al. used different organic compounds to synthesize CDs, with an average particle size of about 2.6–7.9 nm, with many carboxyl and amine groups, proving that the quantum yield varies with functional groups [9]. In 2013, Qu et al. used citric acid and ethylenediamine to synthesize CDs, leading to a quantum yield as high as 60.2% [10]. It can be speculated that naturally synthesized carbon nanomaterials should have higher biocompatibility and lower toxicity. To date, many techniques for synthesizing CDs from natural sources have been developed, including laser ablation, arc discharge, electrochemistry, thermal decomposition, ultrasonic, and microwave methods, among others. In particular, the thermal decomposition method is simple, safe, fast, and effective. Research has found that carbon nanoparticles synthesized by this method have good fluorescent performance, although the quantum yield is not high [11–13]. The optical properties, stability, and biocompatibility are still powerful enough for use in bioapplications. This review introduces and simplifies the current carbon nanocomposite synthesis techniques. We provide a brief introduction to fluorescent CDs, mainly reviewing their structural characteristics, carbon source materials, preparation methods, luminescence mechanism, and applications in the field of biomedicine. Obtaining CDs from natural resources can produce nanomaterials that are more environmentally friendly. The considered technique intends to use available (in a daily sense) materials to synthesize low-toxicity fluorescent nanomaterials. The production process does not require the use of many precursor chemicals to build natural CDs. We hope that developing a safe and straightforward production process can unite fluorescent materials and natural CDs and contribute to understanding and exploring the application of nano-fluorescent materials.

2. Natural Carbon Nanodots

CDs are a new type of carbon nanomaterials with luminescence characteristics, quasi-zero dimension, relatively simple preparation, abundant sources of raw materials, easy surface functionalization, low toxicity, and good biocompatibility. The fluorescence wavelength can be adjusted, and the two-photon absorption area is large. In various literature reports, CDs have also been referred to as carbon quantum dots (CQDs), carbon nanodots (CNDs), Graphene quantum dots (GQDs), carbon nanocrystals, and so on. Due to their excellent performance, CDs show promising potential application prospects in sensing, biology, medicine, food, environment, catalysis, photoelectricity, energy, and so on. Researchers have conducted extensive scientific studies and made significant progress [14]. CDs synthesis methods can be mainly divided into two categories: One uses physical or chemical means to crack larger carbon structures (e.g., carbon nanotubes, graphite rods, graphene, carbon powder) into tiny CDs; these top-down methods include arc discharge methods, laser ablation methods, chemical oxidation methods, and so on. Small organic molecule precursors, such as sugar, citric acid, and amino acids, are used as carbon sources for these methods through functional group coupling to achieve polymerization to prepare the CDs [15–17]. The second category comprises bottom-up methods, such as electrochemical, hydrothermal, pyrolysis, microwave-assisted, and ultrasonic methods [18].

Other methods, such as template methods, neutralization reaction exothermic methods, and micro-fluidized bed technology methods, can also be used to prepare carbon dots [19–21]. The preparation of CDs well-embodies the concept of green chemistry, using cheap, environmentally friendly carbon source precursors and natural renewable, cheap raw materials as carbon sources. The resources for synthesizing CDs can be found in the natural environment, such as eggs, grass, tea leaves, leaves, silk, silkworm pupae, shrimp shells, grapefruit peel, peanut shells, coffee grounds, beer, and other materials. CDs were discovered and debuted for the first time due to their fluorescent luminescence characteristics. The luminescence mechanism of CDs has always been a critical research direction for researchers, considering factors such as the quantum size effect, surface state, functional group mechanisms, electron holes, and radiation. Their rearrangement theory has been studied in various aspects. Although a complete theoretical explanation system of the CDs fluorescence mechanism has not been formed, the absorption and fluorescence of CDs exhibit properties such as photoluminescence, chemiluminescence, electrochemiluminescence, and luminescence. The conversion of photoluminescence, peroxidase-like activity, non-toxicity, and other physical and chemical properties provides a solid and feasible theoretical basis for further research.

The main application research directions of CDs can be divided into the following categories (Figure 1):

1. Imaging: multicolor fluorescent images of mammalian cells, plant cells, and microorganisms, and imaging in mice;
2. Photocatalysis: the degradation of organic molecules, the reduction in CO_2, and water splitting;
3. Optoelectronic devices: LEDs and solar cells (sensitizer/co-sensitizer, transport layer, electrolyte, and/or co-catalyst for counter electrode);
4. Sensors: food quality and safety, drug analysis, environmental pollution determination, immunoassay, and other fields, such as detecting heavy metal ions, anions, pesticides, molecules, small organic molecules, and/or nucleic acids;
5. Electrocatalysis: mainly used in oxidation-reduction reactions, oxygen evolution reactions, hydrogen evolution reactions, and reduction reactions for carbon dioxide, and dual-function catalysts;
6. Biomedicine: photodynamic therapy, photosensitizers for cancer cell destruction, radiotherapy, the tracing and delivery of drugs or genes, drug release, and anticancer drugs.

Figure 1. The main application research directions of CDs. This review focuses on summarizing the biological applications of CDs, such as cell imaging, photocatalysis, optoelectronic devices, molecules sensors, electrocatalysis, and biomedicines, comprehensively. Finally, current challenges, research emphasis, and prospects of this field are also discussed.

Most carbon sources derive from exhausted petroleum resources or non-environmentally friendly manufacturing processes. Therefore, using recycled materials and bio-renewable resources to develop high-performance CDs is a critical green environmental issue. In recent years, the awareness of environmental protection has significantly risen, and, as such, naturally derived CDs have gradually received more attention.

The main factor is that they are derived from renewable and sustainable biological materials; for example, lignocellulose from dead wood, waste wood, rice straw, bagasse, wheat straw, etc. Suppose that such a raw material can be used as a carbon resource and converted into CDs with therapeutic and diagnostic value. In that case, we can imagine the result will not overwhelm the demand of the food supply chain. Still, it can also address waste disposal problems while applying the resource to produce high value-added products such as electronics, energy, and biomedicine. Table 1 summarizes the existing research on converting various natural resources into CDs, and the associated applications and emission spectra are described in detail.

Table 1. Natural CDs and their applications.

Source	Methods	Applied Ex./Emit. (nm)	QY (%)	Application	Applied Concentration	Ref.
Lemon juice	Solvothermal	440–600/575–650	28%	In vivo bioimaging	100 µL (20 µg/mL)	[22]
Taxus leaf	Solvothermal	380–580/673	59%	In vivo bioimaging	20 mg/kg	[23]
Wedelia trilobata	Solvothermal	370–470/483–520, 654	H_2O: 10.52% EtOH: 18.16%	Glutathione sensing and cell imaging	50–400 µg/mL	[24]
Leek	Solvothermal	390–450; 325–385/440–500, 676; 440	DCD: 1.7% SCD: 1.14%	Cell imaging, Cu^{2+} and tetracycline sensing	0.5 mg/ mL	[25]
Lemon juice	Solvothermal	237, 279, 570–670/704	31%	In vivo bioimaging	50 µL (30 µg/mL)	[26]
Plum	Solvothermal	328–418/450–550	0.54%	Doxorubicin sensing	200 µL (1.0 mg/mL)	[27]
F. nucleatum	Hydrothermal	300–400/450–470	9.9%	In vivo bioimaging and Fe^{3+} sensing	10 µL/g (0.7 mg/mL)	[28]
Green pepper	Hydrothermal	310–380/400–460	8.7%	Fe^{3+} sensing and cell imaging	50 mg/mL	[29]
Papaya	Hydrothermal	350–490/445–530	H_2O: 18.98% EtOH: 18.39%	Fe^{3+} sensing and cell imaging	100, 175 µL (0.94 mg/mL) 100, 175 µL (1.17 mg/mL)	[30]
P. avium	Hydrothermal	280–360/411–430	13%	Fe^{3+} sensing and cell imaging	0–40 µL	[31]
Honey	Hydrothermal	320–410/410–475	19.8%	Fe^{3+} sensing and cell imaging	40 µL (1 mg/mL)	[32]
Sweet potato	Hydrothermal	300–410/406–486	8.64%	Fe^{3+} sensing and cell imaging	0–100 µg/mL	[33]
Black tea	Hydrothermal	290–420/398–490	n/a	Fe^{3+} sensing	990 µL (8 µg/mL)	[34]
Fish-scale	Hydrothermal	220–390/425–455	6.9%	Fe^{3+} sensing	5 mg/mL	[35]
Kiwi	Hydrothermal	300–450/432–500	14%/19%	Fe^{3+} sensing	0.5 mL	[36]
Goose feather	Hydrothermal	300–500/410–560	17.1%	Fe^{3+} sensing	1 mL	[37]
Cranberry	Hydrothermal	300–500/410–540	10.85%	Fe^{3+} sensing	n/a	[38]
Potato	Hydrothermal	323/405	15%	Fe^{3+} sensing	n/a	[39]

Table 1. *Cont.*

Source	Methods	Applied Ex./Emit. (nm)	QY (%)	Application	Applied Concentration	Ref.
Boswellia ovalifoliolata bark	Hydrothermal	275–440/400–535	10.2%	Fe^{3+} sensing	20 µg/mL	[40]
Rosin	Hydrothermal	290–370/425–475	1.22%	Fe^{3+} sensing and cell imaging	1.25–160 µg/mL	[41]
Coriander leaf	Hydrothermal	320–480/400–510	6.48%	Fe^{3+} sensing, cell imaging, and antioxidant	0–1.0 mg/mL	[42]
Mint leaf	Hydrothermal	330–420/410–500	7.64%	Fe^{3+} and ascorbic acid sensing	n/a	[43]
Leek	Hydrothermal	300–460/449–534	n/a	DDVP sensing and cell imaging	0–300 mg/mL	[44]
Peach gum	Hydrothermal	330–450/327–505	28.46%	Au^{3+} sensing and cell imaging	0.5 mL (20 mg/mL)	[45]
Tomato	Hydrothermal	367/440	n/a	Carcinoembryonic antigen and aptamer sensing	1 µg/mL	[46]
Bean pod and onion	Hydrothermal	310–380/410–450	5.55%	Ag^{1+} sensing and cell imaging	200 µg/mL	[47]
D. Salina	Hydrothermal	310–400/400–475	8%	Hg^{2+} and Cr^{4+} sensing and cell imaging	0–75 µg/mL	[48]
Chinese yam	Hydrothermal	280–440/400–525	9.3%	6-mercaptopurine and Hg^{2+} sensing	n/a	[49]
Pomelo peel	Hydrothermal	365/444	6.9%	Hg^{2+} sensing	n/a	[50]
Strawberry	Hydrothermal	344–440/427–500	6.3%	Hg^{2+} sensing	75 µL	[51]
Cucumber	Hydrothermal	418–518/514–571	3.25%	Hg^{2+} sensing	n/a	[52]
Highland barley	Hydrothermal	340–480/450–525	14.4%	Hg^{2+} sensing	0.05 mg/mL	[53]
Lemon peel	Hydrothermal	300–540/441–605	14%	Cr^{6+} sensing	n/a	[54]
Elaeagnus angustifolia	Hydrothermal	310–410/290–450	16.8%	Tartrazine sensing	n/a	[55]
Aloe	Hydrothermal	370–480/443–525	10.37%	Tartrazine sensing	n/a	[56]

Table 1. *Cont.*

Source	Methods	Applied Ex./Emit. (nm)	QY (%)	Application	Applied Concentration	Ref.
Coconut water	Hydrothermal	340–450/430–500	2.8%	Thiamine sensing and cell imaging	n/a	[57]
Lentil	Hydrothermal	310–390/400–460	10%	Thioridazine hydrochloride sensing	200 μL	[58]
Pomegranate juice	Hydrothermal	280–350/350–600	4.8%	Cephalexin sensing	30 μL (1.0 mg/mL)	[59]
Bamboo leaf	Hydrothermal	365–525/440–540	7.1%	Cu^{2+} sensing	n/a	[60]
Pipe tobacco	Hydrothermal	310–430/425–515	3.2%	Cu^{2+} sensing	n/a	[61]
Apple juice	Hydrothermal	300–540/465–565	4.27%	Cell imaging	10 μg/mL	[62]
Hylocereus undatus	Hydrothermal	275–380/400–450	n/a	Cell imaging	0–50 μL/mL	[63]
Saccharum officinarum	Hydrothermal	300–540/450–550	5.76%	Cell imaging	0–400 mg/mL	[64]
Linseed	Hydrothermal	350–450/503	61%	Cell imaging	0.04 mg/mL	[65]
Shiitake mushroom	Hydrothermal	330–450/410–500	5.5%	Cell imaging and pH sensing	2 mg/mL	[66]
Citrus	Hydrothermal	360–500/460–554	1.1%	Cell imaging	30 μL (1.0 mg/mL)	[67]
Carrot	Hydrothermal	360–520/442–565	5.16%	Cell imaging	700 μg/mL	[68]
Dwarf banana	Hydrothermal	310–460/395–505	23%	Cell imaging	0–200 μg/mL	[69]
Bagasse	Hydrothermal	330–510/450–550	12.3%	Cell imaging and biolabeling	100 μg/mL	[70]
Cabbage	Hydrothermal	276, 320/432–584	16.5%	Cell imaging	100 μL (20–1000 μg/mL)	[71]
Alkali lignin	Hydrothermal	280–450/410–510	21%	Cell imaging	0–100 μg/mL	[72]
Shrimp	Hydrothermal	360–530/430–550	54%	Cell imaging and drug delivery	10–500 μg/mL	[73]
Wheat bran	Hydrothermal	360–540/460–600	33.23%	Cell imaging and drug delivery	2 mg/mL	[74]
Milk	Hydrothermal	360/450	n/a	Cell imaging and anticancer drug delivery	100–600 μg/mL	[75]

Table 1. *Cont.*

Source	Methods	Applied Ex./Emit. (nm)	QY (%)	Application	Applied Concentration	Ref.
Chlorhexidine gluconate	Hydrothermal	360–560/480–600	s-CGCDs: 2.6% m-CGCDs: 11.3% l-CGCDs: 8.0%	Antibacterial and cell imaging	0–150 μg/mL	[76]
Turmeric leaf	Hydrothermal	310–470/429–520	n/a	Anti-bacterial	0 1.0 mg/mL	[77]
Rosemary	Hydrothermal	332–422/424–500	n/a	Anti-bacterial	12 μg/mL	[78]
Osmanthus leaf, tea leaf, and milk vetch	Hydrothermal	450/530	n/a	Antibacterial and cell imaging	0–1000 μg/mL	[79]
Mushroom	Hydrothermal	300–500/372–545	n/a	Anti-bacterial, anti-cancer, and Pb^{2+} sensing	0–25 μg/mL	[80]
Watermelon	Hydrothermal	808/900–1200	0.4%	Photothermal therapy and cell imaging	0–20 mg/mL	[81]
Hypocrella Bambusa	Hydrothermal	540–590/600–650	n/a	Photodynamic and photothermal therapy	0–200 μg/mL	[82]
Camellia japonica	Hydrothermal	360/400–700	n/a	Photodynamic and photothermal therapy	45 μg/mL	[83]
Ginger	Hydrothermal	325–445/400–500	13.4%	Cancer inhibition and cell imaging	440 μg	[84]
Garlic	Hydrothermal	320–580/380–600	17.5%	Cell imaging	0–1 mg/mL	[85]
Starch	Hydrothermal	340–500/452–545	21.7%	Cell imaging	0.078–1.250 mg/mL	[86]
Orange juice	Hydrothermal	360–450/441–510	26%	Cell imaging	0–200 μg/mL	[87]
Bee pollen	Hydrothermal	340–450/425–505	c-CDs: 8.9% l-CDs: 6.1%	Cell imaging	0.5 mg/mL	[88]
Gelatin	Hydrothermal	300–500/430–580	31.6%	Cell imaging	5.o mL (0.8 mg/mL)	[89]
Papaya	Hydrothermal	300–500/450–550	7.0%	Cell imaging	16.2–500 μg/mL	[90]

Table 1. *Cont.*

Source	Methods	Applied Ex./Emit. (nm)	QY (%)	Application	Applied Concentration	Ref.
Oatmeal	Hydrothermal	280–460/410–504	37.4%	Cell imaging	1 mg/mL	[91]
Egg white	Hydrothermal	290–450/415–540	61%	Cell imaging	0.04 mg/mL	[92]
Corn flour	Hydrothermal	320–500/401–553	7.7%	Cell imaging and Cu^{2+} sensing	0–640 µg/mL	[93]
Humic acid	Hydrothermal	320–520/440–540	5.7%	Cell imaging	0.2 mg/mL	[94]
Durian	Hydrothermal	400–560/605	79%	Cell imaging	0–500 µg/mL	[95]
Gooseberry	Hydrothermal	300–500/406–545	13.5%	*C. elegans* bioimaging	50 µg/mL	[96]
Rice husk	Hydrothermal	310–340/360–440	8.1%	Cell imaging	50 µg/mL	[97]
Ayurvedic	Chemical ablation	430/518	n/a	Cell imaging and phototherapy	0.5 mg/mL	[98]
Coffee bean shell	Chemical ablation	280–520/368–557	n/a	In vivo bioimaging and antioxidant	0–400 µg/mL	[99]
Muskmelon	Chemical ablation	342, 415, 425/432, 515, 554	7.07%/26.9%/14.3%	Hg^{2+} sensing and Cell imaging	0.25–1.00 mg/mL	[100]
Cow manure	Chemical ablation	320–450/400–530	0.65	Cell imaging	2.5 mg/mL	[101]
Food waste	Ultrasound irradiation	330–405/400–470	2.85%	Cell imaging	0–4 mg/mL	[102]
Citrus limone juice	Ultrasound irradiation	230–450/325–538	12.1%/15%	Cell imaging	2–100 mM	[103]
Crab shell	Ultrasound irradiation	330–390/410–450	14.5%	Cell imaging	0–1000 µg/mL	[104]
Silkworm	Microwave	300–400/350–550	46%	Cell imaging	0–15 mg/mL	[105]
Algal bloom	Microwave	300–500/400–550	13%	Cell imaging	10–1000 µg/mL	[106]
Eggshell	Microwave	275/450	14%	Glutathione sensing	n/a	[107]
Flour	Microwave	360–500/438–550	5.4%	Hg^{2+} sensing	4 µg/mL	[108]
Protein	Microwave	300–420/380–480	14%	Ag^{+} sensing	n/a	[109]

Table 1. *Cont.*

Source	Methods	Applied Ex./Emit. (nm)	QY (%)	Application	Applied Concentration	Ref.
Rose flower	Microwave	330–410/390–435	13.45%	Tetracycline sensing	n/a	[110]
Onion peel	Microwave	300–470/520	n/a	Skin wound healing	1.5 mg/mL	[111]
Lychee	Pyrolysis	365/440	10.6%	Cell imaging	0–1000 μg/mL	[112]
Coffee	Pyrolysis	350–500/400–600	3.8%	Cell imaging	1.2 mg/mL	[113]
Urine	Pyrolysis	275–625/450–650	14%	Cell imaging	0.05–1.5 mg/mL	[114]
Watermelon peel	Pyrolysis	310–550/490–580	7.1%	Cell imaging	n/a	[115]
Konjac flour	Pyrolysis	400–700/575–760	13%	Fe^{3+} and L-lysine sensing and cell imaging	200 μg/mL	[116]
Soybean and broccoli	Pyrolysis	300–460/425–500	12.8%	Cu^{2+} sensing and cell imaging	0–300 μg/mL	[117]
Borassus flabellifer	Pyrolysis	300–400/350–403	11.73%/13.97%/10.83%	Fe^{3+} sensing	n/a	[118]
Peanut shell	Pyrolysis	262–402/413–500	10.58%	Cu^{2+}sensing	n/a	[119]
Assam tea	Pyrolysis	340/446	n/a	Dopamine and ascorbic acid sensing	n/a	[120]
Peanut shell	Pyrolysis	320–480/441–524	9.91%	Cell imaging	0–1.2 mg/mL	[121]
Roast duck	Pyrolysis	300–400/400	10.53%/38.05%	*C. elegans* bioimaging	15 mg/mL	[122]
Artemisia argyi leaf	Pyrolysis	360–440/450–480	n/a	Antibacterial and cell imaging	0–150 μg/mL	[123]
Sugarcane bagasse	Pyrolysis	405/550	n/a	Drug delivery	n/a	[124]
Silkworm cocoon	Pyrolysis	378/459	6.32%	Anti-inflammatory	1.4 mg/mL;	[125]
Lychee exocarp	Pyrolysis	365/423	n/a	Drug delivery and cell imaging	0–15 μg/mL	[126]
Bamboo leaf	Pyrolysis	300–400/425–475	n/a	Cell imaging and anticancer drug delivery	0–400 μg/mL	[127]
Walnut shell	Pyrolysis	360–540/500–560	n/a	Cell imaging	100 μg/mL	[128]

3. Toxicity Evaluation of Natural Carbon Nanodots

CDs can be derived from a wide range of synthetic raw materials, low-cost, stable chemical properties, and non-toxic materials. The application of CDs in medicine and pharmacy was recently extensively studied. One of the most promising nanomaterials is carbon quantum dots (CQDs). For this purpose, the toxicity of CQDs was investigated in cells and living systems (Table 2). In 2019, L. Janus used human dermal fibroblasts to conduct a cytotoxicity study of N-doped chitosan-based CDs [129]. As shown in Figure 2, after 48 h, the cell viability was recorded as 94%. CDs were synthesized by utilizing non-toxic raw materials and removing unreactive residues during purification.

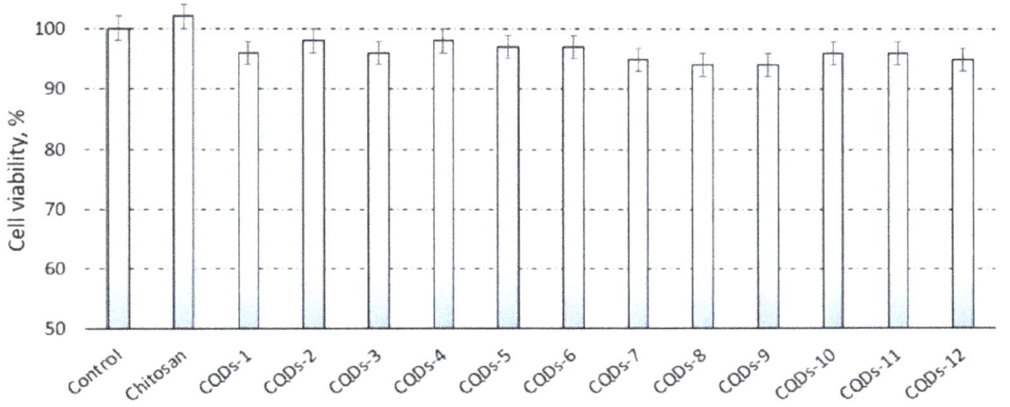

Figure 2. TXT assay study on human dermal fibroblasts. Adapted from [129], MDPI, 2019.

First, we start with the selection of carbon source and the adjustment of dopants that synthesize carbon dots, and at the same time list the test models of carbon dots. Secondly, according to the selective response of the carbon dot to the biological models, the corresponding toxicity was constructed, and the actual application was investigated. As mentioned in Table 2, CQDs are usually doped with nitrogen to enhance their fluorescence quantum yield and optical performance [130,131]. CDs doped with nitrogen (such as carbon nanotubes, graphene, hollow spheres, etc.) have unique properties. They can inject electrons into the carbon substrate to change the electron and transport characteristics such as sensors, nanogenerators, etc., and are widely used and have become a research trend. Moreover, green methods that utilize natural biomass/biowaste and micro-organisms without introducing toxic chemicals as CDs precursors have been widely used to synthesize CQDs [132].

Table 2. Toxicity evaluation methods for CDs.

Material	Sources	Concentration	QY (%)	Cells or Animal Models	Toxicity	Ref.
Carbon quantum dots	Medicinal mulberry leaves	500 ug/mL	9.7	Human normal hepatic stellate cell line LX-2 cells and human HCC cell line HepG2 cells	Almost non-cytotoxic	[133]
Carbon dots	Mango peel	500 ug/L	8.5	A549 cells	Remained above 90% Low toxicity	[134]
Nitrogen-doped carbon quantum dots	Watermelon juice	300 ug/mL	10.6	HepG2 cells	Remained 90% Low cytotoxicity	[135]

Table 2. *Cont.*

Material	Sources	Concentration	QY (%)	Cells or Animal Models	Toxicity	Ref.
N-doped carbon quantum dots	Bio-waste lignin	100 mg/mL	8.1	Mouse macrophage cells	Remained 96.8% Low toxicity	[136]
Carbon dots	Roast duck	1 mg/mL	38.05	PC12 cells and *C. elegans*	Remained 91.19% Low toxicity	[122]
Nitrogen-doped carbon dots	*P. acidus* fruit juice	200 ug/mL	12.5	Cells and *C. elegans*	Remained 93% Low cytotoxicity	[137]
Carbon quantum dots	*Salvia hispanica L.* seeds	250 ug/mL	17.8	HEK293 cell line	Remained 91.7% Low toxicity	[138]
Carbon dots	Wheat straw	0.8 mg/mL	7.5	HeLa cells	Negligible cytotoxicity	[139]
Carbon dots	*Malus floribunda* fruit	200 ug/mL	19	Cells and *C. elegans*	Remained 93% and low toxicity	[140]
Carbon quantum dots	Banana peel waste	200 ug/mL	20	*C. elegans*	Low toxicity	[141]
Nitrogen and sulfur dual-doped carbon quantum dots	Fungus fibers	400 ug/mL	28.11	HepG2 cells	Remained over 95% Low cytotoxicity	[142]
Carbon dots	Sweet lemon peel	500 ug/mL	n/a	MDA-MB231 cells	Remained above 75% Low cytotoxicity	[143]
Carbon dots	Lychee waste	1.2 mg/mL	23.5	Skin melanoma cells	Remained above 89% Low cytotoxicity	[144]
Nitrogen-doped carbon quantum dots	Citrus lemon	2 mg/mL	31	Human breast adenocarcinoma cells	Remained above 88% low cytotoxicity,	[145]
Carbon dots	*Daucus carota subsp. sativus* roots	1 mg/mL	7.6	MCF-7 cells	Remained above 95% Low toxicity	[146]
Carbon nanodots	Custard apple peel waste biomass	100 ug/mL	n/a	HeLa and L929 cells	Remained above 85% Low toxicity	[147]
Carbon quantum dots	Pineapple peel	1 mg/mL	42	HeLa and MCF-7 cells	Remained 84% Low toxicity	[148]
N-carbon dots	Jackfruit seeds	2 mg/mL	17.91	A549 cells	Remained 96% and less toxic	[149]

S. Cong et al. [122] used PC12 cells for a cytotoxicity study of CQDs obtained from a roast duck. After 36 h, the cell viability of PC12 was recorded as 91.19% at the concentration of 1 mg/mL. In addition, using CDs at a concentration of 15 mg/mL for treating nematodes did not lead to any death for 24 h. These results indicate the low toxicity of CDs, even after a long period of exposure at high concentrations. In 2021, R. Atchudan et al. [141] used *C. elegans* as an animal model for their toxicity evaluation of CQDs. As displayed in Figure 3, the CQDs synthesized from banana peel were shown to have low nematode toxicity, even at a high concentration of 200 ug/mL. These results can be explained by their utilizing non-toxic raw materials without adding any passivates or additives.

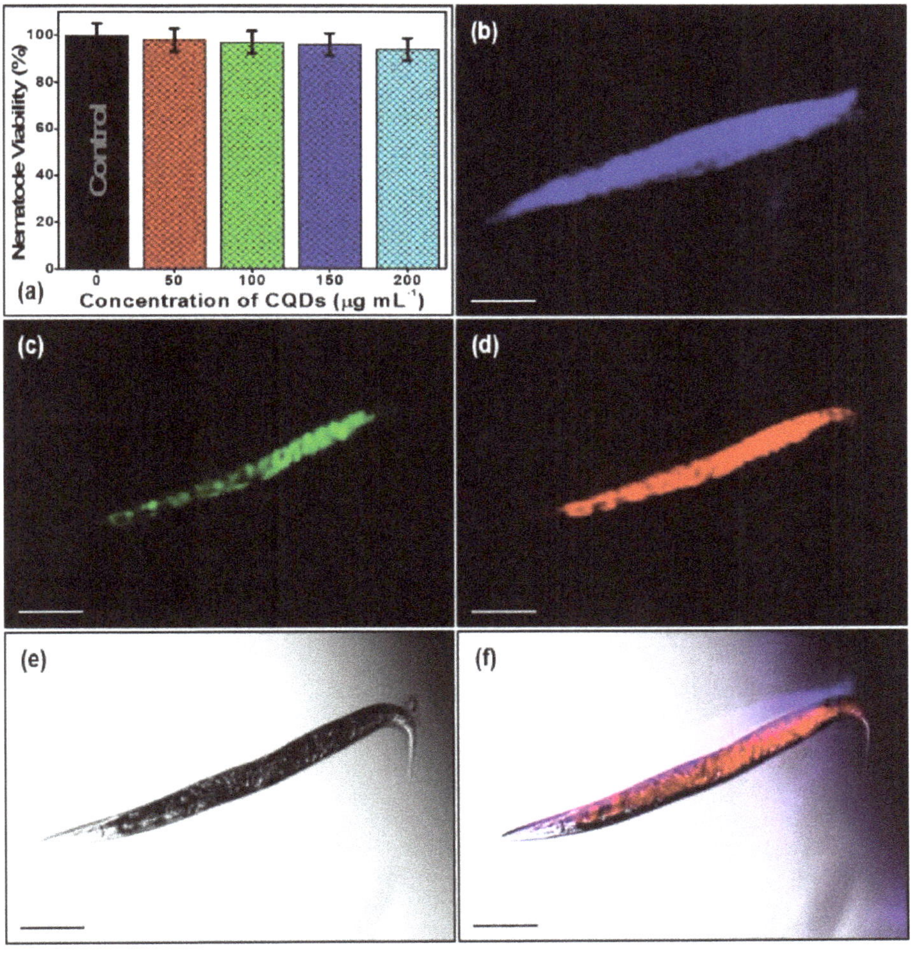

Figure 3. (**a**) Toxicity assay of nematode incubation under different concentrations of synthesized CQDs. Multicolor imaging of in vivo model nematode incubated with CQDs (100 μg/mL) under the excitation wavelengths of (**b**) 400 nm, (**c**) 470 nm, (**d**) 550 nm, (**e**) bright-field (BF), and (**f**) merge (overlap). Live nematodes were immobilized using 0.05% sodium azide (NaN_3) for imaging under fluorescence filters. Adapted with permission from [141], Elsevier, 2021.

4. Theranostic Application of Natural Carbon Nanodots

Therefore, fluorescent materials are expected to be critical in biological applications. CDs have become the focus of attention as new nanomaterials due to their excellent light stability, fluorescence, good biocompatibility, and low toxicity. Surgery, radiotherapy, and chemotherapy are inevitable treatments for most cancer patients, but these processes have considerable side effects on the human body. However, fluorescent nanomaterials have the advantages of high fluorescence stability, low biotoxicity, and good biocompatibility. The most important thing is that they can be used to perform non-invasive targeted therapy on human lesions, exploiting their characteristics to avoid harm to the human body caused by the abovementioned treatments.

This paragraph covers the luminescence mechanism of CDs and their applications in biology, focusing on applying natural CDs in biological diagnosis and treatment. We discuss the combination of CDs with specific targeting molecules to form CD-based probes for detecting fluorescent signals. With the help of advanced optical imaging technology, real-time dynamic monitoring of molecules in cells and organisms can be carried out, and rapid immunofluorescence analysis of primary infectious disease sources can be carried out. They can provide new technologies and methods for disease occurrence, diagnosis, and treatment research [150].

4.1. Bioimaging

The discrete and diverse microstates of CDs lead to broad excitation and emission ranges [151,152]. CDs have many unique properties, and their excellent photostability can provide fluorescent information in a biological environment. The surface modification of functional groups can make CDs more helpful in applying biomarkers (Table 3). In this section, we compare bioimaging from cells to living animals, emphasizing the biological diversity to prove the generalized safety of CDs. In cell imaging applications, CDs are usually applied to HepG2 cells, HeLa cells, T24 cells, and so on [27,102,120,124]. In addition, normal cell lines have been used in trials, showing good cell compatibility [24,71]. Alam et al. treated HaCaT cells with cabbage-derived CDs and showed that, at 500 μg/mL of CDs, the cell viability was 100%. Furthermore, CDs' tunable excitation and emission show promise for normal cell imaging under the irradiation of confocal fluorescence microscopy [71]. In a cell imaging experiment, Mehta et al. considered CDs originating from apple juice and three fungal (*M. tuberculosis*, *P. aeruginosa*, and *M. oryzae*) sources. Cell compatibility was shown by feeding more than 100 mg/mL of CDs. The germination of *M. oryzae* spores also strongly indicated how CDs are good biocompatible nanocomposites at high concentrations (i.e., 400 mg/mL). The fungi appeared red, green, and blue in confocal laser microscopic images [62].

Kasibabu et al. provided pictures of *Bacillus subtilis* and *Aspergillus aculeatus* after incubating with CDs derived from papaya juice for 1–6 h. The uptake ability was shown by observing well-dispersed CDs in the cytoplasm of the fungi [90]. In vivo tests are critical standards for investigating the potential and toxicity of CDs in animals. Atchudan et al. explored colorful nitrogen-doped CDs (NCDs) derived from gooseberry by hydrothermal methods in *C. elegans* imaging. *C. elegans* presented blue, green, and red in the whole body when excited under 400 nm, 470 nm, and 550 nm. The cell viability was over 97% after incubating *C. elegans* in NCDs for 24 h from 0 to 200 μg/mL (Figure 4a–f) [96]. Cong et al. using roast duck as the source and pyrolyzing at 200 °C, 250 °C, and 300 °C, synthesized single-fluorescence CDs. *C. elegans* were treated with 15 mg/mL of the CDs pyrolyzed at 300 °C (300-CDs) for 24 h. Compared with the wild-type group, the accumulation and uptake of 300-CDs made the intestine appear blue under UV light exposure (Figure 4g–j) [122]. The murine model has also been widely used for determining the efficiency of CDs.

Ding et al. subcutaneously injected 100 μL of red-emitting CDs (R-CDs) into nude mice. Strong fluorescence at 700 nm was detected under an excitation wavelength of 535 nm, indicating the excellent penetration ability from tissues to skin. Furthermore, the mice were still alive after 10 days (Figure 5a) [22]. Liu et al. investigated how the accumulation of carbonized polymer dots (CPDs) varied with time. Most CPDs remained in the lung and liver in the early period, with negligible dispersion to the brain and heart. The metabolism of CPDs was confirmed after 4 h, as their fluorescence decreased sharply (Figure 5b) [23]. Ding et al. compared subcutaneous and intravenous injections of near-infrared emissive CDs (NIR-CDs). Through the use of subcutaneous methods, NIR-CDs were distributed into mouse skin and tissues. As for intravenously injected nude mice, fluorescence was seen in the bladder, indicating NIR-CDs elimination via urine (Figure 5c) [26]. Liu et al. tracked CDs (Fn-CDs) in Kunming mice for different periods.

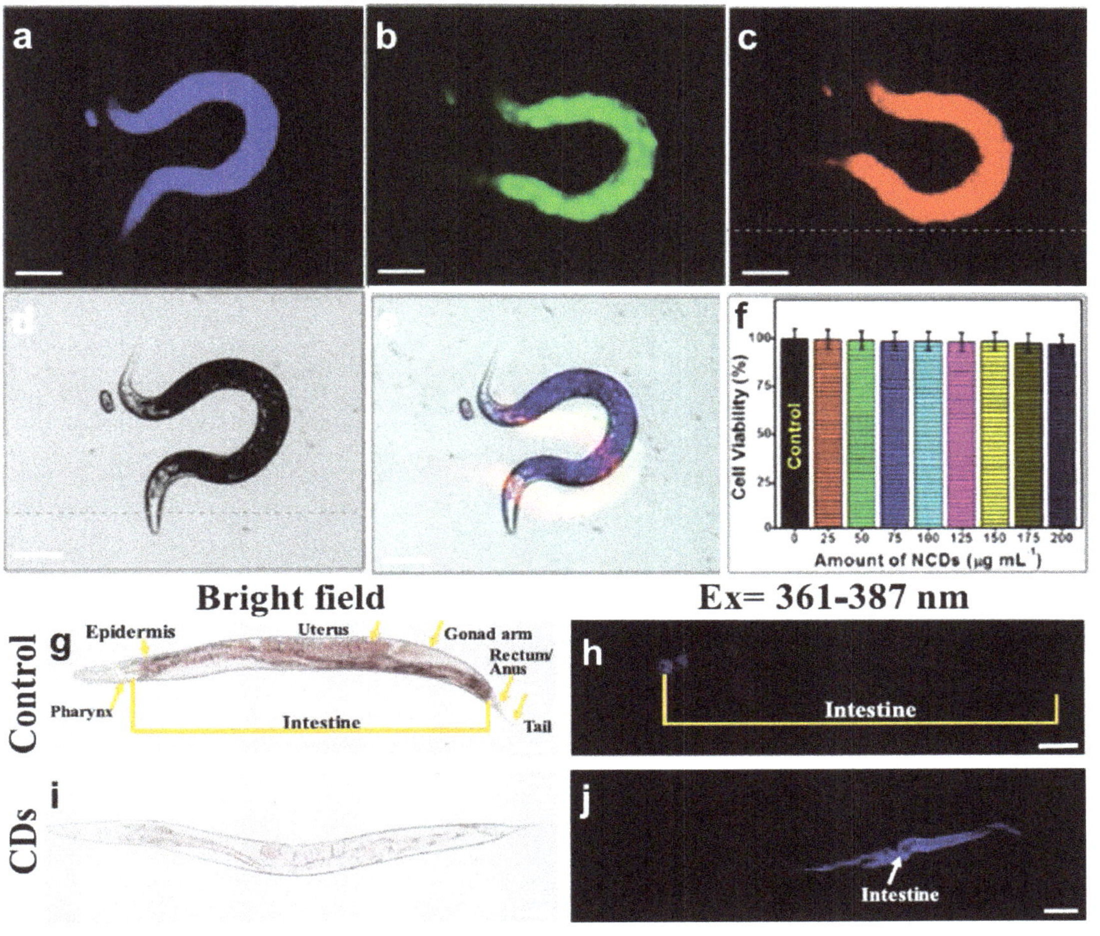

Figure 4. *C. elegans* confocal imaging excited under the wavelengths of (**a**) 400 nm, (**b**) 470 nm, and (**c**) 550 nm; as well as (**d**) bight field; (**e**) the merged image; and (**f**) cell viability test under different concentrations of NCDs, adapted with permission from [96], ACS Publications, 2018. Bright-field (**g**) wild-type *C. elegans* imaging and (**i**) 300-CDs treated *C. elegans* imaging; UV-exposed (**h**) wild-type *C. elegans* imaging and (**j**) 300-CDs treated *C. elegans* imaging, adapted with permission from [122], Elsevier, 2019.

They concluded that, with the assistance of PL, Fn-CDs show strong fluorescence in the bladder and are eliminated after 7 h (Figure 5d) [28]. Therefore, CDs can perform well as imaging nanoparticles and be adapted to different cell lines and living animals. In addition, their multiple emission ranges, low toxicity, and small size confer their high potential in future clinical applications.

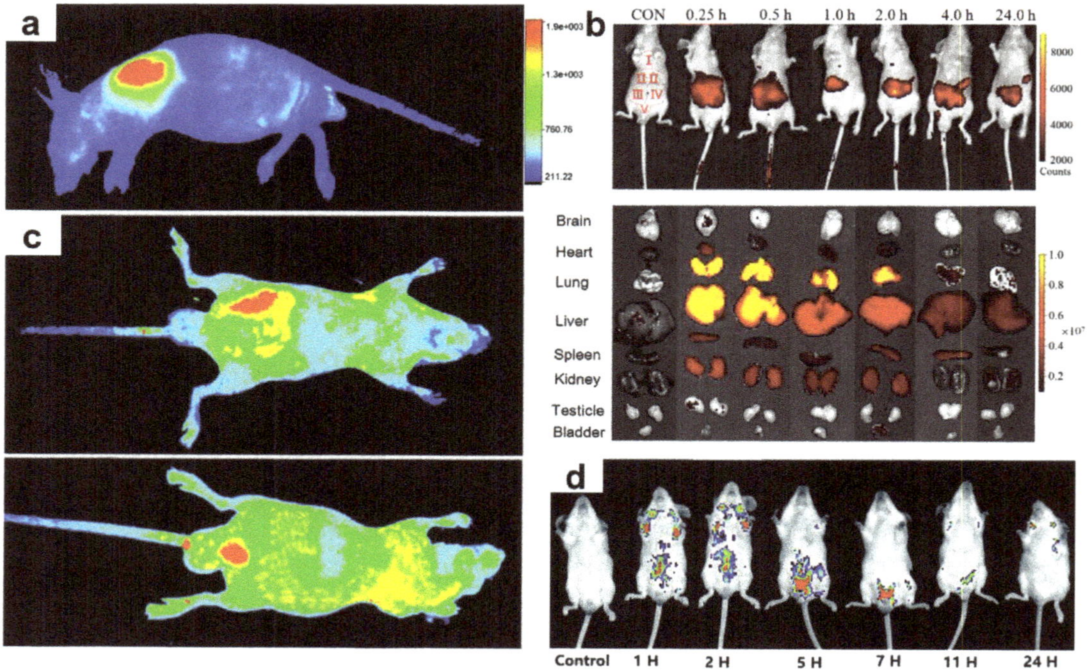

Figure 5. In vivo images and accumulation of CDs: (**a**) in vivo images of R-CDs, adapted with permission from [22], Royal Society of Chemistry, 2017; (**b**) metabolism of CPDs, adapted with permission from [23], Wiley, 2020; (**c**) subcutaneous and intravenous injection of NIR-CDs, adapted with permission from [26], Elsevier, 2019; and (**d**) metabolism of Fn-CDs, adapted with permission from [28], Royal Society of Chemistry, 2021.

4.2. Sensors

Inorganic ions are critical for creatures, not only for enhancing the efficiency of catalytic reactions in bio-systems but also for maintaining our fundamental life functions. However, it is harmful to have too many metal cations, which are highly toxic to human beings. Therefore, developing sensors for inorganic ions is a simple beneficial method to collate the concentration and standard. Owing to the multiple PL of CDs, we can observe the intensity variation and chelation quenching effect at other peaks in the PL image. For natural CDs sensors, Fe^{3+}, Hg^{2+}, and Cu^{2+} are the most examined targets. Shen et al. assessed HepG2 and HeLa cell images after incubating with CDs and Fe^{3+}.

According to the decreased blue fluorescence under 405 nm irradiated light, Fe^{3+} had a practical quenching effect on the CDs (Figure 6a–d) [33]. Hu et al. prepared double–emitted biomass nitrogen co-doped CDs (B-NCdots) for Cu^{2+} probing in T24 cells. Similarly, the quenching effect was still available for Cu^{2+}, causing a decreased intensity of blue and green, as shown in Figure 6e,f [117]. Furthermore, bio-related molecules, including peptides and drug-containing cells, are even more crucial. Some researchers designed chemically sensitive CDs to assist in directly resolving the effects of molecules by monitoring the decrease or increase in fluorescence. Liang et al. added 0.5 mM and 1 mM of glutathione and 150 μg/mL rose-red fluorescence CDs (wCDs) in L929 cells, HeLa cells, and HepG2 cells. An intense quenching effect of glutathione was observed in L929 cells and HeLa cells. However, no apparent variation occurred in HepG2 cells, which implies a distinct response of wCDs to glutathione (GSH) in various cells (Figure 7) [24].

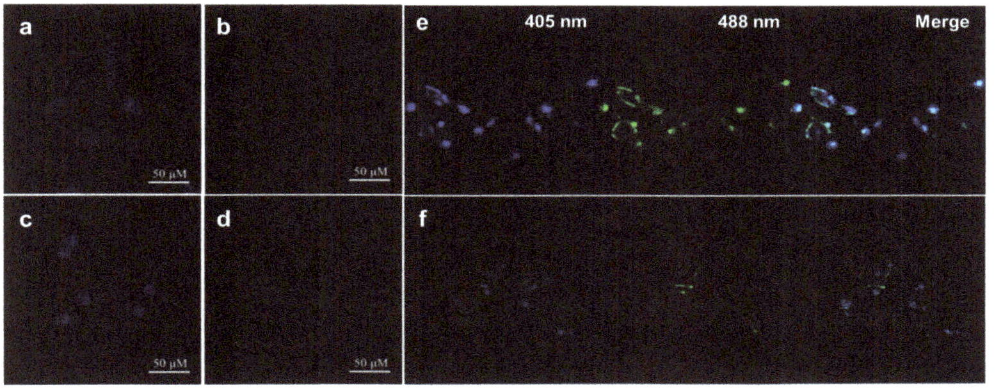

Figure 6. Confocal images collected at 405 nm of (**a**) HeLa cell; (**c**) HepG2 cell incubated with CDs; adding Fe^{3+} and CDs to (**b**) HeLa cell; and (**d**) HepG2 cell, adapted with permission from [33], Elsevier, 2017. Confocal laser images of T24 cells excited at 405 nm and 488 nm, (**e**) treated with B-NCdots and (**f**) treated with Cu^{2+} and B-NCdots, adapted with permission from [117], Springer Nature, 2019.

Wang et al. constructed a glutathione assay composed of eggshell-derived CDs and Cu^{2+}. The authors quantify without other indicators by plotting the fluorescence ratio versus the glutathione concentration (Figure 8a) [107]. Wang et al. examined how the fluorescence intensity of Shiitake mushroom-derived CDs (MCDs) varied with pH in dexamethasone-induced HeLa cell apoptosis. At higher concentrations of dexamethasone, the fluorescence under excitation at 405 nm and 488 nm was stronger, indicating a positive relationship on MCDs with increasing intracellular acidification (Figure 8c–e) [66]. As for drug probing, Zhu et al. analyzed doxorubicin (DOX), an anthracycline-based anticancer medicine, by taking advantage of the PL of plum-based carbon quantum dots (PCQDs). The dual-emitted property at the wavelengths of 491 nm and 591 nm provided a ratiometric calibration curve as a function of the DOX concentration. They also confirmed the accuracy by analyzing urine and serum samples (Figure 8b) [27]. CDs are suitable for detecting different kinds of molecules and ions. The intensity changed at a single emission peak, but the amplitude ratio of two emission peaks is valid for sensing experiments. Due to their intrinsic properties, CDs show promise in the bio-sensing field and are applied to cancer therapy.

4.3. Antibacterial Activity

Bacteria are well-known as the origins of various diseases. Recently, super bacteria have appeared globally, which cause incurable illnesses due to the abuse of antibiotics. In addition, people have come to pay more attention to the side effects of antibiotics and wish to avoid unexpected risks. Nanomedicines, especially CDs, have been taken into consideration as substitute methods. *E. coli* and *S. aureus* are the most common types of bacteria for investigating how nanomedicines or antibiotics work to induce apoptosis in bacteria. Wang et al. used CDs (ACDs) derived from *Artemisia argyi* leaves to treat *E. coli* and *S. aureus* cultures. According to the SEM images (Figure 9a–h), it can be seen that the cell walls of *E. coli* were destroyed; however, there was no distinct difference between treated and untreated *S. aureus*. This means that ACDs are selective to Gram-negative bacteria due to the structural properties of their cell walls [123]. Sun et al. synthesized chlorhexidine gluconate CDs from large to small (l-CGCDs, m-CGCD, and s-CGCDs) to determine the relationship between size and antibacterial activity. From the SEM imagery (Figure 9i), it can be seen that the rigidity of cell walls was the strongest in the control group and decreased from l- to s-CGCDs groups. These results revealed that CGCDs lead to frustration in the walls and membranes of *E. coli* and *S. aureus*. Bacterial death can be controlled by tuning the size of the CGCDs [76]. Ma et al. tested three kinds of CDs, including osmanthus leaves-derived CDs (OCDs), milk vetch-derived CDs (MCDs), and tea leaves-derived CDs (TCDs). In Figure 10, 80% of *E. coli* and *S. aureus* were killed by OCDs at a 1 mg/mL concentration, while 70% of bacteria survived in the MCDs group. In addition, *E. coli* had stronger resistivity than *S. aureus* among these CDs. CDs are internalized into bacteria. The outer surface of bacteria is attached to CDs leading to indirect proliferating inhibition [153–155]. These results prove the natural sources are essential for the synthesis of CDs [79]. As mentioned above, the tunability of raw material and diameters primarily affect the antibacterial efficiency and selectivity of the resultant CDs. Treating the patient's wounds after surgery with CDs with editable properties can tremendously decrease the associated risks.

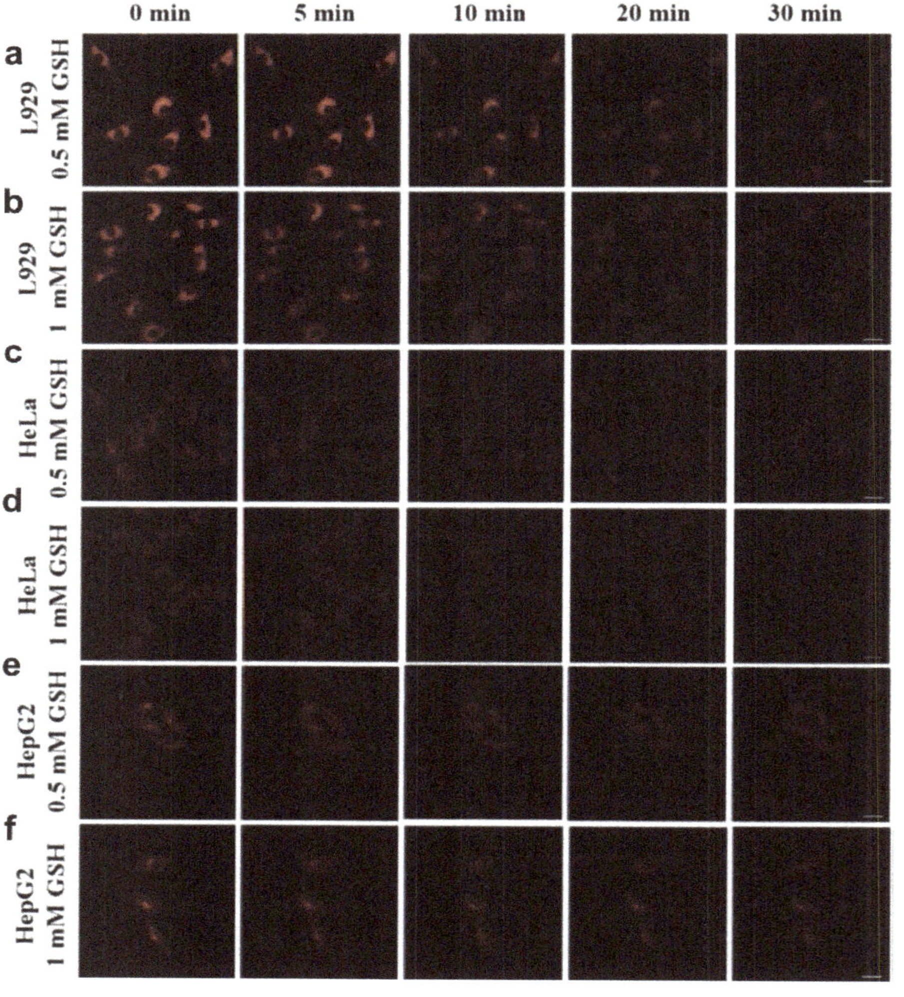

Figure 7. Confocal laser images of: (**a**) L929 cells, (**c**) HeLa cells, and (**e**) HepG2 cells in 0.5 mM GSH and 150 μg/mL wCDs; and (**b**) L929 cells, (**d**) HeLa cells, (**f**) and HepG2 cells in 1 mM GSH and 150 μg/mL wCDs, adapted with permission from [24], Royal Society of Chemistry, 2021.

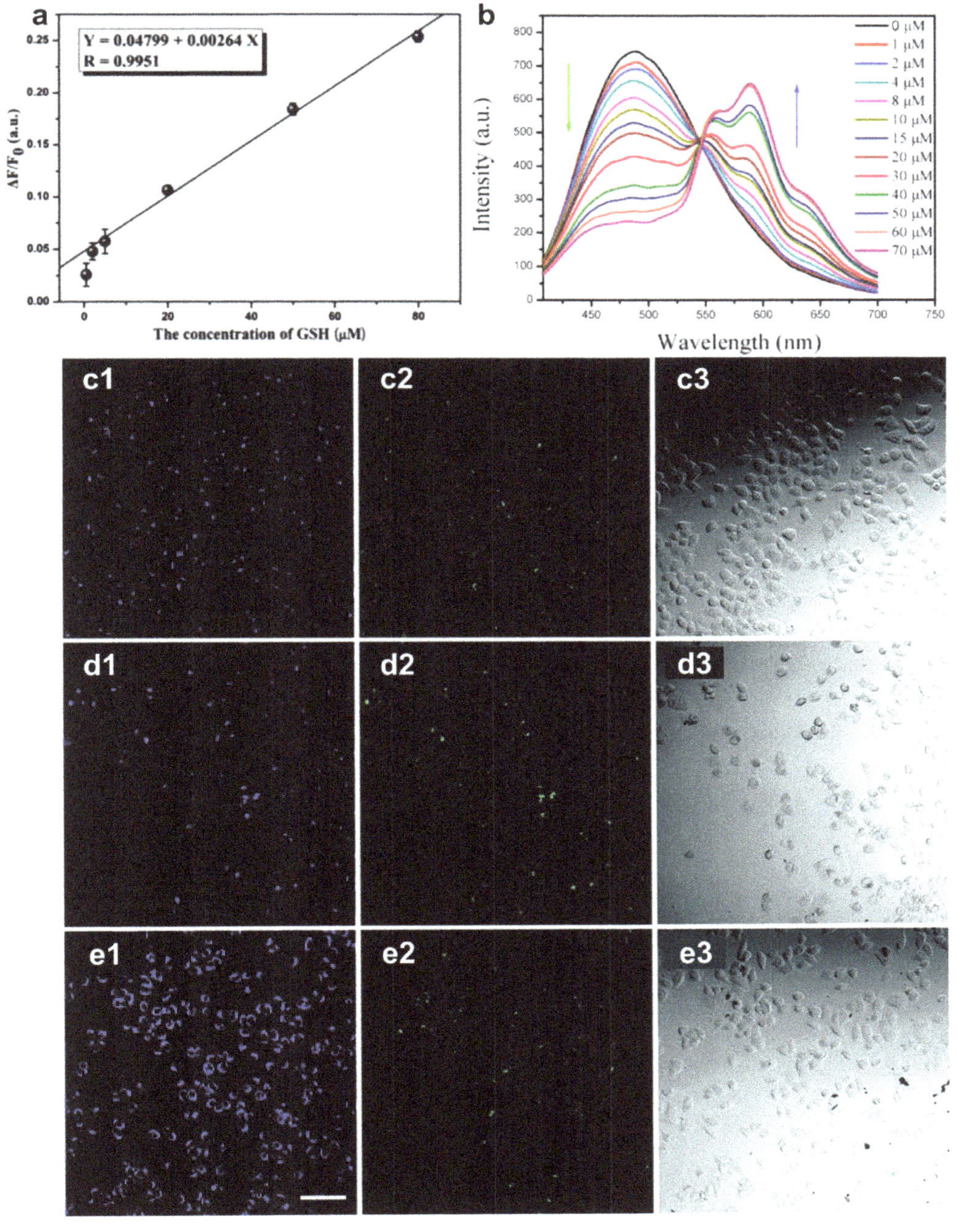

Figure 8. (**a**) Linear calibration of GSH probing, adapted with permission from [107], Royal Society of Chemistry, 2012. (**b**) The fluorescence spectra of PCQDs with varying DOX concentrations, adapted with permission from [27], Elsevier, 2021. (**c**) Confocal images of HeLa cells treated with MCDs, (**d**) adding 10 μM dexamethasone or (**e**) 100 μM dexamethasone, at excitation wavelengths of (**c1**,**d1**,**e1**) 405 nm, (**c2**,**d2**,**e2**) 488 nm, and (**c3**,**d3**,**e3**) bright-field, adapted with permission from [66], Royal Society of Chemistry, 2016.

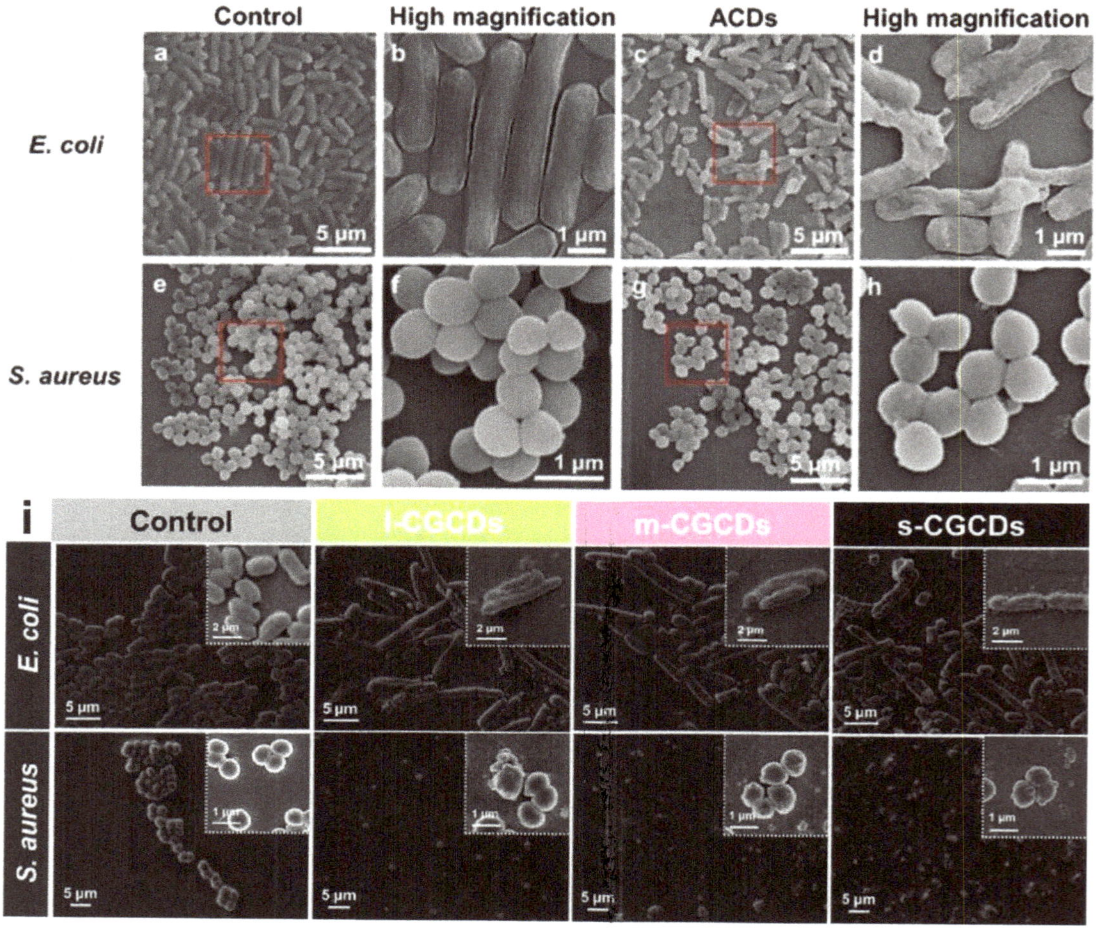

Figure 9. SEM images without ACDs of (**a**) *E. coli*, (**e**) *S. aureus*, and ACDs-treated (**c**) *E. coli*, and (**g**) *S. aureus*. Magnified SEM images in the red square (**b**) *E. coli*, (**f**) *S. aureus*; and ACDs-treated (**d**) *E. coli*, and (**h**) *S. aureus*, adapted with permission from [123], Royal Society of Chemistry, 2020. SEM images of (**i**) *E. coli* and *S. aureus* untreated and treated with 75 μg/mL and 50 μg/mL of s-CGCDs, m-CGCDs, and l-CGCDs in Luria–Bertani broth medium for 6 h, adapted with permission from [76], Elsevier, 2021.

4.4. Anticancer Activity

At present, cancers are prevalent within all age ranges. Cancers may be fatal due to unexpected syndromes as well as the disorder of living functions. Furthermore, conventional cancer therapies are long-term processes. Surgeries are straightforward methods, but recovery typically poses a challenge for patients. Even though chemotherapy seems safer, the currently used drugs lack selectivity and affinities to specific tumors. Some targeted therapies have been developed in recent years. However, they are expensive and only valid for certain types of cancers. CDs can provide great theranostic nanomedicines in cancer treatments. Scientists have attempted to eliminate cancer cells through photothermal therapy (PTT) [81,82] and photodynamic therapy (PDT) [83,87,129] to fulfill tumor targeting. Li et al. tested NIR-II emitted (900–1200 nm) CDs (CDs), adapted for 808 nm laser photothermal therapy. According to the in vivo test (Figure 11), the temperature increased to 50 °C in the intratumoral environment after intravenous injection. Tumor inhibition and volume shrinkage were observed within 6 days, compared with the PBS group. No detectable damage to tissues or weight loss after the treatment confirmed the high biocompatibility of the CDs [156]. Jia et al. prepared red-light absorbing (610 nm) CDs from *Hypocrella Bambusa* (HBCUs). They found that HBCDs highly generate $^1O_2$ under 635 laser irradiation. The reactive radicals induced apoptosis of cancer cells, which is helpful in the hypoxia tumor environment.

As shown in an in vivo experiment (Figure 12), due to the synergistic effect of PDT and PTT, the temperature at the tumor site increased to 56.4 °C in 10 min. Secondly, a good tumor inhibition effect was found after 14 days of therapy, even though the tumor could not be depleted thoroughly. No harmful phenomena were observed in other organs, indicating the safety of the treatment [82]. Xue et al. conducted modification with polyethylene glycol diamine (H_2N-PEG-NH_2), chlorin e6 (Ce6), and transferrin (Tf) on natural biomass CDs (NBCDs) to increase the targeting efficacy.

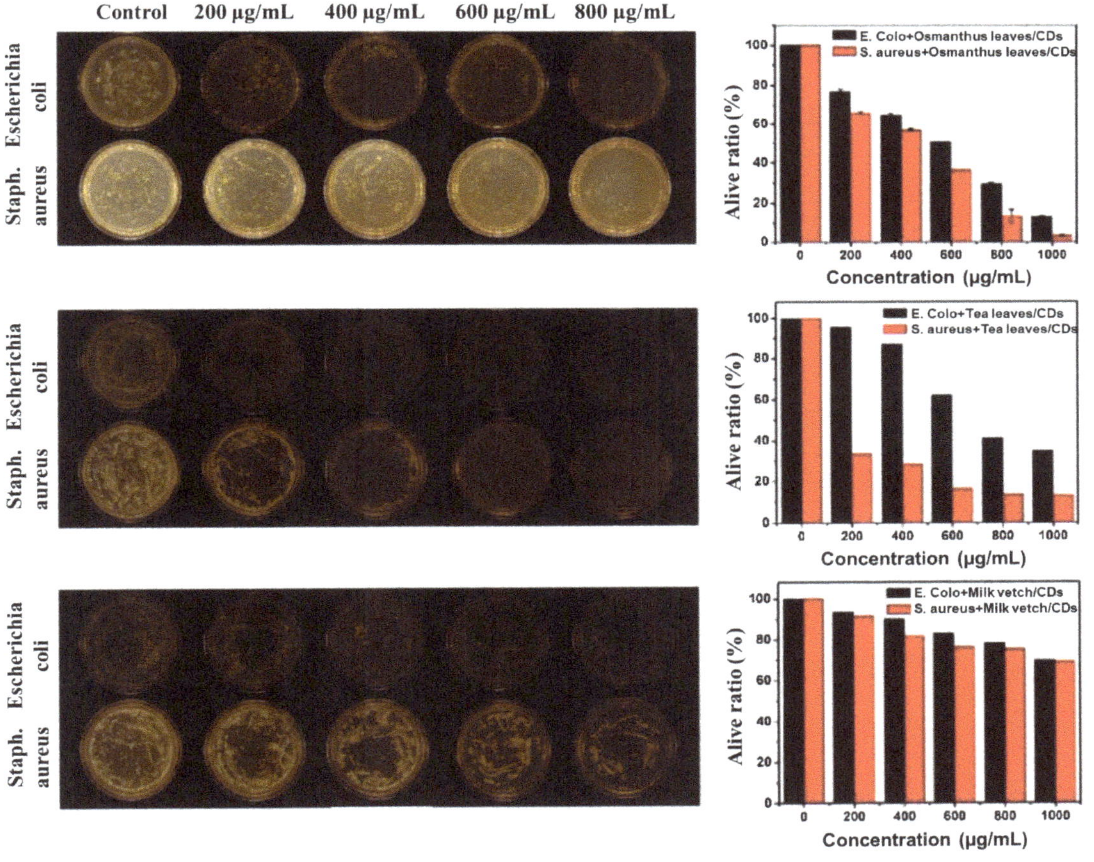

Figure 10. Graphs of *E. coli* and *S. aureus* incubated with varying concentrations of OCDs, TCDs, and MCDs for 24 h; alive ratio of *E. coli* and *S. aureus* calculated using UV–Vis spectroscopy methods, adapted with permission from [79], Elsevier, 2020.

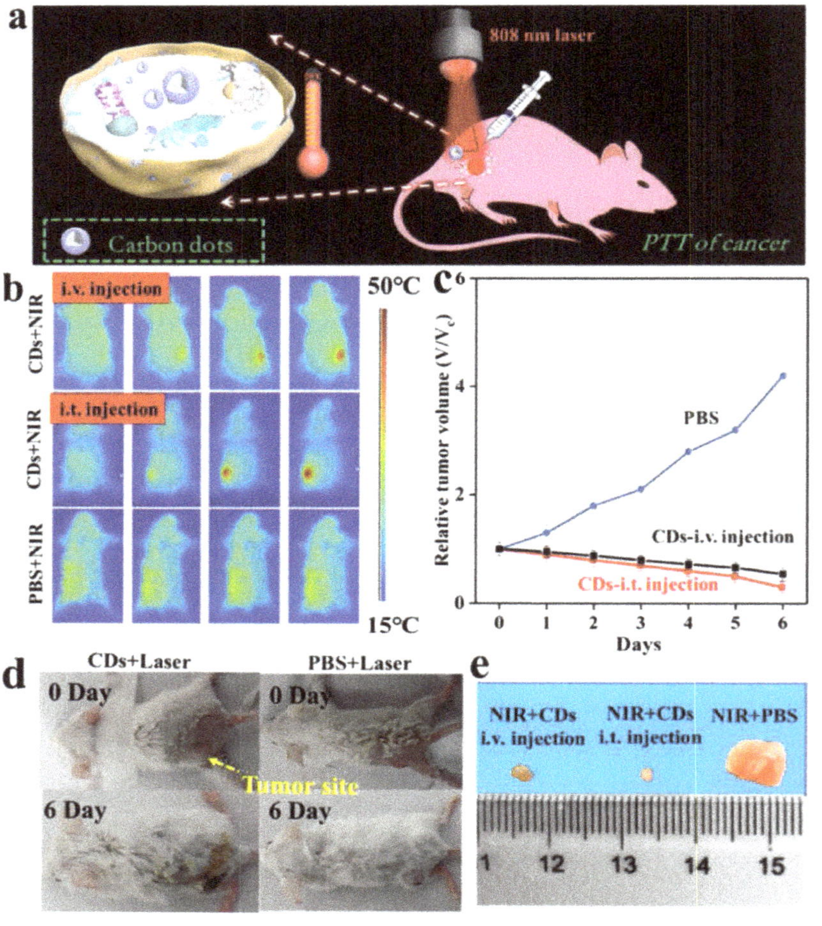

Figure 11. (**a**) Schematic diagram of photothermal therapy of cancer in vivo; (**b**) infrared thermal images of tumor-bearing mice with intravenous (i.v.) or intratumoral (i.t.) injection of CDs (50 μL, 20 mg/mL), and PBS (50 μL); (**c**) tumor volume after treatment; and (**d**,**e**) images of tumor-bearing mice and harvested tumors after 6 days, adapted with permission from [156], ACS publications, 2019.

The resulting products, NBCD-PEG-Ce6-Tf, were shown to remain within the tumor environment for 120 h using a real-time NIR fluorescence image (Figure 13a). The mice were irradiated daily under a 650 nm laser to generate $^1O_2$. During the 21-day process, tumor growth was stopped, and the tumors were ablated, indicating no conflict between the NBCDs and modifications (Figure 13b,c) [126]. Li et al. synthesized reactive oxygen species (ROS)-generating CDs from ginger. The CDs were harvested from HepG2 tumor inoculating mice; next, the tumor regressions were observed in the C-dot (440 μg) treated group; the tumor growth was prominently delayed, which attained only 3.7 ± 0.2 mg. In contrast, tumors in the PBS group grew up to 104 mg [84]. Boobalan et al. added 30 μg/mL of CDs into *P. aeruginosa*. They observed the destruction of cell walls due to ROS attack, in agreement with the results of ROS fluorescence detection using a fluorogenic dye, 2′,7′-dichlorofluorescein diacetate (DCFDA) (Figure 14a–c). MDA-MB-231 breast cancer cells were treated with CDs (3.34 μg/mL). Cell apoptosis staining, acridine orange and ethidium bromide (AO/EtBr), and nuclear staining (Hoechst 33342) were applied to the cells. The presence of orange colors and blue dots indicate cell fragmentation due to CDs (Figure 14d–g) [80]. CDs have a powerful potential in the anticancer field. Their flexibility is because the CDs are modified with various molecules, which can improve the uptake by tumor cells and increase the tumor-killing ability of the nanohybrids.

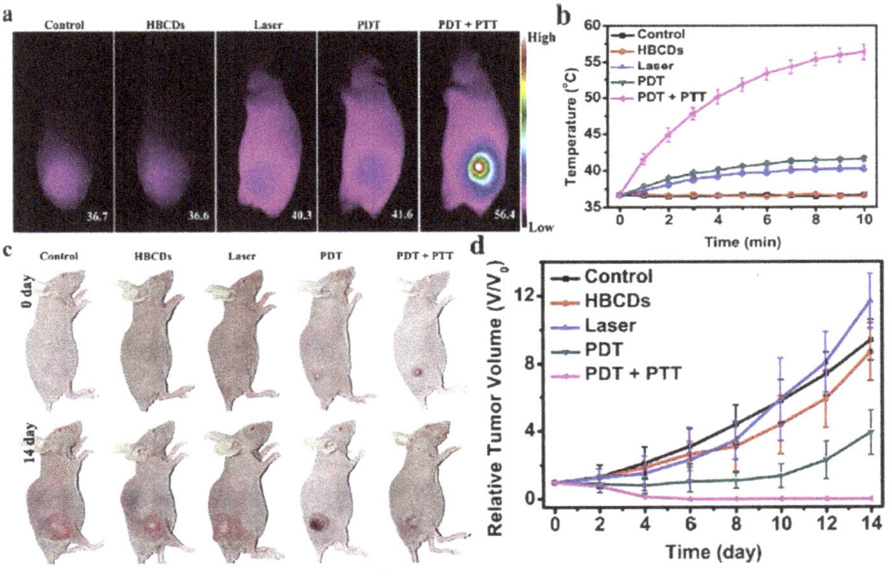

Figure 12. (**a**) IR thermal images post-injection; (**b**) temperature increase trends; (**c**) images of mice with different therapeutic methods; and (**d**) growth of the tumor during 14 days, adapted with permission from [82], Elsevier, 2018.

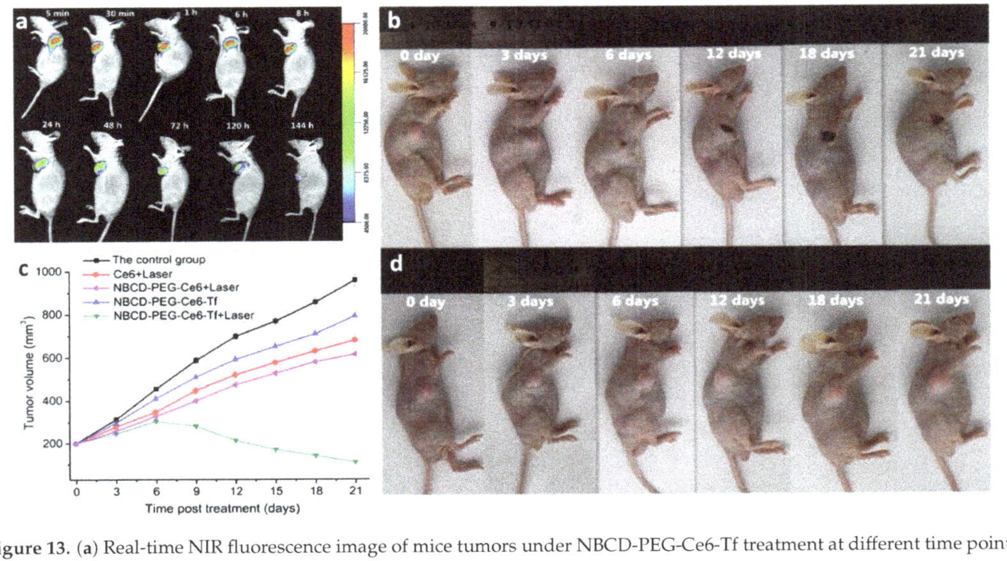

Figure 13. (**a**) Real-time NIR fluorescence image of mice tumors under NBCD-PEG-Ce6-Tf treatment at different time points; tumor propagating images for (**b**) NBCD-PEG-Ce6-Tf + laser-treated and (**d**) control groups; (**c**) time-dependent tumor growth curves after other treatments, adapted with permission from [126], Royal Society of Chemistry, 2018.

Table 3. Modifications of CDs.

CDs Type	Modification	Goal	Ref.
Carbon quantum dots	Ethylene diamine	Nucleoli selection	[101]
Nitrogen-doped carbon dots	Folic acid	Cancer cell targeting	[104]
Carbon dots	Polyethylene glycol diamine; chlorin e6; transferrin	Photosensitizing and cancer cell targeting	
Carbon dots	4-carboxy-benzyl boronic acid	Tumor cell targeting	[127]

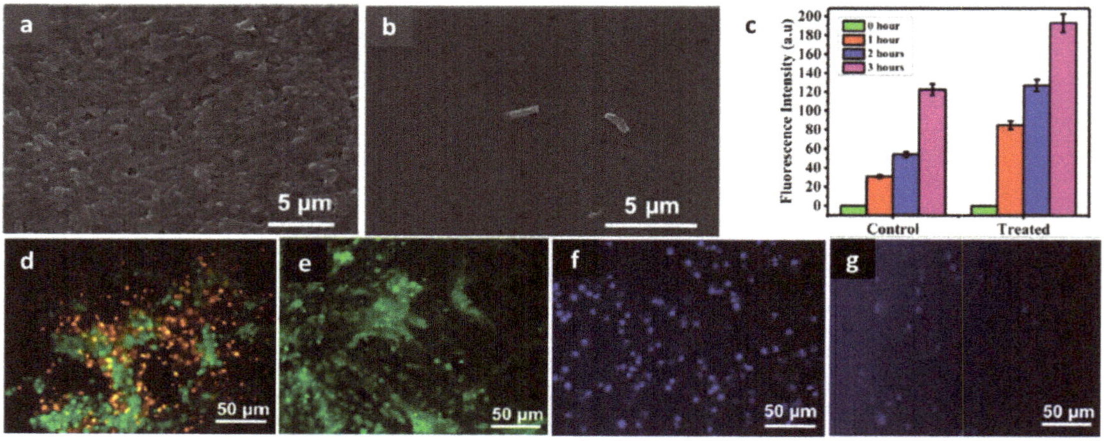

Figure 14. SEM images of (**a**) untreated and (**b**) treated *P. aeruginosa* with 30 µg/mL of CDs; (**c**) DCFDA probing of ROS generation; fluorescent microscopic images of MDA-MB-231 cells (**d**,**f**) CDs (3.34 µg/mL) treated and (**e**,**g**) untreated with nuclear staining assays (AO and Hoechst 33342), adapted with permission from [80], ACS publications, 2020.

5. Discussion and Conclusions

Carbon nanomaterials have been widely used in various scientific, engineering, and commercial fields, due to their high catalytic activity and good stability. Among them, the new "zero-dimensional" carbon nanomaterials, CDs, have unique optical properties, such as stable fluorescence signals, no light scintillation, adjustable excitation and emission wavelength, and low biological toxicity and biocompatibility. These advantages gradually led to the popularity of researching carbon nanomaterials, widely used in bioimaging, natural cell labeling, sensors, photocatalysis, solar cells, and light-emitting elements. This article mainly reviewed the different synthesis methods of CDs (including top-down and bottom-up methods) and their applications. Their luminescence properties can be adjusted through surface modification. They have been applied in many fields and have great potential. The function of CDs can also be modified by using various surface functional groups, allowing them to act, for example, as detectors and cleaners for different heavy metal ions or by doping with other ions. By controlling the surface light energy groups, they can be better used in the required fields.

Green chemistry is a discipline that has gradually received attention in recent years. The core concept focuses on the development of environmentally friendly chemical technologies. At the technical level, chemical technologies and methods are applied to reduce or eliminate the use and generation of hazardous substances in chemical synthesis and analysis, and recovery and reuse technologies are combined with increasing energy and material use efficiency. Green chemistry and nanotechnology have become emerging technology research and development directions in recent years. With the deepening of the concept of sustainability, combining the advantages of the two and accelerating the expansion of their research and development applications has become a top priority. Laboratories are committed to determining the complementary relationships between green chemistry practices and nanotechnology and applying them to materials development, chemical analysis, energy, environmental, and other related fields. The preparation of CDs well embodies the concept of green chemistry. Cheap, environmentally friendly carbon source precursors and natural renewable raw materials as carbon sources for preparation, such as eggs, grass, leaves, silk, coffee grounds, beer, and other materials, have become carbon sources for the synthesis of CDs.

Excellent performance and a unique structure provide natural CDs unlimited charm and various changes. Natural CDs, combined with biological and pharmaceutical molecules of interest through surface modification, seem to be an emerging platform for imaging probes that are both diagnostic and therapeutic. The next generation of nano-molecular probes integrates a variety of fluorescent dyes, drugs, and multifunctional nanomaterials into a single nanoprobe, providing superior signal contrast, controllable transmission, and targeted drug delivery capabilities. However, before this kind of multifunctional imaging probe was used in diagnosis and treatment, there were still many challenges, such as long-term safety, risk-benefit, biocompatibility, and biodistribution, to be evaluated. In the future, we need to solve several critical scientific problems in the research of natural CDs. First of all, the uncertain chemical groups on the surface indicate that natural-synthetic CDs are a kind of unsure material. It means the method of natural mass production of high-quality CDs is still a big challenge. Secondly, due to the different sources of naturally transformed CDs, the luminescence centers of CDs are also dissimilar. Finding a suitable luminescence position is also essential to research content. Third, categorizing different natural CDs and conducting a systematic comparative analysis will be beneficial research methods. Nanotoxicology is the emerging study of potential adverse effects derived from the interaction between nanomaterials and biological systems, and it is bound to become more critical. To further clarify its physical toxicity and adjust the size and structure accordingly, it will significantly improve CDs' application performance while expanding a more comprehensive range of applications. The scale and complexity of biomedical issues have always been an enormous challenge for researchers. It is more necessary to conduct cross-disciplinary research in chemistry, physics, materials, biomedical engineering, toxicology, public health, and clinical medicine. As more research on natural CDs continues to develop, the topic will achieve breakthroughs and progress in a short period.

Author Contributions: Conceptualization, M.-H.C., B.-G.C., R.-S.L. and M.H.; writing—original draft preparation, M.-H.C., B.-G.C. and L.T.N.; writing—review and editing, M.-H.C., B.-G.C., L.T.N., W.-T.H. and C.-H.L.; supervision, R.-S.L. and M.H.; project administration, R.-S.L. and M.H. All authors have read and agreed to the published version of the manuscript.

Funding: This study was supported by Genomics Research Center, Academia Sinica, Taiwan, to Michael Hsiao. This work was also financially supported by the Ministry of Science and Technology in Taiwan (MOST 109-2113-M-002-020-MY3) to Ru-Shi Liu. Academia Sinica Outstanding Postdoctoral Fellowship supported to Ming-Hsien Chan.

Acknowledgments: The authors would like to express their gratitude to support from Genomics Research Center, Academia Sinica, Taiwan.

Conflicts of Interest: The authors declare no conflict of interest.

Abbreviations

AO/EtBr	Acridine orange and ethidium bromide
B-NCdots	Biomass nitrogen co-doped carbon dots
c-CDs	Camellia bee pollen carbon dots
CDs	Carbon dots

CNDs	Carbon Nanodots
CPDs	Carbonized polymer dots
CQDs	Carbon quantum dots
CdSe	Cadmium selenide
CdS	Cadmium sulfide
ACDs	CDs derived from *Artemisia argyi* leaves
Ce6	Chlorin e6
l-CGCDs	Large chlorhexidine gluconate carbon dots
m-CGCDs	Medium chlorhexidine gluconate carbon dots
s-CGCDs	Small chlorhexidine gluconate carbon dots
DCFDA	2′,7′-dichlorofluorescein diacetate
DOX	Doxorubicin
DCDs	Dual emission carbon dots
EtOH	Ethanol
Fn-CDs	*F. nucleatum*-carbon dots
HBCUs	*Hypocrella Bambusa*
HBCDs	*Hypocrella Bambusa* CDs
GSH	Glutathione
GQDs	Graphene quantum dots
l-CDs	Lotus bee pollen carbon dots
MCDs	Milk vetch-derived CDs
NCDs	Nitrogen-doped CDs
NBCDs	Natural biomass CDs
NIR-CDs	Near-infrared emissive CDs
PDT	Photodynamic therapy
PL	Photoluminescence
PTT	Photothermal therapy
PCQDs	Plum-based carbon quantum dots
H_2N-PEG-NH_2	Polyethylene glycol diamine
OCDs	Osmanthus leaves-derived CDs
QY, ϕ	Quantum yield
ROS	Reactive oxygen species
R-CDs	Red-emitting CDs
wCDs	Rose-red fluorescence CDs
SCDs	Single-emission carbon dots
MCDs	Shiitake mushroom derived CDs
TCDs	Tea leaves-derived CDs
Tf	Transferrin

References

1. Wu, D.; Zhou, J.J.; Creyer, M.N.; Yim, W.; Chen, Z.; Messersmith, P.B.; Jokerst, J.V. Phenolic-enabled nanotechnology: Versatile particle engineering for biomedicine. *Chem. Soc. Rev.* **2021**, *50*, 4432–4483. [CrossRef]
2. Mauro, N.; Utzeri, M.A.; Drago, S.E.; Buscarino, G.; Cavallaro, G.; Giammona, G. Carbon Nanodots as Functional Excipient to Develop Highly Stable and Smart PLGA Nanoparticles Useful in Cancer Theranostics. *Pharmaceutics* **2020**, *12*, 1012. [CrossRef] [PubMed]
3. Nair, A.; Haponiuk, J.T.; Thomas, S.; Gopi, S. Natural carbon-based quantum dots and their applications in drug delivery: A review. *Biomed. Pharmacother.* **2020**, *132*. [CrossRef]
4. Zhao, X.Q.; Wang, L.; Ren, S.M.; Hu, Z.; Wang, Y.M. One-pot synthesis of Forsythia@carbon quantum dots with natural anti-wood rot fungus activity. *Mater. Des.* **2021**, *206*. [CrossRef]
5. Song, J.Q.; Zhao, N.; Qu, Y.; Zhao, L.S. Natural deep eutectic solvent-assisted preparation of nitrogen-doped carbon dots for ratiometric determination of pirimicarb and pH. *Dyes Pigment.* **2021**, *193*. [CrossRef]
6. Gagic, M.; Kociova, S.; Smerkova, K.; Michalkova, H.; Setka, M.; Svec, P.; Pribyl, J.; Masilko, J.; Balkova, R.; Heger, Z.; et al. One-pot synthesis of natural amine-modified biocompatible carbon quantum dots with antibacterial activity. *J. Colloid Interf. Sci.* **2020**, *580*, 30–48. [CrossRef]
7. Wang, C.X.; Pan, C.W.; Wei, X.R.; Yang, F.; Wu, W.J.; Mao, L.Q. Emissive carbon dots derived from natural liquid fuels and its biological sensing for copper ions. *Talanta* **2020**, *208*. [CrossRef]

8. Kang, L.X.; Hu, Y.; Liu, L.L.; Wu, J.X.; Zhang, S.C.; Zhao, Q.C.; Ding, F.; Li, Q.W.; Zhang, J. Growth of Close-Packed Semiconducting Single-Walled Carbon Nanotube Arrays Using Oxygen-Deficient TiO_2 Nanoparticles as Catalysts. *Nano Lett.* **2015**, *15*, 403–409. [CrossRef] [PubMed]
9. Crista, D.M.A.; El Mragui, A.; Algarra, M.; da Silva, J.C.G.E.; Luque, R.; da Silva, L.P. Turning Spent Coffee Grounds into Sustainable Precursors for the Fabrication of Carbon Dots. *Nanomaterials* **2020**, *10*, 1209. [CrossRef]
10. Qu, D.; Zheng, M.; Du, P.; Zhou, Y.; Zhang, L.G.; Li, D.; Tan, H.Q.; Zhao, Z.; Xie, Z.G.; Sun, Z.C. Highly luminescent S, N co-doped graphene quantum dots with broad visible absorption bands for visible light photocatalysts. *Nanoscale* **2013**, *5*, 12272–12277. [CrossRef]
11. Gu, S.Y.; Hsieh, C.T.; Gandomi, Y.A.; Li, J.L.; Yue, X.X.; Chang, J.K. Tailoring fluorescence emissions, quantum yields, and white light emitting from nitrogen-doped graphene and carbon nitride quantum dots. *Nanoscale* **2019**, *11*, 16553–16561. [CrossRef]
12. Han, Y.; Tang, B.J.; Wang, L.; Bao, H.; Lu, Y.H.; Guan, C.T.; Zhang, L.; Le, M.Y.; Liu, Z.; Wu, M.H. Machine-Learning-Driven Synthesis of Carbon Dots with Enhanced Quantum Yields. *ACS Nano* **2020**, *14*, 14761–14768. [CrossRef]
13. Chao, D.Y.; Chen, J.X.; Dong, Q.; Wu, W.W.; Qi, D.S.; Dong, S.J. Ultrastable and ultrasensitive pH-switchable carbon dots with high quantum yield for water quality identification, glucose detection, and two starch-based solid-state fluorescence materials. *Nano Res.* **2020**, *13*, 3012–3018. [CrossRef]
14. Manzur, A.; Oluwasanmi, A.; Moss, D.; Curtis, A.; Hoskins, C. Nanotechnologies in Pancreatic Cancer Therapy. *Pharmaceutics* **2017**, *9*, 39. [CrossRef]
15. Messina, M.M.; Barrionuevo, S.D.; Coustet, M.E.; Kreuzer, M.P.; Saccone, F.D.; Claro, P.C.D.; Ibanez, F.J. Graphene and Carbon Dots for Photoanodes with Enhanced Performance. *ACS Appl. Nano Mater.* **2021**, *4*, 7309–7318. [CrossRef]
16. Akbar, K.; Moretti, E.; Vomiero, A. Carbon Dots for Photocatalytic Degradation of Aqueous Pollutants: Recent Advancements. *Adv. Opt. Mater.* **2021**, *9*. [CrossRef]
17. Shen, J.L.; Chen, W.F.; Yang, Z.H.; Lv, G.; Cao, J.; Li, D.Y.; Liu, X. A Critical Review of Graphene Quantum Dots: Synthesis and Application in Biosensors. *Nano* **2021**, *16*. [CrossRef]
18. De, B.; Karak, N. A green and facile approach for the synthesis of water soluble fluorescent carbon dots from banana juice. *RSC Adv.* **2013**, *3*, 8286–8290. [CrossRef]
19. Michaud, V.; Pracht, J.; Schilfarth, F.; Damm, C.; Platzer, B.; Haines, P.; Harreiss, C.; Guldi, D.M.; Spiecker, E.; Peukert, W. Well-separated water-soluble carbon dots via gradient chromatography. *Nanoscale* **2021**, *13*, 13116–13128. [CrossRef]
20. Kurniawan, D.; Chiang, W.H. Microplasma-enabled colloidal nitrogen-doped graphene quantum dots for broad-range fluorescent pH sensors. *Carbon* **2020**, *167*, 675–684. [CrossRef]
21. Ru, Y.; Ai, L.; Jia, T.T.; Liu, X.J.; Lu, S.Y.; Tang, Z.Y.; Yang, B. Recent advances in chiral carbonized polymer dots: From synthesis and properties to applications. *Nano Today* **2020**, *34*. [CrossRef]
22. Ding, H.; Ji, Y.; Wei, J.-S.; Gao, Q.-Y.; Zhou, Z.-Y.; Xiong, H.-M. Facile synthesis of red-emitting carbon dots from pulp-free lemon juice for bioimaging. *J. Mater. Chem. B* **2017**, *5*, 5272–5277. [CrossRef] [PubMed]
23. Liu, J.; Geng, Y.; Li, D.; Yao, H.; Huo, Z.; Li, Y.; Zhang, K.; Zhu, S.; Wei, H.; Xu, W.; et al. Deep Red Emissive Carbonized Polymer Dots with Unprecedented Narrow Full Width at Half Maximum. *Adv. Mater.* **2020**, *32*, 1906641. [CrossRef] [PubMed]
24. Liang, C.; Xie, X.; Zhang, D.; Feng, J.; Lu, S.; Shi, Q. Biomass carbon dots derived from Wedelia trilobata for the direct detection of glutathione and their imaging application in living cells. *J. Mater. Chem. B* **2021**, *9*, 5670–5681. [CrossRef]
25. Wu, L.; Long, R.; Li, T.; Tang, C.; Tong, X.; Guo, Y.; Shi, S.; Xiang, H.; Tong, C. One-pot fabrication of dual-emission and single-emission biomass carbon dots for Cu^{2+} and tetracycline sensing and multicolor cellular imaging. *Anal. Bioanal. Chem.* **2020**, *412*, 7481–7489. [CrossRef] [PubMed]
26. Ding, H.; Zhou, X.; Qin, B.; Zhou, Z.; Zhao, Y. Highly fluorescent near-infrared emitting carbon dots derived from lemon juice and its bioimaging application. *J. Lumin.* **2019**, *211*, 298–304. [CrossRef]
27. Zhu, J.; Chu, H.; Shen, J.; Wang, C.; Wei, Y. Green preparation of carbon dots from plum as a ratiometric fluorescent probe for detection of doxorubicin. *Opt. Mater.* **2021**, *114*, 110941. [CrossRef]
28. Liu, L.; Zhang, S.; Zheng, X.; Li, H.; Chen, Q.; Qin, K.; Ding, Y.; Wei, Y. Carbon dots derived from Fusobacterium nucleatum for intracellular determination of Fe^{3+} and bioimaging both in vitro and in vivo. *Anal. Methods* **2021**, *13*, 1121–1131. [CrossRef]
29. Wang, W.; Chen, J.; Wang, D.; Shen, Y.; Yang, L.; Zhang, T.; Ge, J. Facile synthesis of biomass waste-derived fluorescent N, S, P co-doped carbon dots for detection of Fe^{3+} ions in solutions and living cells. *Anal. Methods* **2021**, *13*, 789–795. [CrossRef]
30. Wang, N.; Wang, Y.; Guo, T.; Yang, T.; Chen, M.; Wang, J. Green preparation of carbon dots with papaya as carbon source for effective fluorescent sensing of Iron (III) and Escherichia coli. *Biosens. Bioelectron.* **2016**, *85*, 68–75. [CrossRef]
31. Edison, T.N.J.I.; Atchudan, R.; Shim, J.-J.; Kalimuthu, S.; Ahn, B.-C.; Lee, Y.R. Turn-off fluorescence sensor for the detection of ferric ion in water using green synthesized N-doped carbon dots and its bio-imaging. *J. Photochem. Photobiol. B Biol.* **2016**, *158*, 235–242. [CrossRef]
32. Yang, X.; Zhuo, Y.; Zhu, S.; Luo, Y.; Feng, Y.; Dou, Y. Novel and green synthesis of high-fluorescent carbon dots originated from honey for sensing and imaging. *Biosens. Bioelectron.* **2014**, *60*, 292–298. [CrossRef] [PubMed]
33. Shen, J.; Shang, S.; Chen, X.; Wang, D.; Cai, Y. Facile synthesis of fluorescence carbon dots from sweet potato for Fe^{3+} sensing and cell imaging. *Mater. Sci. Eng. C* **2017**, *76*, 856–864. [CrossRef]
34. Song, P.; Zhang, L.; Long, H.; Meng, M.; Liu, T.; Yin, Y.; Xi, R. A multianalyte fluorescent carbon dots sensing system constructed based on specific recognition of Fe(iii) ions. *RSC Adv.* **2017**, *7*, 28637–28646. [CrossRef]

35. Liu, W.; Zhang, R.; Kang, Y.; Zhang, X.Y.; Wang, H.J.; Li, L.H.; Diao, H.P.; Wei, W.L. Preparation of nitrogen-doped carbon dots with a high fluorescence quantum yield for the highly sensitive detection of Cu^{2+} ions, drawing anti-counterfeit patterns and imaging live cells. *New Carbon Mater.* **2019**, *34*, 390–401. [CrossRef]
36. Atchudan, R.; Edison, T.N.J.I.; Perumal, S.; Vinodh, R.; Sundramoorthy, A.K.; Babu, R.S.; Lee, Y.R. Leftover Kiwi Fruit Peel-Derived Carbon Dots as a Highly Selective Fluorescent Sensor for Detection of Ferric Ion. *Chemosensors* **2021**, *9*, 166. [CrossRef]
37. Liu, R.; Zhang, J.; Gao, M.; Li, Z.; Chen, J.; Wu, D.; Liu, P. A facile microwave-hydrothermal approach towards highly photoluminescent carbon dots from goose feathers. *RSC Adv.* **2015**, *5*, 4428–4433. [CrossRef]
38. Zulfajri, M.; Gedda, G.; Chang, C.-J.; Chang, Y.-P.; Huang, G.G. Cranberry Beans Derived Carbon Dots as a Potential Fluorescence Sensor for Selective Detection of Fe^{3+} Ions in Aqueous Solution. *ACS Omega* **2019**, *4*, 15382–15392. [CrossRef]
39. Xu, J.; Zhou, Y.; Liu, S.; Dong, M.; Huang, C. Low-cost synthesis of carbon nanodots from natural products used as a fluorescent probe for the detection of ferrum(iii) ions in lake water. *Anal. Methods* **2014**, *6*, 2086. [CrossRef]
40. Venkatesan, G.; Rajagopalan, V.; Chakravarthula, S.N. Boswellia ovalifoliolata bark extract derived carbon dots for selective fluorescent sensing of Fe^{3+}. *J. Environ. Chem. Eng.* **2019**, *7*, 103013. [CrossRef]
41. Zhou, J.; Ge, M.; Han, Y.; Ni, J.; Huang, X.; Han, S.; Peng, Z.; Li, Y.; Li, S. Preparation of Biomass-Based Carbon Dots with Aggregation Luminescence Enhancement from Hydrogenated Rosin for Biological Imaging and Detection of Fe^{3+}. *ACS Omega* **2020**, *5*, 11842–11848. [CrossRef] [PubMed]
42. Sachdev, A.; Gopinath, P. Green synthesis of multifunctional carbon dots from coriander leaves and their potential application as antioxidants, sensors and bioimaging agents. *Analyst* **2015**, *140*, 4260–4269. [PubMed]
43. Raveendran, V.; Suresh Babu, A.R.; Renuka, N.K. Mint leaf derived carbon dots for dual analyte detection of Fe(iii) and ascorbic acid. *RSC Adv.* **2019**, *9*, 12070–12077. [CrossRef]
44. Hu, Y.; Li, J.; Li, X. Leek-derived co-doped carbon dots as efficient fluorescent probes for dichlorvos sensitive detection and cell multicolor imaging. *Anal. Bioanal. Chem.* **2019**, *411*, 7879–7887. [CrossRef] [PubMed]
45. Liao, J.; Cheng, Z.; Zhou, L. Nitrogen-Doping Enhanced Fluorescent Carbon Dots: Green Synthesis and Their Applications for Bioimaging and Label-Free Detection of Au^{3+} Ions. *ACS Sustain. Chem. Eng.* **2016**, *4*, 3053–3061. [CrossRef]
46. Miao, H.; Wang, L.; Zhuo, Y.; Zhou, Z.; Yang, X. Label-free fluorimetric detection of CEA using carbon dots derived from tomato juice. *Biosens. Bioelectron.* **2016**, *86*, 83–89. [CrossRef]
47. Lu, H.; Li, C.; Wang, H.; Wang, X.; Xu, S. Biomass-Derived Sulfur, Nitrogen Co-Doped Carbon Dots for Colorimetric and Fluorescent Dual Mode Detection of Silver (I) and Cell Imaging. *ACS Omega* **2019**, *4*, 21500–21508. [CrossRef]
48. Singh, A.K.; Singh, V.K.; Singh, M.; Singh, P.; Khadim, S.R.; Singh, U.; Koch, B.; Hasan, S.H.; Asthana, R.K. One pot hydrothermal synthesis of fluorescent NP-carbon dots derived from Dunaliella salina biomass and its application in on-off sensing of Hg (II), Cr (VI) and live cell imaging. *J. Photochem. Photobiol. A* **2019**, *376*, 63–72. [CrossRef]
49. Tsai, S.R.; Yin, R.; Huang, Y.Y.; Sheu, B.C.; Lee, S.C.; Hamblin, M.R. Low-level light therapy potentiates NPe6-mediated photodynamic therapy in a human osteosarcoma cell line via increased ATP. *Photodiagn. Photodyn.* **2015**, *12*, 123–130. [CrossRef] [PubMed]
50. Lu, W.; Qin, X.; Liu, S.; Chang, G.; Zhang, Y.; Luo, Y.; Asiri, A.M.; Al-Youbi, A.O.; Sun, X. Economical, Green Synthesis of Fluorescent Carbon Nanoparticles and Their Use as Probes for Sensitive and Selective Detection of Mercury(II) Ions. *Anal. Chem.* **2012**, *84*, 5351–5357. [CrossRef]
51. Huang, H.; Lv, J.-J.; Zhou, D.-L.; Bao, N.; Xu, Y.; Wang, A.-J.; Feng, J.-J. One-pot green synthesis of nitrogen-doped carbon nanoparticles as fluorescent probes for mercury ions. *RSC Adv.* **2013**, *3*, 21691. [CrossRef]
52. Wang, C.; Sun, D.; Zhuo, K.; Zhang, H.; Wang, J. Simple and green synthesis of nitrogen-, sulfur-, and phosphorus-co-doped carbon dots with tunable luminescence properties and sensing application. *RSC Adv.* **2014**, *4*, 54060–54065. [CrossRef]
53. Xie, Y.; Cheng, D.; Liu, X.; Han, A. Green Hydrothermal Synthesis of N-doped Carbon Dots from Biomass Highland Barley for the Detection of Hg^{2+}. *Sensors* **2019**, *19*, 3169. [CrossRef]
54. Tyagi, A.; Tripathi, K.M.; Singh, N.; Choudhary, S.; Gupta, R.K. Green synthesis of carbon quantum dots from lemon peel waste: Applications in sensing and photocatalysis. *RSC Adv.* **2016**, *6*, 72423–72432. [CrossRef]
55. Ghereghlou, M.; Esmaeili, A.A.; Darroudi, M. Green Synthesis of Fluorescent Carbon Dots from Elaeagnus angustifolia and its Application as Tartrazine Sensor. *J. Fluoresc.* **2021**, *31*, 185–193. [CrossRef] [PubMed]
56. Xu, H.; Yang, X.; Li, G.; Zhao, C.; Liao, X. Green Synthesis of Fluorescent Carbon Dots for Selective Detection of Tartrazine in Food Samples. *J. Agric. Food Chem.* **2015**, *63*, 6707–6714. [CrossRef]
57. Purbia, R.; Paria, S. A simple turn on fluorescent sensor for the selective detection of thiamine using coconut water derived luminescent carbon dots. *Biosens. Bioelectron.* **2016**, *79*, 467–475. [CrossRef]
58. Amjadi, M.; Hallaj, T.; Mayan, M.A. Green synthesis of nitrogen-doped carbon dots from lentil and its application for colorimetric determination of thioridazine hydrochloride. *RSC Adv.* **2016**, *6*, 104467–104473. [CrossRef]
59. Akhgari, F.; Samadi, N.; Farhadi, K.; Akhgari, M. A green one-pot synthesis of nitrogen and sulfur co-doped carbon quantum dots for sensitive and selective detection of cephalexin. *Can. J. Chem.* **2017**, *95*, 641–648. [CrossRef]
60. Liu, Y.; Zhao, Y.; Zhang, Y. One-step green synthesized fluorescent carbon nanodots from bamboo leaves for copper(II) ion detection. *Sens. Actuators B Chem.* **2014**, *196*, 647–652. [CrossRef]
61. Sha, Y.; Lou, J.; Bai, S.; Wu, D.; Liu, B.; Ling, Y. Hydrothermal synthesis of nitrogen-containing carbon nanodots as the high-efficient sensor for copper(II) ions. *Mater. Res. Bull.* **2013**, *48*, 1728–1731. [CrossRef]

62. Mehta, V.N.; Jha, S.; Basu, H.; Singhal, R.K.; Kailasa, S.K. One-step hydrothermal approach to fabricate carbon dots from apple juice for imaging of mycobacterium and fungal cells. *Sens. Actuators B Chem.* **2015**, *213*, 434–443. [CrossRef]
63. Arul, V.; Edison, T.N.J.I.; Lee, Y.R.; Sethuraman, M.G. Biological and catalytic applications of green synthesized fluorescent N-doped carbon dots using Hylocereus undatus. *J. Photochem. Photobiol. B Biol.* **2017**, *168*, 142–148.
64. Mehta, V.N.; Jha, S.; Kailasa, S.K. One-pot green synthesis of carbon dots by using Saccharum officinarum juice for fluorescent imaging of bacteria (*Escherichia coli*) and yeast (*Saccharomyces cerevisiae*) cells. *Mater. Sci. Eng. C* **2014**, *38*, 20–27. [CrossRef] [PubMed]
65. Song, Y.; Yan, X.; Li, Z.; Qu, L.; Zhu, C.; Ye, R.; Li, S.; Du, D.; Lin, Y. Highly photoluminescent carbon dots derived from linseed and their applications in cellular imaging and sensing. *J. Mater. Chem. B* **2018**, *6*, 3181–3187. [CrossRef]
66. Wang, W.-J.; Xia, J.-M.; Feng, J.; He, M.-Q.; Chen, M.-L.; Wang, J.-H. Green preparation of carbon dots for intracellular pH sensing and multicolor live cell imaging. *J. Mater. Chem. B* **2016**, *4*, 7130–7137. [CrossRef] [PubMed]
67. Zhao, X.J.; Zhang, W.L.; Zhou, Z.Q. Sodium hydroxide-mediated hydrogel of citrus pectin for preparation of fluorescent carbon dots for bioimaging. *Colloids Surf. B Biointerfaces* **2014**, *123*, 493–497. [CrossRef]
68. Liu, Y.; Liu, Y.; Park, M.; Park, S.-J.; Zhang, Y.; Akanda, M.R.; Park, B.-Y.; Kim, H.Y. Green synthesis of fluorescent carbon dots from carrot juice for in vitro cellular imaging. *Carbon Lett.* **2017**, *21*, 61–67.
69. Atchudan, R.; Edison, T.N.J.I.; Perumal, S.; Muthuchamy, N.; Lee, Y.R. Hydrophilic nitrogen-doped carbon dots from biowaste using dwarf banana peel for environmental and biological applications. *Fuel* **2020**, *275*, 117821. [CrossRef]
70. Du, F.; Zhang, M.; Li, X.; Li, J.; Jiang, X.; Li, Z.; Hua, Y.; Shao, G.; Jin, J.; Shao, Q. Economical and green synthesis of bagasse-derived fluorescent carbon dots for biomedical applications. *Nanotechnology* **2014**, *25*, 315702. [CrossRef]
71. Alam, A.-M.; Park, B.-Y.; Ghouri, Z.K.; Park, M.; Kim, H.-Y. Synthesis of carbon quantum dots from cabbage with down- and up-conversion photoluminescence properties: Excellent imaging agent for biomedical applications. *Green Chem.* **2015**, *17*, 3791–3797. [CrossRef]
72. Ding, Z.; Li, F.; Wen, J.; Wang, X.; Sun, R. Gram-scale synthesis of single-crystalline graphene quantum dots derived from lignin biomass. *Green Chem.* **2018**, *20*, 1383–1390. [CrossRef]
73. D'Souza, S.L.; Deshmukh, B.; Bhamore, J.R.; Rawat, K.A.; Lenka, N.; Kailasa, S.K. Synthesis of fluorescent nitrogen-doped carbon dots from dried shrimps for cell imaging and boldine drug delivery system. *RSC Adv.* **2016**, *6*, 12169–12179. [CrossRef]
74. John, T.S.; Yadav, P.K.; Kumar, D.; Singh, S.K.; Hasan, S.H. Highly fluorescent carbon dots from wheat bran as a novel drug delivery system for bacterial inhibition. *Luminescence* **2020**, *35*, 913–923. [CrossRef] [PubMed]
75. Yuan, Y.; Guo, B.; Hao, L.; Liu, N.; Lin, Y.; Guo, W.; Li, X.; Gu, B. Doxorubicin-loaded environmentally friendly carbon dots as a novel drug delivery system for nucleus targeted cancer therapy. *Colloids Surf. B Biointerfaces* **2017**, *159*, 349–359. [CrossRef] [PubMed]
76. Sun, B.; Wu, F.; Zhang, Q.; Chu, X.; Wang, Z.; Huang, X.; Li, J.; Yao, C.; Zhou, N.; Shen, J. Insight into the effect of particle size distribution differences on the antibacterial activity of carbon dots. *J. Colloid Interface Sci.* **2021**, *584*, 505–519. [CrossRef]
77. Saravanan, A.; Maruthapandi, M.; Das, P.; Luong, J.H.T.; Gedanken, A. Green Synthesis of Multifunctional Carbon Dots with Antibacterial Activities. *Nanomaterials* **2021**, *11*, 369. [CrossRef]
78. Eskalen, H.; Çeşme, M.; Kerli, S.; Özğan, Ş. Green synthesis of water-soluble fluorescent carbon dots from rosemary leaves: Applications in food storage capacity, fingerprint detection, and antibacterial activity. *J. Chem. Res.* **2021**, *45*, 428–435. [CrossRef]
79. Ma, Y.; Zhang, M.; Wang, H.; Wang, B.; Huang, H.; Liu, Y.; Kang, Z. N-doped carbon dots derived from leaves with low toxicity via damaging cytomembrane for broad-spectrum antibacterial activity. *Mater. Today Commun.* **2020**, *24*, 101222. [CrossRef]
80. Boobalan, T.; Sethupathi, M.; Sengottuvelan, N.; Kumar, P.; Balaji, P.; Gulyás, B.; Padmanabhan, P.; Selvan, S.T.; Arun, A. Mushroom-Derived Carbon Dots for Toxic Metal Ion Detection and as Antibacterial and Anticancer Agents. *ACS Appl. Nano Mater.* **2020**, *3*, 5910–5919. [CrossRef]
81. Castaing, V.; Sontakke, A.D.; Xu, J.; Fernandez-Carrion, A.J.; Genevois, C.; Tanabe, S.; Allix, M.; Viana, B. Persistent energy transfer in ZGO:Cr^{3+},Yb^{3+}: A new strategy to design nano glass-ceramics featuring deep red and near infrared persistent luminescence. *Phys. Chem. Chem. Phys.* **2019**, *21*, 19458–19468. [CrossRef]
82. Jia, Q.; Zheng, X.; Ge, J.; Liu, W.; Ren, H.; Chen, S.; Wen, Y.; Zhang, H.; Wu, J.; Wang, P. Synthesis of carbon dots from Hypocrella bambusae for bimodel fluorescence/photoacoustic imaging-guided synergistic photodynamic/photothermal therapy of cancer. *J. Colloid Interface Sci.* **2018**, *526*, 302–311. [CrossRef] [PubMed]
83. Kim, D.; Jo, G.; Chae, Y.; Subramani, S.; Lee, B.Y.; Kim, E.J.; Ji, M.-K.; Sim, U.; Hyun, H. Bioinspired Camellia japonica carbon dots with high near-infrared absorbance for efficient photothermal cancer therapy. *Nanoscale* **2021**, *13*, 14426–14434. [CrossRef] [PubMed]
84. Li, C.-L.; Ou, C.-M.; Huang, C.-C.; Wu, W.-C.; Chen, Y.-P.; Lin, T.-E.; Ho, L.-C.; Wang, C.-W.; Shih, C.-C.; Zhou, H.-C.; et al. Carbon dots prepared from ginger exhibiting efficient inhibition of human hepatocellular carcinoma cells. *J. Mater. Chem. B* **2014**, *2*, 4564. [CrossRef] [PubMed]
85. Zhao, S.; Lan, M.; Zhu, X.; Xue, H.; Ng, T.-W.; Meng, X.; Lee, C.-S.; Wang, P.; Zhang, W. Green Synthesis of Bifunctional Fluorescent Carbon Dots from Garlic for Cellular Imaging and Free Radical Scavenging. *ACS Appl. Mater. Inter.* **2015**, *7*, 17054–17060. [CrossRef]
86. Chen, W.; Li, D.; Tian, L.; Xiang, W.; Wang, T.; Hu, W.; Hu, Y.; Chen, S.; Chen, J.; Dai, Z. Synthesis of graphene quantum dots from natural polymer starch for cell imaging. *Green Chem.* **2018**, *20*, 4438–4442. [CrossRef]

87. Sahu, S.; Behera, B.; Maiti, T.K.; Mohapatra, S. Simple one-step synthesis of highly luminescent carbon dots from orange juice: Application as excellent bio-imaging agents. *Chem. Commun.* **2012**, *48*, 8835. [CrossRef] [PubMed]
88. Zhang, J.; Yuan, Y.; Liang, G.; Yu, S.-H. Scale-Up Synthesis of Fragrant Nitrogen-Doped Carbon Dots from Bee Pollens for Bioimaging and Catalysis. *Adv. Sci.* **2015**, *2*, 1500002. [CrossRef]
89. Liang, Q.; Ma, W.; Shi, Y.; Li, Z.; Yang, X. Easy synthesis of highly fluorescent carbon quantum dots from gelatin and their luminescent properties and applications. *Carbon* **2013**, *60*, 421–428. [CrossRef]
90. Kasibabu, B.S.B.; D'Souza, S.L.; Jha, S.; Kailasa, S.K. Imaging of Bacterial and Fungal Cells Using Fluorescent Carbon Dots Prepared from Carica papaya Juice. *J. Fluoresc.* **2015**, *25*, 803–810. [CrossRef]
91. Yu, C.; Xuan, T.; Chen, Y.; Zhao, Z.; Sun, Z.; Li, H. A facile, green synthesis of highly fluorescent carbon nanoparticles from oatmeal for cell imaging. *J. Mater. Chem. C* **2015**, *3*, 9514–9518. [CrossRef]
92. Zhang, Z.; Sun, W.; Wu, P. Highly Photoluminescent Carbon Dots Derived from Egg White: Facile and Green Synthesis, Photoluminescence Properties, and Multiple Applications. *ACS Sustain. Chem. Eng.* **2015**, *3*, 1412–1418. [CrossRef]
93. Wei, J.; Zhang, X.; Sheng, Y.; Shen, J.; Huang, P.; Guo, S.; Pan, J.; Feng, B. Dual functional carbon dots derived from cornflour via a simple one-pot hydrothermal route. *Mater. Lett.* **2014**, *123*, 107–111. [CrossRef]
94. Shi, W.; Fan, H.; Ai, S.; Zhu, L. Preparation of fluorescent graphene quantum dots from humic acid for bioimaging application. *New J. Chem.* **2015**, *39*, 7054–7059. [CrossRef]
95. Wang, G.; Guo, Q.; Chen, D.; Liu, Z.; Zheng, X.; Xu, A.; Yang, S.; Ding, G. Facile and Highly Effective Synthesis of Controllable Lattice Sulfur-Doped Graphene Quantum Dots via Hydrothermal Treatment of Durian. *ACS Appl. Mater. Inter.* **2018**, *10*, 5750–5759. [CrossRef]
96. Atchudan, R.; Jebakumar Immanuel Edison, T.N.; Perumal, S.; Lee, Y.R. Indian Gooseberry-Derived Tunable Fluorescent Carbon Dots as a Promise for In Vitro/In Vivo Multicolor Bioimaging and Fluorescent Ink. *ACS Omega* **2018**, *3*, 17590–17601. [CrossRef]
97. Hou, J.; Wang, W.; Zhou, T.Y.; Wang, B.; Li, H.Y.; Ding, L. Synthesis and formation mechanistic investigation of nitrogen-doped carbon dots with high quantum yields and yellowish-green fluorescence. *Nanoscale* **2016**, *8*, 11185–11193. [CrossRef]
98. Meena, R.; Singh, R.; Marappan, G.; Kushwaha, G.; Gupta, N.; Meena, R.; Gupta, J.P.; Agarwal, R.R.; Fahmi, N.; Kushwaha, O.S. Fluorescent carbon dots driven from ayurvedic medicinal plants for cancer cell imaging and phototherapy. *Heliyon* **2019**, *5*, e02483. [CrossRef]
99. Zhang, X.; Wang, H.; Ma, C.; Niu, N.; Chen, Z.; Liu, S.; Li, J.; Li, S. Seeking value from biomass materials: Preparation of coffee bean shell-derived fluorescent carbon dots via molecular aggregation for antioxidation and bioimaging applications. *Mater. Chem. Front.* **2018**, *2*, 1269–1275. [CrossRef]
100. Desai, M.L.; Jha, S.; Basu, H.; Singhal, R.K.; Park, T.-J.; Kailasa, S.K. Acid Oxidation of Muskmelon Fruit for the Fabrication of Carbon Dots with Specific Emission Colors for Recognition of Hg^{2+} Ions and Cell Imaging. *ACS Omega* **2019**, *4*, 19332–19340. [CrossRef]
101. D'Angelis do ES Barbosa, C.; Corrêa, J.R.; Medeiros, G.A.; Barreto, G.; Magalhães, K.G.; De Oliveira, A.L.; Spencer, J.; Rodrigues, M.O.; Neto, B.A.D. Carbon Dots (C-dots) from Cow Manure with Impressive Subcellular Selectivity Tuned by Simple Chemical Modification. *Chem. Eur. J.* **2015**, *21*, 5055–5060. [CrossRef] [PubMed]
102. Park, S.Y.; Lee, H.U.; Park, E.S.; Lee, S.C.; Lee, J.-W.; Jeong, S.W.; Kim, C.H.; Lee, Y.-C.; Huh, Y.S.; Lee, J. Photoluminescent Green Carbon Nanodots from Food-Waste-Derived Sources: Large-Scale Synthesis, Properties, and Biomedical Applications. *ACS Appl. Mater. Inter.* **2014**, *6*, 3365–3370. [CrossRef]
103. Oza, G.; Oza, K.; Pandey, S.; Shinde, S.; Mewada, A.; Thakur, M.; Sharon, M.; Sharon, M. A Green Route Towards Highly Photoluminescent and Cytocompatible Carbon dot Synthesis and its Separation Using Sucrose Density Gradient Centrifugation. *J. Fluoresc.* **2015**, *25*, 9–14. [CrossRef] [PubMed]
104. Dehvari, K.; Liu, K.Y.; Tseng, P.-J.; Gedda, G.; Girma, W.M.; Chang, J.-Y. Sonochemical-assisted green synthesis of nitrogen-doped carbon dots from crab shell as targeted nanoprobes for cell imaging. *J. Taiwan Inst. Chem. Eng.* **2019**, *95*, 495–503. [CrossRef]
105. Feng, J.; Wang, W.-J.; Hai, X.; Yu, Y.-L.; Wang, J.-H. Green preparation of nitrogen-doped carbon dots derived from silkworm chrysalis for cell imaging. *J. Mater. Chem. B* **2016**, *4*, 387–393. [CrossRef] [PubMed]
106. Ramanan, V.; Thiyagarajan, S.K.; Raji, K.; Suresh, R.; Sekar, R.; Ramamurthy, P. Outright Green Synthesis of Fluorescent Carbon Dots from Eutrophic Algal Blooms for In Vitro Imaging. *ACS Sustain. Chem. Eng.* **2016**, *4*, 4724–4731. [CrossRef]
107. Wang, Q.; Liu, X.; Zhang, L.; Lv, Y. Microwave-assisted synthesis of carbon nanodots through an eggshell membrane and their fluorescent application. *Analyst* **2012**, *137*, 5392. [CrossRef] [PubMed]
108. Qin, X.; Lu, W.; Asiri, A.M.; Al-Youbi, A.O.; Sun, X. Microwave-assisted rapid green synthesis of photoluminescent carbon nanodots from flour and their applications for sensitive and selective detection of mercury(II) ions. *Sens. Actuators B Chem.* **2013**, *184*, 156–162. [CrossRef]
109. Liu, X.; Li, T.; Hou, Y.; Wu, Q.; Yi, J.; Zhang, G. Microwave synthesis of carbon dots with multi-response using denatured proteins as carbon source. *RSC Adv.* **2016**, *6*, 11711–11718. [CrossRef]
110. Feng, Y.; Zhong, D.; Miao, H.; Yang, X. Carbon dots derived from rose flowers for tetracycline sensing. *Talanta* **2015**, *140*, 128–133. [CrossRef] [PubMed]
111. Bankoti, K.; Rameshbabu, A.P.; Datta, S.; Das, B.; Mitra, A.; Dhara, S. Onion derived carbon nanodots for live cell imaging and accelerated skin wound healing. *J. Mater. Chem. B* **2017**, *5*, 6579–6592. [CrossRef]

112. Xue, M.; Zou, M.; Zhao, J.; Zhan, Z.; Zhao, S. Green preparation of fluorescent carbon dots from lychee seeds and their application for the selective detection of methylene blue and imaging in living cells. *J. Mater. Chem. B* **2015**, *3*, 6783–6789. [CrossRef]
113. Hsu, P.-C.; Shih, Z.-Y.; Lee, C.-H.; Chang, H.-T. Synthesis and analytical applications of photoluminescent carbon nanodots. *Green Chem.* **2012**, *14*, 917. [CrossRef]
114. Essner, J.B.; Laber, C.H.; Ravula, S.; Polo-Parada, L.; Baker, G.A. Pee-dots: Biocompatible fluorescent carbon dots derived from the upcycling of urine. *Green Chem.* **2016**, *18*, 243–250. [CrossRef]
115. Zhou, J.; Sheng, Z.; Han, H.; Zou, M.; Li, C. Facile synthesis of fluorescent carbon dots using watermelon peel as a carbon source. *Mater. Lett.* **2012**, *66*, 222–224. [CrossRef]
116. Teng, X.; Ma, C.; Ge, C.; Yan, M.; Yang, J.; Zhang, Y.; Morais, P.C.; Bi, H. Green synthesis of nitrogen-doped carbon dots from konjac flour with "off–on" fluorescence by Fe^{3+} and l-lysine for bioimaging. *J. Mater. Chem. B* **2014**, *2*, 4631. [CrossRef] [PubMed]
117. Hu, Y.; Chen, Z.; Lai, F.; Li, J. Biomass-codoped carbon dots: Efficient fluorescent probes for isocarbophos ultrasensitive detection and for living cells dual-color imaging. *J. Mater. Sci.* **2019**, *54*, 8627–8639. [CrossRef]
118. Murugan, N.; Sundramoorthy, A.K. Green synthesis of fluorescent carbon dots from Borassus flabellifer flowers for label-free highly selective and sensitive detection of Fe^{3+} ions. *New J. Chem.* **2018**, *42*, 13297–13307. [CrossRef]
119. Ma, X.; Dong, Y.; Sun, H.; Chen, N. Highly fluorescent carbon dots from peanut shells as potential probes for copper ion: The optimization and analysis of the synthetic process. *Mater. Today Chem.* **2017**, *5*, 1–10. [CrossRef]
120. Baruah, U.; Gogoi, N.; Konwar, A.; Jyoti Deka, M.; Chowdhury, D.; Majumdar, G. Carbon Dot Based Sensing of Dopamine and Ascorbic Acid. *J. Nanoparticles* **2014**, *2014*, 1–8. [CrossRef]
121. Xue, M.; Zhan, Z.; Zou, M.; Zhang, L.; Zhao, S. Green synthesis of stable and biocompatible fluorescent carbon dots from peanut shells for multicolor living cell imaging. *New J. Chem.* **2016**, *40*, 1698–1703. [CrossRef]
122. Cong, S.; Liu, K.; Qiao, F.; Song, Y.; Tan, M. Biocompatible fluorescent carbon dots derived from roast duck for in vitro cellular and in vivo C. elegans bio-imaging. *Methods* **2019**, *168*, 76–83. [CrossRef]
123. Wang, H.; Zhang, M.; Ma, Y.; Wang, B.; Shao, M.; Huang, H.; Liu, Y.; Kang, Z. Selective inactivation of Gram-negative bacteria by carbon dots derived from natural biomass: Artemisia argyi leaves. *J. Mater. Chem. B* **2020**, *8*, 2666–2672. [CrossRef]
124. Chung, H.K.; Wongso, V.; Sambudi, N.S.; Isnaeni. Biowaste-derived carbon dots/hydroxyapatite nanocomposite as drug delivery vehicle for acetaminophen. *J. Sol-Gel Sci. Technol.* **2020**, *93*, 214–223. [CrossRef]
125. Wang, X.; Zhang, Y.; Kong, H.; Cheng, J.; Zhang, M.; Sun, Z.; Wang, S.; Liu, J.; Qu, H.; Zhao, Y. Novel mulberry silkworm cocoon-derived carbon dots and their anti-inflammatory properties. *Artif. Cells Nanomed. Biotechnol.* **2020**, *48*, 68–76. [CrossRef] [PubMed]
126. Xue, M.; Zhao, J.; Zhan, Z.; Zhao, S.; Lan, C.; Ye, F.; Liang, H. Dual functionalized natural biomass carbon dots from lychee exocarp for cancer cell targetable near-infrared fluorescence imaging and photodynamic therapy. *Nanoscale* **2018**, *10*, 18124–18130. [CrossRef] [PubMed]
127. Fahmi, M.Z.; Haris, A.; Permana, A.J.; Nor Wibowo, D.L.; Purwanto, B.; Nikmah, Y.L.; Idris, A. Bamboo leaf-based carbon dots for efficient tumor imaging and therapy. *RSC Adv.* **2018**, *8*, 38376–38383. [CrossRef]
128. Cheng, C.; Shi, Y.; Li, M.; Xing, M.; Wu, Q. Carbon quantum dots from carbonized walnut shells: Structural evolution, fluorescence characteristics, and intracellular bioimaging. *Mater. Sci. Eng. C* **2017**, *79*, 473–480. [CrossRef]
129. Janus, Ł.; Piątkowski, M.; Radwan-Pragłowska, J.; Bogdał, D.; Matysek, D. Chitosan-based carbon quantum dots for biomedical applications: Synthesis and characterization. *Nanomaterials* **2019**, *9*, 274. [CrossRef]
130. Qu, D.; Zheng, M.; Zhang, L.; Zhao, H.; Xie, Z.; Jing, X.; Haddad, R.E.; Fan, H.; Sun, Z. Formation mechanism and optimization of highly luminescent N-doped graphene quantum dots. *Sci. Rep.* **2014**, *4*, 1–11. [CrossRef]
131. Zhang, R.; Chen, W. Nitrogen-doped carbon quantum dots: Facile synthesis and application as a "turn-off" fluorescent probe for detection of Hg^{2+} ions. *Biosens. Bioelectron.* **2014**, *55*, 83–90. [CrossRef] [PubMed]
132. Kurian, M.; Paul, A. Recent trends in the use of green sources for carbon dot synthesis-A short review. *Carbon Trends* **2021**, 100032. [CrossRef]
133. Lai, Z.; Guo, X.; Cheng, Z.; Ruan, G.; Du, F. Green synthesis of fluorescent carbon dots from cherry tomatoes for highly effective detection of trifluralin herbicide in soil samples. *ChemistrySelect* **2020**, *5*, 1956–1960.
134. Jiao, X.-Y.; Li, L.-s.; Qin, S.; Zhang, Y.; Huang, K.; Xu, L. The synthesis of fluorescent carbon dots from mango peel and their multiple applications. *Colloids Surfaces A Physicochem. Eng. Asp.* **2019**, *577*, 306–314. [CrossRef]
135. Lu, M.; Duan, Y.; Song, Y.; Tan, J.; Zhou, L. Green preparation of versatile nitrogen-doped carbon quantum dots from watermelon juice for cell imaging, detection of Fe^{3+} ions and cysteine, and optical thermometry. *J. Mol. Liq.* **2018**, *269*, 766–774. [CrossRef]
136. Shi, Y.; Liu, X.; Wang, M.; Huang, J.; Jiang, X.; Pang, J.; Xu, F.; Zhang, X. Synthesis of N-doped carbon quantum dots from bio-waste lignin for selective irons detection and cellular imaging. *Int. J. Biol. Macromol.* **2019**, *128*, 537–545. [CrossRef]
137. Atchudan, R.; Edison, T.N.J.I.; Perumal, S.; Selvam, N.C.S.; Lee, Y.R. Green synthesized multiple fluorescent nitrogen-doped carbon quantum dots as an efficient label-free optical nanoprobe for in vivo live-cell imaging. *J. Photochem. Photobiol. A* **2019**, *372*, 99–107. [CrossRef]
138. Marouzi, S.; Darroudi, M.; Hekmat, A.; Sadri, K.; Oskuee, R.K. One-pot hydrothermal synthesis of carbon quantum dots from Salvia hispanica L. seeds and investigation of their biodistribution, and cytotoxicity effects. *J. Environ. Chem. Eng.* **2021**, *9*, 105461. [CrossRef]

139. Liu, S.; Liu, Z.; Li, Q.; Xia, H.; Yang, W.; Wang, R.; Li, Y.; Zhao, H.; Tian, B. Facile synthesis of carbon dots from wheat straw for colorimetric and fluorescent detection of fluoride and cellular imaging. *Spectrochim. Acta A Mol. Biomol. Spectrosc.* **2021**, *246*, 118964.
140. Atchudan, R.; Edison, T.N.J.I.; Perumal, S.; Vinodh, R.; Lee, Y.R. Multicolor-emitting carbon dots from Malus floribunda and their interaction with Caenorhabditis elegans. *Mater. Lett.* **2020**, *261*, 127153. [CrossRef]
141. Atchudan, R.; Edison, T.N.J.I.; Shanmugam, M.; Perumal, S.; Somanathan, T.; Lee, Y.R. Sustainable synthesis of carbon quantum dots from banana peel waste using hydrothermal process for in vivo bioimaging. *Phys. E Low-Dimens. Syst. Nanostruct.* **2021**, *126*, 114417. [CrossRef]
142. Shi, C.; Qi, H.; Ma, R.; Sun, Z.; Xiao, L.; Wei, G.; Huang, Z.; Liu, S.; Li, J.; Dong, M. N, S-self-doped carbon quantum dots from fungus fibers for sensing tetracyclines and for bioimaging cancer cells. *Mater. Sci. Eng. C* **2019**, *105*, 110132. [CrossRef] [PubMed]
143. Ghosh, S.; Ghosal, K.; Mohammad, S.A.; Sarkar, K. Dendrimer functionalized carbon quantum dot for selective detection of breast cancer and gene therapy. *Chem. Eng. J.* **2019**, *373*, 468–484. [CrossRef]
144. Sahoo, N.K.; Jana, G.C.; Aktara, M.N.; Das, S.; Nayim, S.; Patra, A.; Bhattacharjee, P.; Bhadra, K.; Hossain, M. Carbon dots derived from lychee waste: Application for Fe^{3+} ions sensing in real water and multicolor cell imaging of skin melanoma cells. *Mater. Sci. Eng. C* **2020**, *108*, 110429. [CrossRef] [PubMed]
145. Tadesse, A.; Hagos, M.; RamaDevi, D.; Basavaiah, K.; Belachew, N. Fluorescent-nitrogen-doped carbon quantum dots derived from citrus lemon juice: Green synthesis, mercury (II) ion sensing, and live cell imaging. *ACS omega* **2020**, *5*, 3889–3898.
146. D'souza, S.L.; Chettiar, S.S.; Koduru, J.R.; Kailasa, S.K. Synthesis of fluorescent carbon dots using Daucus carota subsp. sativus roots for mitomycin drug delivery. *Optik* **2018**, *158*, 893–900. [CrossRef]
147. Babu, P.J.; Saranya, S.; Singh, Y.D.; Venkataswamy, M.; Raichur, A.M.; Doble, M. Photoluminescence carbon nano dots for the conductivity based optical sensing of dopamine and bioimaging applications. *Opt. Mater.* **2021**, *117*, 111120. [CrossRef]
148. Vandarkuzhali, S.A.A.; Natarajan, S.; Jeyabalan, S.; Sivaraman, G.; Singaravadivel, S.; Muthusubramanian, S.; Viswanathan, B. Pineapple peel-derived carbon dots: Applications as sensor, molecular keypad lock, and memory device. *ACS Omega* **2018**, *3*, 12584–12592.
149. Raji, K.; Ramanan, V.; Ramamurthy, P. Facile and green synthesis of highly fluorescent nitrogen-doped carbon dots from jackfruit seeds and its applications towards the fluorimetric detection of Au^{3+} ions in aqueous medium and in in vitro multicolor cell imaging. *New J. Chem.* **2019**, *43*, 11710–11719. [CrossRef]
150. Kumari, S.; Sharma, N.; Sahi, S.V. Advances in Cancer Therapeutics: Conventional Thermal Therapy to Nanotechnology-Based Photothermal Therapy. *Pharmaceutics* **2021**, *13*, 1174. [CrossRef]
151. Molaei, M.J. Principles, mechanisms, and application of carbon quantum dots in sensors: A review. *Anal. Methods* **2020**, *12*, 1266–1287. [CrossRef]
152. Xu, Q.; Zhou, Q.; Hua, Z.; Xue, Q.; Zhang, C.; Wang, X.; Pan, D.; Xiao, M. Single-Particle Spectroscopic Measurements of Fluorescent Graphene Quantum Dots. *ACS Nano* **2013**, *7*, 10654–10661. [CrossRef] [PubMed]
153. Wang, L.L.; Hu, C.; Shao, L.Q. The antimicrobial activity of nanoparticles: Present situation and prospects for the future. *Int. J. Nanomed.* **2017**, *12*, 1227–1249. [CrossRef] [PubMed]
154. Dong, X.L.; Liang, W.X.; Meziani, M.J.; Sun, Y.P.; Yang, L.J. Carbon Dots as Potent Antimicrobial Agents. *Theranostics* **2020**, *10*, 671–686. [CrossRef] [PubMed]
155. Ghirardello, M.; Ramos-Soriano, J.; Galan, M.C. Carbon Dots as an Emergent Class of Antimicrobial Agents. *Nanomaterials* **2021**, *11*, 1877. [CrossRef] [PubMed]
156. Li, Y.B.; Bai, G.X.; Zeng, S.J.; Hao, J.H. Theranostic Carbon Dots with Innovative NIR-II Emission for in Vivo Renal-Excreted Optical Imaging and Photothermal Therapy. *ACS Appl. Mater. Inter.* **2019**, *11*, 4737–4744. [CrossRef] [PubMed]

pharmaceutics

MDPI

Review

RNAi-Based Approaches for Pancreatic Cancer Therapy

Min Ju Kim [1,2], Hyeyoun Chang [1], Gihoon Nam [1,2], Youngji Ko [1], Sun Hwa Kim [2], Thomas M. Roberts [1,*] and Ju Hee Ryu [2,*]

[1] Department of Cancer Biology, Dana-Farber Cancer Institute, Boston, MA 02115, USA; simple5336@gmail.com (M.J.K.); chang.hyeyoun@gmail.com (H.C.); ghnam@shiftbio.net (G.N.); YoungJ_Ko@dfci.harvard.edu (Y.K.)

[2] Center for Theragnosis, Biomedical Research Institute, Korea Institute of Science and Technology (KIST), Seoul 02792, Korea; sunkim@kist.re.kr

* Correspondence: Thomas_Roberts@dfci.harvard.edu (T.M.R.); jhryu@kist.re.kr (J.H.R.); Tel.: +1-617-632-3049 (T.M.R.); +82-2-958-5942 (J.H.R.)

Abstract: Pancreatic cancer is one of the most lethal forms of cancer, predicted to be the second leading cause of cancer-associated death by 2025. Despite intensive research for effective treatment strategies and novel anticancer drugs over the past decade, the overall patient survival rate remains low. RNA interference (RNAi) is capable of interfering with expression of specific genes and has emerged as a promising approach for pancreatic cancer because genetic aberrations and dysregulated signaling are the drivers for tumor formation and the stromal barrier to conventional therapy. Despite its therapeutic potential, RNA-based drugs have remaining hurdles such as poor tumor delivery and susceptibility to serum degradation, which could be overcome with the incorporation of nanocarriers for clinical applications. Here we summarize the use of small interfering RNA (siRNA) and microRNA (miRNA) in pancreatic cancer therapy in preclinical reports with approaches for targeting either the tumor or tumor microenvironment (TME) using various types of nanocarriers. In these studies, inhibition of oncogene expression and induction of a tumor suppressive response in cancer cells and surrounding immune cells in TME exhibited a strong anticancer effect in pancreatic cancer models. The review discusses the remaining challenges and prospective strategies suggesting the potential of RNAi-based therapeutics for pancreatic cancer.

Keywords: RNA interference; siRNA; miRNA; pancreatic cancer; nanocarrier

Citation: Kim, M.J.; Chang, H.; Nam, G.; Ko, Y.; Kim, S.H.; Roberts, T.M.; Ryu, J.H. RNAi-Based Approaches for Pancreatic Cancer Therapy. *Pharmaceutics* **2021**, *13*, 1638. https://doi.org/10.3390/pharmaceutics13101638

Academic Editors: Sophia Hatziantoniou

Received: 20 August 2021
Accepted: 27 September 2021
Published: 8 October 2021

1. Introduction

Pancreatic cancer is one of the most aggressive malignancies, ranking as the fourth leading cause of cancer-related deaths with a five-year survival rate of 8% [1]. If outcomes are not improved, pancreatic cancer is predicted to be the second leading cause of cancer-related mortality by 2025 [2]. With a minimal increase in patient survival rate in past four decades, pancreatic cancer is often discovered and diagnosed at advanced stage where metastasis have already taken asymptomatically, and limited treatment options are left [3]. Because of aggressive perineural and vascular local growth, chemotherapy, radiotherapy, and molecularly target therapies are better treatment options than surgical resection. Typically, chemotherapy elongates the life span only by 8 to 16 weeks, and there is an urgent need for new therapeutic options for pancreatic cancer patients [4]. In particular, pancreatic cancer is characterized both by complicated genetic mutations and epigenetic alterations and by development of a dense tumor microenvironment that together result in a low survival rate of just ~15–25% of patients who undergo surgical resection [5]. One of the unique histological features of pancreatic cancer is a dense fibrotic stroma surrounding the tumor that comprises around 80% of the whole tumor mass [6]. In addition to the early metastasis and robustly immunosuppressive TME, this protective stromal barrier and distorted tumor blood vessels contribute greatly to poor survival by increasing the difficulty of drug delivery and distribution [7–10]. In addition, the abnormal vasculature induces hypoxic environment in TME which contributes to resistance towards chemotherapy and immunotherapy [11,12]. Desmoplastic features with an abundance extracellular matrix are known to be developed by the cancer-associated pancreatic stellate cells (PSCs) in communication with the surrounding cells in the TME [13]. The PSCs that are normally quiescent in the healthy pancreas are recruited upon cancer cell communication to fuel the aggressive proliferation of tumors [14–16]. Given the clear importance of TME in tumorigenesis, approaches targeting specific features within TME have been investigated in addition to the focus on directly targeting tumor cells.

In recent decades, studies have delineated the genetic mutations unique to pancreatic cancer. For example, 94% of pancreatic tumors feature mutations along with mutations in the *TP53* (64%), *SMAD4* (21%), and *CDKN2A* (17%) tumor suppressor genes [17]. The largest challenge with the genetic aberrations of pancreatic cancer is that the *KRAS* oncogene has been 'undruggable' using the now conventional treatment approach of targeted chemical inhibitors [18]. Undruggable targets represent proteins that is pharmacologically incapable of being targeted due to elusive factors from molecular shapes and mechanism of action. For this reason, the potential to selectively inhibit undruggable genes with RNAi represents a promising strategy to halt pancreatic cancer progression. RNAi is a post-transcriptional mechanism that involves the inhibition of specific gene expression through promoting cleavage of the relevant messenger RNA (mRNA). After decades of investigation, in 2018 the first RNAi-based drug, Patisiran (ONPATTRO®) finally proved the potential of siRNA therapeutics as demonstrated by FDA approval and market entry [19]. In the following years, Givosiran (GIVLAARI®) and Lumasiran (OXLUMO®) were FDA approved in 2019 and 2020, embarking the era of RNAi-based oligo therapeutics. Thus, RNAi-based technology has presented an opportunity to regulate a set of tumor-promoting genes in cancer cells and other targets in TME. However, the clinical application of RNAi therapeutics is still hindered by poor delivery to the target organs and inefficient internalization into target cells. In this review, we discuss the prospects and challenges in RNAi-based approaches using various strategies of nanomedicine to target key genes in cancer cells or TME of pancreatic cancer.

2. RNAi Targets for Pancreatic Cancer Therapy

RNAi is a regulatory mechanism using small regulatory RNAs to prevent translation of mRNA and regulate protein expression levels. As an intrinsic mechanism for post-transcription editing, RNAi comprises two major different types of RNA molecules, siRNA, and miRNA, generated via similar mechanisms involving the dicer protein (Figure 1). Both siRNA and miRNA are short non-coding RNA strands that can silence the target mRNA in sequence-specific manner, therefore offering the advantage of greater selectivity and expanding the inhibitory action for undruggable targets. Despite numerous challenges involving delivery, safety, and potency, RNAi-based drugs have shown therapeutic potential in rare genetic diseases and complex diseases such as cancer [20–24].

siRNA is a 20 to 25 base pairs double-stranded non-coding RNA that plays a key role in RNAi machinery. Once internalized in a cellular system, the siRNA forms an activated RNA-induced silencing complex to guide pairing with a target mRNA of complementary sequence resulting in cleavage and degradation [25,26]. The ability of siRNA to silence specific gene expression involved in diseases has led to a significant effort to use it as an alternative treatment option. Despite the breakthroughs in the generation of siRNA therapeutics beyond preclinical studies, siRNA delivery into cells at the disease site has remained the major hurdle in various clinical applications. The challenges include the negatively charged nature of siRNA, immediate degradation by the serum proteins, and fast clearance from the reticuloendothelial system [27]. Furthermore, the dense fibrotic stroma and vascular barriers of pancreatic cancer add an extra layer of difficulty in successfully delivering siRNA to the tumor and surrounding cells in TME. We will use this forum to introduce siRNA-based approaches to pancreatic cancer therapy using various reagents and strategies.

miRNAs are 19–25 nucleotides long, endogenous non-coding RNAs found in living organisms as one of the key players in the RNAi machinery as shown in Figure 1 [28]. Similar to siRNA, miRNA interfere with the expression of specific proteins by binding to the mRNA strand. miRNA is a long bimolecular RNA duplex that partially binds to the untranslated region of multiple genes to interfere with the post-transcriptional expression. Contrary to the direct and specific inhibition of siRNA, miRNA often hinders the translation of several mRNAs via closely related binding sites in the 3′ regions of target mRNAs [29]. Compelling evidence has demonstrated that miRNA is involved in the cellular machinery and signaling contributing to the dysregulation of cancer cell growth [30]. miRNA may function as oncogenes or tumor suppressors interfering with a set of specific protein expression. The dysregulation of miRNA is known to be prevalent across cancer types, contributing to sustaining proliferative signaling, evading growth suppressors, resisting cell death, activating invasion, and inducing metastasis and angiogenesis. For this reason, miRNA-based therapeutics that inhibit the levels of oncogenic signaling or elevate tumor suppressor functionality have enormous potential [31].

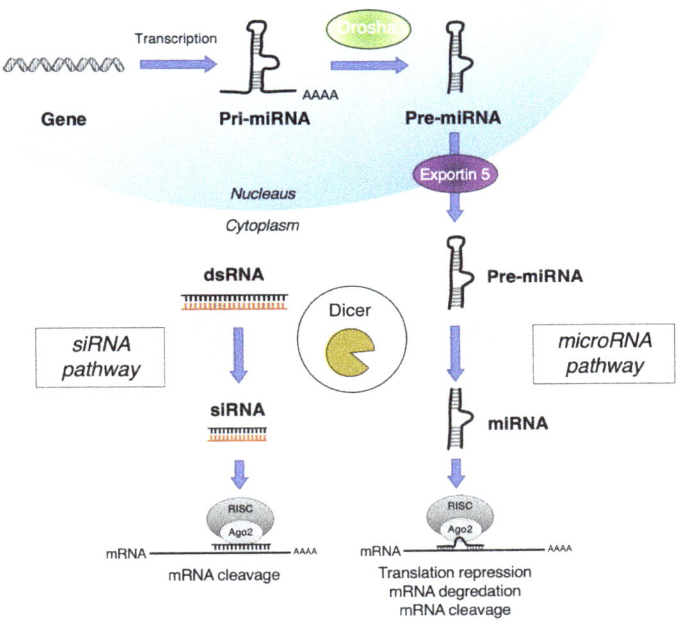

Figure 1. Gene silencing mechanism of siRNA and miRNA of RNAi.

2.1. *Pancreatic Tumor Targets*

Conventional treatment approaches are focused on the eradication of pancreatic cancer using chemotherapies. The potency of a of given procedure is related to its ability to effectively remove cancerous cells; however, many therapies often harm the normal organs causing major side effects in patients [32]. To overcome these drawbacks, targeted therapies have been advanced as the next generation of cancer therapies. The carcinogenesis of pancreatic ductal adenocarcinoma (PDAC), comprising 90% of the pancreatic cancers, involves progressive accumulation of key driver mutations, usually leading to the activation of the KRAS oncogene and loss of the tumor suppressor gene TP53, contributing to the histological evolution of pancreatic intraepithelial neoplasia to PDAC [33–35]. These genetic aberrations induce invasive adenocarcinoma and remodel the surrounding stromal structure into one more favorable for tumor formation. Because of the unique stromal feature and frequent oncogenic mutations, wide range of pancreatic cancer models exist for investigation for varying purposes as summarized in Table 1.

Often, the genes discovered to be the novel cancer-related oncogenes that promote tumor progression are found to be 'undruggable,' since they are difficult to inhibit using small drug molecules [36]. Because of lack of suitable ligand binding sites or the presence of large protein-protein interaction interfaces, various oncogenes are relatively intractable using small-molecule inhibitors [37]. A notable undruggable target, mutationally activated KRAS, is present in >90% of PDAC and represents the most frequent (>90%) and the earliest genetic alternation, found even in low-grade pancreatic intraepithelial neoplasia 1A lesions [38,39]. Moreover, KRAS activation is also a poor prognostic factor resulting resistance to traditional treatments [5,39,40]. Genetically engineered mouse models also support the initiating role of KRAS mutation in PDAC and its involvement in progression to the malignancy [41,42]. When coupled with mutational inactivation of TP53 (TP53R175H), CDKN2A, or SMAD4, pancreatic intraepithelial neoplasia formation rapidly evolves into high-frequency occurrence of metastatic PDAC [41,43,44]. In recent decades, KRAS has been an intractable challenge as the therapeutic target with no targeted inhibitors. Recently a breakthrough was made with the generation of $KRAS^{G12C}$ inhibitors which have proved effective in non-small cell lung cancer patients in phase I clinical trials; however, only 2% of the PDAC carry $KRAS^{G12C}$ specific mutations rather than the common $KRAS^{G12D}$ mutation. Thus, while $KRAS^{G12C}$ inhibitors are a great proof of concept, their existence only increases the urgency to develop alternative KRAS-targeting moieties such as RNAi [45]. In addition to the prominent undruggable KRAS mutation, other aberrations in gene expression in PDAC also emphasize the need for RNAi mediated approach for therapy. Table 2 summarizes the reported siRNA therapeutics used in pancreatic cancer therapy by targeting various 'undruggable' genes. Most importantly, the incorporation of delivery agents such as nanoparticles and extracellular vesicles have facilitated the delivery of siRNA to cancer cells.

For miRNA, there are numerous evident data from in vitro models as well as from PDAC patient samples proving dysregulated levels of miRNA in pancreatic cancer [46]. From analyzing 95 miRNAs expression levels in pancreatic cancer tissues and cell lines, eight miRNAs—miR-196a, miR-190, miR-186, miR-221, miR-222, miR-200b, miR-15b, and miR-95—were significantly up-regulated in most pancreatic cancer samples. On the other hand, miR-148a, miR-217, miR-34a and miR-375 tumor suppressor miRNAs are frequently down-regulated in PDAC [47,48]. Over the past decade, extensive research efforts have been dedicated to investigating the underlying mechanism of chemoresistance. Along with dysregulation of various genes involved in key signaling pathways, such as phosphoinositide-3-kinase (PI3K), NF-κB, and hedgehog, miRNAs have also been demonstrated to modulate the expression associated with chemoresistance in PDAC [49]. For this reason, miRNA has gained attention as a useful biomarker for PDAC prognosis and as a target for therapeutic applications. Table 3 summarizes the miRNA-based approaches for pancreatic cancer therapy now in preclinical studies.

2.2. Targets in the Pancreatic Tumor Stroma and Immunosuppressive Microenvironment

The prominent stromal structure of PDAC consists of several cellular and acellular components, which together play a critical part in disease progression [83–85]. Although substantial research studies have focused on developing small-molecule or biological inhibitors directly targeting the cancer cells interest has more recently turned to agents targeting the tumor microenvironment. The key players of PDAC microenvironment are PSCs and cancer-associated fibroblasts responsible for the desmoplasia as well as various immune cells including immunosuppressive myeloid-derived suppressor cells, tumor-associated macrophages, and regulatory T cells [86]. The unique stromal barrier is often the cause of drug resistance while tumor induced alterations of the stroma have been suggested to facilitate tumor progression and proliferation. Thus, the concept of stomal reprogramming holds considerable promise [87].

Table 1. Various models of pancreatic cancer for preclinical studies.

	In Vitro		In Vivo		
Models	Cell Line	Patient-Derived Organoid	Cell Line Xenograft	Patient-Derived Xenograft	Genetically Modified Mouse Model (GEMM)
TME	-	+++	+	++	++
Immune system	-	++	+	+	+++
Pros	• Fast and easy growth • Rapid drug screening	• Mimic complex TME and stroma • Orthotopically transplantable	• Useful for testing drug efficacy and safety	• Mirror patient response to drug • Personalized drug regimen	• Useful tool for oncogenic mutation investigation and biomarker discovery
Cons	• Genetically uniform	• New model and needs further analysis	• Limited TME and immune system	• Engraftment difficulties • Long duration of growth • Selection for aggressive tumors	• Species discrepancy • High cost and complexity of GEMM generation

Parameters are appreciated as abundant (+++), moderate (++), minimal (+) and none (-).

Table 2. siRNA-based therapeutics in preclinical pancreatic cancer models.

Category	Delivery Vehicle	siRNA Target	Tumor Model	Drug Route	Combination Therapy	Reference
Lipid-based Nanoparticles	Lipid nanoparticle (LNPK15)	KRAS	MIA PaCa-2 s.c.	I.V.		[50]
	Lipoplex	KRAS	PANC1 s.c.	I.V.		[51]
	Lipoplex (Atu027)	PKN3	DanG orthotopic	I.V.		[52]
	Liposome	KRAS	PANC-1 s.c.	I.V.	Gemcitabine	[53]
Polymer-based Nanoparticles	Gold nanocluster siRNA (GNC-siRNA)	Nerve growth factor	PANC-1 s.c., orthotopic and PDX	I.T.		[54]
	Superparamagnetic iron oxide nanoparticles (siPLK1-StAv-SPIONs)	PLK1	6606PDA orthotopic [55]	I.V.		[56]
	Star polymeric nanoparticles different lengths of cationic PDMAEMA side-arms and varied amounts of POEGMA	βIII-tubulin	MiaPaCa-2 and HPAF-II orthotopic	I.T.		[57,58]
	Poly(ethylene glycol) and charge-conversional polymer (PEG-CCP)	VEGF	L1-Luc/TAg transgenic [59]	I.V.		[60,61]
	Local Drug EluteR, LODER (PLGA)	KRAS	PANC1-Luc or Capan1-Luc s.c., synograft, and orthotopic	I.T.		[62]
	PLGA/poloxamer	EPAS1	BxPC3 s.c.	I.T.		[63]
	Cholesterol-modified polymeric CXCR4 antagonist (PCX) nanoparticles	NCOA3	CD18/HPAF orthotopic	I.V.		[64]
	Cholesterol-modified polymeric CXCR4 antagonist (PCX) nanoparticles	KRAS	KPC8060 orthotopic	I.V. and I.P.		[65]
	BCPV	KRAS	MiaPaCa-2 s.c.	Peritumoral		[66]
	Folic acid (FA)-modified PEG-chitosan oligosaccharide lactate (COL) nanoparticles	ARHGEF4, CCDC88A, LAMTOR2, mTOR, NUP85, and WASF2	S2-013 orthotopic	I.V.		[67]
	PEGylated iRGD-guided tumor-penetrating nanocomplexes (TPN)	KRAS	KP D8-175 orthotopic from Pdx1-Cre; Krasþ/LSL-G12D; Trp53fl/ (KPC) mouse	I.V.		[68]
	poly(ethylene glycol)-block-poly-L-lysine (PEG-PLL)	KRAS	PANC-1 (mutant KrasGGT_GAT), BXPC-3 (KrasWT) s.c.	I.V.	Arsenic therapy	[69]
	Peptide-conjugated PSPG (PSPGP)	TR3	PANC-1 s.c.	I.V.	Paclitaxel	[70]
	Magnetic nanocarrier	PD-L1	PANC-02 syngeneic	I.V.	Gemcitabine	[71]

Table 2. *Cont.*

Category	Delivery Vehicle	siRNA Target	Tumor Model	Drug Route	Combination Therapy	Reference
Extracellular vesicle	Exosome	KRAS	PANC-1 or BxPC-3/1) PANC-1, BxPC-3, or KPC689 orthotopic tumor mice model 2) KTC (Ptf1acre/+;LSL-Kras^{G12D}/+;Tgfbr2lox/lox) genetically engineered mouse	I.P.		[72]
	Extracellular vesicle	Galectin-9	PANC-02 orthotopic	I.V.	Oxaliplatin	[73]

Abbreviations: PDMAEMA, Poly[2 (Dimethylamino)ethyl methacrylate]; POEGMA, Polyoligo(ethylene glycol) methyl ether methacrylate; PLGA, Poly lactic-co-glycolic acid; PSPG, G2 dendrimers through disulfide linkages; PEG, polyethylene glycol; RGD, arginylglycylaspartic acid; CXCR4, C-X-C chemokine receptor type 4; BCPV, Biodegradable charged polyester-based vector; PKN3, protein kinase N3; NGF, nerve growth factor; PLK1, polo-like kinase 1; VEGF, vascular endothelial growth factor; EPAS1, endothelial PAS domain protein 1; NCOA3, nuclear receptor coactivator 3; ARHGEF4, rho guanine nucleotide exchange factor 4; CCDC88A, coiled-coil domain containing 88a; LAMTOR2, late endosomal/lysosomal adaptor, MAPK, and mTOR activator 2; mTOR, mechanistic target of rapamycin; NUP85, nucleoporin 85; WASF2, Wiskott–Aldrich syndrome protein family member 2; PD-L1, programed death-ligand 1; s.c., subcutaneous; PDX, patient-derived xenograft; I.V., intravenous; I.P., intraperitoneal; I.T., intratumoral.

There have been numerous siRNA targets in TME and stromal or immune cells to seek therapeutic effect for pancreatic cancer. One of the unique drivers of the dense stroma in PDAC are the activated PSCs. The PSCs are normally in a quiescent state functioning as a storage of vitamin A-rich lipid droplets [88]. During the initial development of PDAC, the PSCs are switched to an activated state losing the ability to produce lipid droplets and instead, producing an excessive amount of extracellular matrix components and proinflammatory chemokines [89]. The extracellular matrix production disrupts the normal parenchyma, leading to hypovascularity, and hypoxic conditions, which in turn further activates PSCs maintaining a hypoxia-fibrosis cycle [90]. Toll-like receptors are proteins that recognize foreign pathogen-associated molecular patterns and play a key role in the innate immune system. Recently, it has been shown that PSCs express a variety of toll-like receptors contributing to the activation of proinflammatory signaling pathways [91,92]. In particular, the toll-like receptor 4 is heavily involved in the stromal development in pancreatic cancer and has been advanced as a therapeutic target in PSCs [92]. Based on this, vitamin A-coupled liposomes carrying toll-like receptor 4-silencing small hairpin RNAs were used to target PSCs, reducing fibrosis by inducing mitochondrial apoptosis [93]. Ishiwatari et al. similarly used vitamin A-coupled liposomes (VA-lip) containing siRNA to induce collagen secretion by activated PSCs [94]. The siRNA targeting the collagen specific chaperone protein gp46 in VA-lip successfully inhibited the secretion of collagen and tissue collagenases, which dissolves the existing collagen matrix. Furthermore, Han et al. used a TME-responsive nanosystem to re-educate the PSCs to return to a quiescent state [95]. The nanosystem consists of polyethylene glycol-modified polyethyleneimine-coated gold nanoparticles to co-deliver all-*trans*-retinoic acid and siRNA targeting heat shock protein 47 (hsp47). The combination strategy successfully restored homeostatic stromal function and improved the efficacy of chemotherapy in stroma-rich tumor Another molecular target in the TME is the tumor-associated macrophages, of which depletion or reprogramming can induce immunogenic attack of the tumor cells [96]. Although various targets have been explored to reprogram or deplete tumor-associated macrophages, inhibition of the colony stimulating factor 1 (CSF1) or C-C Motif Chemokine Ligand 2 (CCL2) signaling have been particularly reported [97,98]. As one of the most abundant tumor-infiltrating leukocytes in all types of cancer, tumor-associated macrophages tend to stay in an alternative M2 state rather than the classically activated M1 state [99]. The M2 phase macrophage exhibit numerous tumor-promoting properties such as angiogenic signaling and restraining adaptive immune responses arising from the interactions of dendritic cells and CD^{8+} T cells. To deplete M2 macrophage, Qian et al. developed a dual-targeting nanoparticle (M2NPs), whose structure and function are controlled by two targeting moieties, an α-peptide coupled with an M2 macrophage binding protein for extremely specific targeting of M2 macrophage [100]. Upon loading of an anti-colony stimulating factor-1 receptor targeted siRNA, M2NPs showed higher affinity to M2-like tumor-associated macrophages than to tissue-resident macrophages resulting in the elimination of M2 macrophage, thus decreasing tumor size in cancer bearing mice. In a later study, colony stimulating factor-1 receptor silencing siRNA was co-delivered with PI3K inhibitor, BEZ-235 using an M2 tumor-associated macrophage targeting nanomicelle to elicit therapeutic immune responses against pancreatic cancer cells both in vitro and in tumor models [101]. The combination treatment switched the macrophage state from M2 to M1, thus, reducing tumor infiltration of myeloid-derived suppressor cells. The reprogramming of M2 tumor-associated macrophage activated antitumor immune responses and enhanced anti-pancreatic tumor effects in a Pan-02 pancreatic xenograft model.

Table 3. miRNAs used for preclinical pancreatic cancer therapy.

miRNA	Target	PC Cell Line/Model	Combination Therapy	Reference
miR-150	MUC4 and HER2	Colo-357 and HPAF cells		[74]
miR-155	SOD2, CAT, and DCK	MiaPaCa and Colo-357 cells	Gemcitabine	[75]
miR-199a ASO	RPS18, Acta-2, Collagen1α1, PDGFR-β, and mTOR	hPSCs		[76]
miR-21 and miR-221 ASO	CDK6, IRAK3, NRP1, SMAD7, SOCS6, C5ORF41, KLF12, MAPK10, EFNA1	L3.6plGres-SP orthotopic		[77]
miR-21 ASO	PDCD4 and PTEN	MIA PaCa-2 s.c.	Gemcitabine	[78]
miR-212	USP9X	PDX	Doxorubicin	[79]
miR-345	SHH, Gli-1, MUC4, and Ki67	Capan-1 and CD18/HDAF s.c.	Gemcitabine	[80]
miR-34a	E2F3, Bcl-2, c-myc, and cyclin D1	PANC-1 s.c.		[81]
miR-34a and miR-143/145	SIRT1, CD44, aldehyde dehydrogenase, KRAS2, and RREB1	MiaPaCa-2 s.c. and orthotopic		[82]

Abbreviations: miR, microRNA; ASO, antisense oligonucleotide; MUC4, mucin 4; HER2, human epidermal growth factor receptor 2; SOD2, superoxide dismutase 2; CAT, catalase; DCK, deoxycytidine kinase; RPS18, ribosomal protein S18; Acta-2, actin alpha 2; PDGFR, platelet-derived growth factor receptor; mTOR, mechanistic target of rapamycin; CDK6, cell division protein kinase 6; IRAK3, interleukin 1 receptor associated kinase 3; NRP1, neuropilin 1; SMAD7, SMAD family member 7; SOCS6, suppressor of cytokine signaling 6; C5ORF41, chromosome 5 open reading frame 41; KLF12, Krueppel-like factor 12; MAPK10, mitogen-activated protein kinase 10; EFNA1, Ephrin-A1 precursor; PDCD4, programed cell death protein 4; PTEN, phosphatase, and tensin homolog; USP9X, ubiquitin specific peptidase 9 x-linked; SHH, sonic hedgehog; Bcl-2, B-cell lymphoma 2; SIRT1, sirtuin 1; RREB1, ras responsive element binding protein 1; s.c., subcutaneous; PDX, patient-derived xenograft.

3. RNAi Delivery Strategies for Pancreatic Cancer Therapy

3.1. Nanocarrier-Mediated RNAi Therapy

Over the last several decades of nanotechnology advancement, several applications and products containing nanomaterials have become available in the field of medicine [102]. Nanotechnology for medical purposes has been termed nanomedicine and is defined as the use of nanomaterials for diagnosis, monitoring, prevention, and treatment of diseases [103]. Nanomaterials can be applied in nanomedicine largely in three areas: diagnosis, drug delivery, and regenerative medicine. As discussed earlier, the direct injection of naked, unmodified siRNA was not found to be effective as siRNA will faces challenges of RNA degradation, short half-life, and circulation time, and poor biodistribution on systemic injection. Direct chemical modifications on the 2′ carbon of the RNA ribose ring have been shown to prolong the half-life of siRNA without altering its functionality [104]. However, the modified siRNA has still shown discouraging results in clinical trials due to poor delivery to target tissues. For this reason, nanocarrier-mediated strategies have emerged as tool for therapy by acting as a protective delivery agent for siRNA. To provide the best opportunity to penetrate and accumulate within the cancer cells of the tumor, the size of delivery agent should range from 10 to 200 nm [105]. This size enables the nanoparticle to passively localize in the tumor sites via 'enhanced permeability and retention effect' (EPR effect) which arises from the leaky tumor vascular structure [106] (Figure 2). Nanoparticles that are over 10 nm in size are too large to penetrate healthy vessels and tight-gap junctions [107]. Several passively targeted nanomedicines—including Doxil™, Abraxane™, Marqibo™, DaunoXome™, and Onivyde™—are currently in clinical use, suggesting that the EPR effect can drive therapeutic outcomes [108]. Although EPR effect is the underlying mechanism for nanomedicine, active targeting of nanoparticles with ligands and modalities are also being used as strategies to increase the target ability of RNAi therapeutics. Active targeting can significantly increase the quantity of drug delivered to the target cell compared to free drug or passively target nanosystems [109]. Ligands that target transferrin receptors or nicotinic acetylcholine receptors or even antibodies greatly enhance targeting of nanomedicine. In addition to nanoparticles, naturally occurring extracellular vesicles have also gained attention as a possible RNAi delivery agent also capable of biological communication with the tumor by harnessing cytokines and signaling proteins. Here, we will discuss some recent studies have demonstrated the potential of siRNA therapeutics using nanoparticles and extracellular vesicles for effective siRNA delivery and release into the tumor cell.

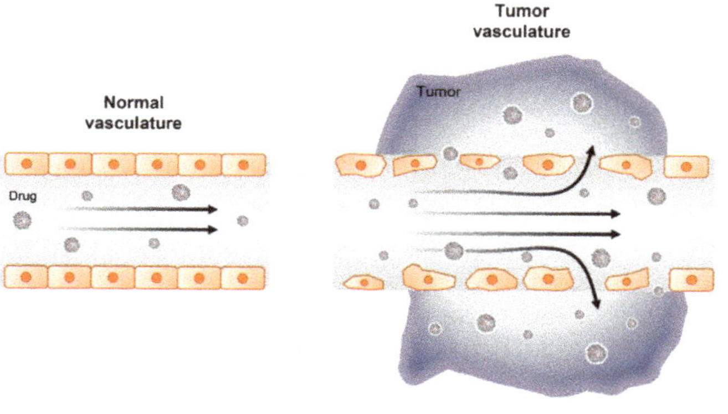

Figure 2. Drug delivery in vasculature of normal and tumor tissue.

3.1.1. Lipid-Based Nanoparticle Delivery

Lipid-based nanoparticles are siRNA delivery agents consisting of pH-sensitive cationic lipid components. The lipids are amphiphilic and capable of self-assembly into micelles or liposomes in conjunction with siRNA via electronic charge interactions under acidic conditions [110]. Lipid-based nanoparticles such as liposomes are a suitable carrier for both small-molecule drug and nucleic acid delivery because of their excellent biocompatibility, biodegradability, low toxicity and immunity, structural flexibility, and ease of mass-production [111]. Decades after the discovery of RNAi, Patisiran (ONTATTRO™), which is a lipid-based nanoparticle formulated with siRNA targeting transthyretin, was FDA approved in 2018 as the first siRNA drug for the treatment of transthyretin-type amyloid polyneuropathy [112]. Because *KRAS* is so frequently mutated and active in cancer, especially in pancreatic cancer, *KRAS* is often the molecular target of siRNA therapeutics.

In 2019, Sasayama et al. generated a lipid nanoparticle formulation (LNPK15) to deliver KRAS-targeting siRNA to MIA PaCa-2 cells in vitro and in tumors in vivo achieving food protein knock down suggesting antitumor efficacy [50]. The group developed a PEGylated lipid-based nanoparticle named LNPK15 using two novel cationic lipids, SST-01 and SST-31, encapsulating KRAS siRNA that showed long circulation and efficient siRNA delivery to tumor sites. The *KRAS*-targeting LNPK15 showed strong antitumor efficacy in tumor xenograft mice of MIA PaCa-2 pancreatic cancer although the in vitro knock down activity of lipid-based nanoparticle was rather weak compared to its effect in tumors.

Although PEGylation with 3.3% 1,2-distearoyl-sn-glycero-3-phosphatidylethanolamine-N-(polyethylene glycol-2000) (PEG-DSPE) provides benefit in terms of long durations, it could be disadvantageous for intracellular delivery and endosomal escape of siRNA, which explains the lower activity in cell lines. The in vitro activity was rescued with lipase treatment for lipid nanoparticle hydrolysis for enhanced knock down of *KRAS*.

Rao et al. have demonstrated *KRAS* down-regulation by a lipoplex encoding a short hairpin RNA via systemic injection in a pancreatic cancer model [51]. The lipoplex contained a plasmid encoding a short hairpin RNA that executes selective silencing of the mutated KRAS gene with tumor cell entry facilitated by facilitated by fusogenic liposomes. The bifunctional short hairpin RNA design is designed for more rapid silencing, higher efficacy, and greater durability of siRNA action compared to conventional short hairpin RNA [113]. The lipoplex containing the *KRAS*-targeting bi-short hairpin RNA was systemically injected in female athymic Nu/Nu mice bearing PANC1 xenografts and showed successful delivery. In addition to the *KRAS* mutation as therapeutic target for RNAi, other genes such as protein kinase N3 have been investigated as targets for siRNA in pancreatic cancer. Kaufmann's group generated a liposomal formulation Atu027) for protein kinase N3 siRNA delivery from neutral fusogenic and polyethylene glycol-modified lipid components (for improved pharmacokinetic properties, cellular uptake, and efficient siRNA release in cells [52]. The siRNA-lipoplexes were capable of accumulating in tumor endothelial cells in vivo and halted tumor growth in two different xenograft mouse models [114]. As protein kinase N3 has previous been identified as a downstream effector of the PI3K signaling pathway, reduction of protein kinase N3 expression with Atu027 resulted in changes in tumor lymphatic vasculature, suggesting an anti-lymph angiogenesis therapeutic strategy for pancreatic cancer [115].

In addition to the demonstration of therapeutic potential of siRNA, miRNA has also been investigated in pancreatic cancer therapy. Pramanik et al. selected miR-34a and miR-143/145 for PDAC therapy based on their involvement in cancer stem cell survival and expression of KRAS and effectors. The miRNAs were formulated in lipid-based nanovector and intravenously injected in MiaPaCa-2 subcutaneous and orthotopic models where they effected tumor growth inhibition and increased apoptosis [82]. The success of miRNA-based therapeutics in preclinical studies has led some candidates to enter human clinical trials for cancer and other diseases [116–118]. The first ever miRNA-based clinical trial was with the miR-122 targeting called drug Miravirsen, which features the use of an LNA-based miRNA inhibitor, and has gone on to Phase II with for Hepatitis C virus infected liver cells [119]. Another miRNA mimicking drug MRX34, a lipid-based nanoparticle, was tested in clinical trials to treat advanced solid tumors [120]. However, the clinical trials involving miRNA-based drugs have largely been terminated because of severe adverse events arising from immune responses after treatments [121]. The two major discontinued miRNA-drug clinical trial results indicated that more than 30 genes in immunogenic and cytokine signaling pathways were identified to be affected by the miRNA machinery [122]. This results in the immune-related adverse events in phase I studies of miRNA-based drug clinical trials.

3.1.2. Metal Nanoparticle USE for Delivery

For centuries, the noble metal gold has been studied first as a bulk material and now impacting nanomedicine [123]. In particular, gold nanoparticles (AuNPs) have shown important roles in the vaccine field as both an adjuvant and a carrier, by reducing toxicity, and increasing stability for storage [124]. Outside of vaccine applications, Lei et al. have shown that AuNPs are capable of siRNA delivery in vivo as nanocarriers for pancreatic tumor models [54]. As the nervous microenvironment has been recognized as a novel niche for PDAC progression and metastasis, nerve growth factor (NGF) and brain derived growth factor (BDGF) are rising targets for pancreatic cancer [125]. The gold nanocluster-assisted delivery of NGF targeting siRNA (GNC-siRNA) allowed efficient NGF gene silencing and pancreatic tumor treatment in subcutaneous, orthotopic, and patient-derived xenograft models. To add the biomimetic property, the AuNP nanoflowers (Au@BSA) were coated with bovine serum albumin (BSA) and conjugated with survivin-targeting siRNA for efficient delivery and inhibition of gene expression in BxPC-3 pancreatic cancer cells [126]. Other than AuNPs, iron oxide nanoparticles have also gained attention as a stable agent for thermal and immunotherapies for cancer [127]. Mahajan et al. developed superparamagnetic iron oxide nanoparticles coupled with siRNA directed against polo-like kinase-1 (PLK1), serving a dual purpose by both delivering siRNA to the tumor and facilitating non-invasive assessment of targeting efficiency via molecular imaging [56]. With uMUC1-specific ligand, EPPT1 and the myristoylated polyarginine peptide ligand and dextran, superparamagnetic iron oxide nanoparticle uptake was specific to pancreatic cancer and induced a drastic decrease in tumor volume in a syngeneic orthotopic model driven by PLK1 silencing. In addition to published reports of siRNA delivery using AuNPs, other inorganic nanoparticles such as graphene oxide and carbon nanotubes use were demonstrated in pancreatic cancer cells and models. Feng et al. developed a multi-functionalized monolayer graphene oxide as delivery agent for HDAC1 and KRAS siRNA in MIA PaCa-2 pancreatic cancer cells [128]. The dual gene silencing induced apoptosis, proliferation inhibition and cell cycle arrest and when treated in combination with near infrared light thermotherapy showed tumor volume growth inhibition for pancreatic in vivo models. Furthermore, Anderson et al. formulated a functionalized single walled carbon nanotubes as a siRNA carrier for in vitro gene therapy [129]. With KRAS siRNA in the carbon nanotube, PANC-1 cells showed great knockdown of KRAS gene expression.

3.1.3. Polymer-Based Nanoparticle Delivery

McCarroll et al. has developed a star-shaped polymeric nanoparticle for delivery of siRNA into pancreatic cancer cells in vitro and in vivo. The star polymers are cost-effective nanoparticles that self-assemble with siRNA to produce a small (<50 nm) delivery system [58]. These star polymers, containing different lengths of cationic poly(dimethylaminoethyl methacrylate) side-arms and varied amounts of poly[oligo(ethylene glycol) methyl ether methacrylate], delivered siRNA targeting βIII-tubulin to orthotopic pancreatic tumors [57]. Using this approach βIII-tubulin expression was silenced by 80% and tumor growth progression was delayed when star polymers were administered in orthotopic models. Often, polyethylene glycol coating offers nanoparticles increased compatibility with aqueous environments. prolonging circulation time and reducing immunogenicity.

However, there has been biosafety issues from PEG recognizing antibodies contributing to immune responses and mitigation of therapeutic efficacy of PEGylated drugs. Despite these concerns, these nanoparticles have shown improved efficacy with polyethylene glycol fabrication when systemically administered [130]. Furthermore, Kataoka's group have developed a polyethylene glycol coated calcium phosphate hybrid micelle system for siRNA delivery in pancreatic cancer [60,61]. First, the PEG-*b*-charge-conversional polymer/calcium phosphate hybrid micelles were systemically administered to EL1-Luc/TAg transgenic mice carrying luciferase targeting siRNA to ensure RNAi activity. As vascular endothelial growth factor (VEGF) is a key pro-angiogenic molecule in tumor progression, VEGF specific siRNA was then therapeutically administered with the help of the hybrid micelle. After systemic administration of purified nanoparticles, the reduction of expression level VEGF mRNA was evaluated in tumor tissues and found to be correlated with the pancreatic tumor growth rates [60]. Because of major issues concerning toxicity, safety, and clearance of inorganic nanoparticles, biodegradable and biocompatible organic materials have been sought out as a nanocarrier option for various drugs [131]. Poly (lactic-co-glycolic acid) (PLGA) is one of the most investigated biodegradable polymeric nanoparticles, and already has won FDA approval for medical applications [132]. Zorde Khvalevsky et al. synthesized a local prolonged siRNA delivery system, siG12D-LODER™, with PLGA for protection and gradual release of drug over time [62]. An siRNA against the mutated $KRAS^{G12D}$ was encapsulated in LODER and implanted in a pancreatic xenograft and an orthotopic model, leading to decreased KRAS levels, as well as inhibition of cell proliferation and epithelial-mesenchymal transition over 155-day period. The group later injected the siRNA encapsulated LODER into tumors of locally advanced pancreatic cancer patients in combination with chemotherapy in a clinical trial study. Pan et al. also designed a PLGA/poloxamer nanoparticle loaded with EPAS1 siRNA and tested it on BxPC-3 pancreatic cancer cells in a mouse model of pancreatic cancer [133]. The endothelial PAS domain protein, also known as hypoxia-inducible factor-2, is a transcription factor known to regulate tumor cell adaption to hypoxic microenvironments and to accelerate metastatic spread [63]. The study demonstrated that the PLGA/poloxamer nanoparticle successfully inhibited VEGF and CD34 expression, therefore halting the growth of pancreatic tumors.

Oupicky's group developed a different type of polyplex for nucleic acid delivery using cholesterol-modified polymeric CXCR4 antagonist nanoparticles to improve pancreatic cancer therapy [64,65]. The nuclear receptor coactivator-3 (NCOA3) is a critical modulator of musin expression that is critical in pancreatic cancer progression. The siRNA targeting NCOA3 was formulated along with a cholesterol-modified polymeric CXCR4 antagonist in a dual-function polyplex for delivery to tumors to inhibit CXCR4 chemokine receptor to treat metastatic pancreatic cancer [64]. Additionally, triple targeted nanoparticles featuring the CXCR4 antagonist for cancer stromal blockade action, as well as anti-miR-210 and siRNA targeting mutated *KRAS* were developed. The triple action-nanoparticle modulated the desmoplastic TME via inactivating PSCs and promoting infiltration of cytotoxic T cells [65].

Another organic material used for siRNA delivery is a biodegradable charged polyester-based vector which has demonstrated no significant toxicities on blood vessels, tissues, or organs even at high dosage [134]. When *KRAS*-targeting siRNA is formulated in this delivery vector and peritumorally injected in pancreatic xenograft models, significant decreases in tumor growth and local invasion were observed [66]. Many polysaccharide-derived nanoparticles such as chitosan nanoparticles are being investigated for nucleic acid drug delivery across diseases [135]. In particular, chitosan oligosaccharide lactate can be assembled with folic acid and polyethylene glycol into nanoparticles that are capable of encapsulating siRNA and intravenously localizing in tumors in pancreatic orthotopic models [67]. This formulated nanoparticle containing siRNA silencing genes *LAMTOR2*, *mTOR*, and *NUP85* have shown clinical potential by inhibiting retroperitoneal invasion and dissemination, therefore improving the prognosis of the mice. Overall, these studies indicate promising strategies of nanoparticle usage for siRNA therapeutics for pancreatic cancer therapy.

Other than the lipid and polymeric nanoparticles, peptides have also been investigated as siRNA delivery materials due to their biocompatibility and biodegradability for minimal toxicity [136]. In addition, peptides also have intrinsic property to act as ligands for targeting specific receptors. An arginine-glycine-aspartic acid-peptide is well-known to specifically bind to the integrin receptors overexpressed on the surface of cancerous cells [137]. Bhatia's group formulated arginine-glycine-aspartic acid peptides with polyethylene glycol to encapsulate siRNA to generate a tumor-penetrating nanocomplex [68]. When a KRAS siRNA was intravenously delivered by this tumor-penetrating nanocomplex into a genetically engineered PDAC mouse model, the growth of pancreatic tumors arising from allografted KP D8-175 cells were significantly delayed. Similarly, pancreatic tumor inhibition was seen when miR-34a was delivered in nanocomplexes formulated with a tumor-targeting and penetrating bifunctional CC9 peptide [81]. The nanoparticle facilitated the cellular uptake of miR-34a and induced cell cycle arrest and blocked angiogenesis in xenograft model using PANC-1 cells. Another type of tumor suppressor miRNA, miR-150 was administered to pancreatic cancer cells in a PLGA-based nanoformulation incorporating polyethyleneimine. The cells showed a significant decrease in MUC4 expression and downstream signaling proteins such as human epidermal growth factor receptor 2 (hEGFR2), therefore reducing tumor growth, clonogenicity, motility, and invasion [74]. Due to these and other successes, the application of various nanoparticles is still being investigated in preclinical and clinical trials for pancreatic cancer therapy.

3.1.4. Extracellular Vesicle-Mediated Delivery

Exosomes or extracellular vesicles are roughly 100 nm spherical vesicles, which are secreted by all cells. Initially, exosomes were considered to be garbage bags for cells [138]. However, recent findings demonstrating that exosomes contain a variety of bioactive molecules reflecting parental cells and play a crucial role in cell-cell signaling, have drawn much attention to exosome research [139]. In addition, exosomes exhibit many advantages as delivery systems including low immunogenicity, multi-functionality, and disease-targetability [140]. Notably, exosomes can be internalized by cells via a fusion mechanism, leading to efficient delivery of drugs into the target cell cytoplasm [141]. Some reports have indicated that exosomes expressing viral fusogen fused significantly with the target cell membrane in acidic conditions compared with native exosomes [142]. Notably, Liu et al. showed that viral fusogen decorated and programed death-ligand 1 siRNA-loaded, macrophage-derived extracellular vesicles substantially elevated tumor-specific CD8$^+$ T cell immune responses in CT26 tumor-bearing mice through efficient programed death-ligand 1 gene silencing [143]. Although packing drug candidates into extracellular vesicles is still challenging, these findings indicate that the exosome has great potential as a drug delivery platform due to strengths mentioned above [144].

Several studies suggested that extracellular vesicles can effectively transfer RNAi into the cancer cells, eliciting substantial antitumor effects. For example, extracellular vesicles containing miRNA-134 enhanced the therapeutic effects of anti-heat shock protein 90 drugs in breast cancer [145]. Similarly, glucose-regulated protein 78 (GRP78) siRNA-loaded extracellular vesicles could sensitize resistant tumor cells to sorafenib [146]. Another study found that extracellular vesicle delivery of let-7a miRNA had potent antitumor effects on breast cancer cells breast cancer cells, and that extracellular vesicles expressing GE11 or AS1411, to target epidermal growth factor or nucleolin, respectively, on breast cancer cells, could enhance the effects of let-7a miRNA by enhancing delivery of their payload to the tumor cells [117,147]. Furthermore, taking advantage of fact that extracellular vesicles can penetrate the blood-brain barrier via transcytosis, systemic administration of extracellular vesicles harboring miRNA-21, miRNA-164b, miRNA-124a, or vesicular endothelial growth factor siRNA significantly retarded brain tumor growth, improving survival [148–151].

There have also been attempts to treat PDAC using exosomal siRNA. Kalluri et al. reported that mesenchymal stem cell-derived exosomes loaded with KRAS^{G12D} siRNA (termed iExosomes) provoked effective antitumor responses in pancreatic cancer models [72]. They also found that iExosomes showed a long half-life due to high expression of exosomal CD47, facilitating evasion of immunological clearance. The production protocol of iExosomes has been successfully established to generate clinical grade-exosomes that have entered phase 1 clinical trials for the treatment of PDAC [ClinicalTrials.gov identifier: NCT03608631] [152]. Zhou et al. recently developed engineered exosomes loaded with galectin-9 siRNA to alleviate immunosuppressive properties of tumor-associated macrophages [73]. Co-delivery of galectin-9 siRNA and the immunogenic chemotherapy oxaliplatin by exosomes elicited successful tumor control via immune activation. Collectively, these studies indicate that exosome-mediated gene silencing could be attractive for the treatment of PDAC.

3.2. Combination Therapy

Many studies have shown that the combination of siRNA and other therapeutic agents can exhibit synergistic effects to overcome multiple drug resistance by enhancing both silencing and chemotherapy activity [153,154]. Continuous administration of inhibitors often leads to the development of resistance via acquired mutations or epigenetic alterations. In combination therapy, the approach is designed to reverse the acquired genetic or epigenetic mechanism with the RNAi drugs instead of additional chemical inhibitors which result in increased toxicity. There have been numerous reports on therapeutic efficacy when existing therapeutic agents are combined with the concomitant use of RNAi drugs [155].

There is also evidence indicating that a combination strategy can also reduce the side effects of individual therapeutics by enhancing the therapeutic index [156].

As one of the most common cancer types with very low survival rate, pancreatic cancer is uniquely identified by its dense surrounding tumor microenvironment that limits the therapeutic effect of existing drugs. Both the dense stroma and acquired resistance to conventional therapies are triggered from dysregulated expression of cancer-associated genes. To target these key genes, RNAi-based approaches have gained attention both to elicit direct antitumor effects and to alter the nature of tumor microenvironment to remove barriers for chemotherapeutic action. Thus, several studies have sought to achieve antitumor therapeutic effects using KRAS siRNA in combination with chemotherapies and other small-molecule drugs. Zeng et al. reported that the combination of KRAS siRNA and arsenic trioxide effectively inhibited pancreatic tumor growth by exerting a synergistic effect of siRNA-induced cell cycle blockage and arsenic-induced apoptosis [69]. Wang et al. showed the combination of KRAS siRNA and gemcitabine exhibited improved inhibition of cell proliferation, cell cycle arrest, increased apoptosis, and suppression of tumor progression without any toxicity. In addition, they demonstrated that the anticancer effect of KRAS siRNA contributes to the low IC50 value of siRNA/gemcitabine-loaded liposomes [53]. Golan et al. tested that a biodegradable implant termed siG12D-LODER™ loaded with siRNA drug against $KRAS^{G12D}$ and gemcitabine for pancreatic cancer. Of the 15 patients, 12 patients did not show tumor progression, 10 out of 12 patients showed stable disease, and 2 patients had partial response, yielding an increased average patient survival rate [157]. In addition, several attempts have been made to knock down several other genes highly associated with pancreatic cancer. Li et al. used a peptide-binding nanovehicle for co-delivery of paclitaxel and siRNA targeting TR3, an orphan nuclear receptor, finding that the co-delivery system exhibited a higher loading efficiency of siRNA, markedly inhibited tumor growth, and induced cancer cell apoptosis [70]. McCarroll et al. also identified βIII-tubulin as a novel mediator for chemoresistance and metastases in pancreatic cancer and used siRNA to increase chemosensitivity [158]. When βIII-tubulin expression is inhibited in orthotopic pancreatic cancer model in the presence of gemcitabine, tumor growth, and metastases were reduced in vivo. Additionally, siRNAs have been used to alter the immunosuppressive microenvironment and to reduce the stromal physical barrier to improve antibody or chemotherapeutic drug actions [159]. Yoo et al. combined an siRNA targeting the programed death-ligand 1 in a magnetic nanocarrier with gemcitabine to activate immune responses against tumor cells while noninvasively monitoring the therapeutic response via magnetic resonance imaging [71].

Currently, miRNA-based drugs are also being investigated in combination with chemotherapeutic drugs to enhance therapeutic efficacy in pancreatic cancer [160]. Although the chemotherapeutic gemcitabine remains a cornerstone of the PDAC treatment, patients acquire resistance within weeks of treatment [161]. To overcome the resistance, several novel therapeutic approaches as well as new entrapment designs with nanoparticles are currently being investigated. Patel et al. encapsulated miR-155 in extracellular vesicles from gemcitabine-resistant pancreatic cancer cells to reduce deoxycytidine kinase (gemcitabine-metabolizing gene) and increase superoxide dismutase 2 and catalase (ROS-detoxifying genes) expression in the presence of chemotherapy [75]. The deoxycytidine kinase reduction by miR-155 delivered by the extracellular vesicles led to abrogation of acquired resistance of gemcitabine in pancreatic cancer cells. Another miRNA that was combined with gemcitabine for pancreatic cancer treatment is miR-345 which was delivered in polymeric dual delivery nanoscale device [80]. The temperature and pH-responsive pentablock copolymer system enhanced/restored the miR-345 levels, making xenograft pancreatic cancer tumors more susceptible to gemcitabine. Immunohistochemical analysis revealed that the miR-345 induced significant down-regulation of desmoplastic reaction, as well as improving SHH, Gli-1, MUC4, and Ki67 levels. Because oncogenic miRNAs are responsible for triggering the cancerous features, antisense oligonucleotides are used to inhibit the activity of commonly dysregulated miRNAs [162]. Among numerous oncogenic miRNAs, miR-21 is one of the earliest identified and most abundant cancer-promoting miRNAs and hence has been targeted in cancer [163]. Li et al. developed a combined therapy using an miR-21 antisense oligonucleotide and gemcitabine co-delivered by a nanoparticle made from polyethylene glycol-polyethyleneimine-magnetic iron oxide [78]. With a coating of anti-CD44v6 single-chain variable fragment for targeted delivery, the miR-21 antisense oligonucleotide and gemcitabine dual-treatment nanoparticle up-regulated programed cell death protein 4 and phosphatase and tensin homolog (PTEN) expression, thus suppressing epithelial-mesenchymal transition, cell proliferation, migration, and invasion of pancreatic cancer cells. This combination therapy also showed a synergistic antitumor effect in a MIA PaCa-2 pancreatic xenograft model.

Other than gemcitabine, doxorubicin, and the hedgehog inhibitor, GDC-0944 have been combined with miRNA for pancreatic cancer therapy. Chen et al. developed a new chimeric peptide composed of plectin-1 for PDAC-specific targeting coupled with arginine-rich RNA-binding motifs for miR-212 delivery [79]. The reduction of USP9X expression by miR-212 enhanced the therapeutic effect of doxorubicin to induce apoptosis and autophagy of PDAC cells in vitro as well as in patient-derived pancreatic xenograft model. Kumar et al. selected miR-let7b as a therapeutic target because of inherently low expression miR-let7b in PDAC tumors with elevated hedgehog levels [164]. Using mPEG-block-poly(2-methyl-2-carboxyl-propylenecarbonate-graft-dodecanol-graft-tetraethylene-pentamine copolymer, miR-let7b and the hedgehog inhibitor, GDC-0944 were co-formulated into micelles, which inhibited tumor growth of an ectopic xenograft model by triggering apoptosis. Instead of combination with chemotherapy, miR-21 and miR-221 antisense oligonucleotides were used to target stem-cell-like cancer cells [77]. The miR-21 and miR-221 antisense oligonucleotides were given to sorted side population cells from gemcitabine-resistant cells and inhibited the growth and metastasis when both miRNAs were given. Schnittert et al. demonstrated a novel peptide-based nanocomplex for miR-199a antisense oligonucleotide delivery into human-derived PSCs [76]. The miR-199a antisense oligonucleotide inhibited PSC differentiation into cancer-associated fibroblasts and reduced the size of 3D hetero-spheroids made of PSCs and cancer cells. Taken together, these studies have shown that the combination of siRNA and chemotherapy not only has excellent anticancer effects, but also reduces toxicity and inhibits resistance.

4. Challenges and Future Prospects

Even with numerous efforts and ongoing investigations of RNAi-based therapeutics, pancreatic cancer remains as one of the most challenging with devastating malignancies with an extremely poor prognosis [165]. The successful delivery of degradation-prone nucleotide-based drugs is still the challenging goal in the field of RNAi therapeutics, requiring alternative strategies and reagents. One of the most promising mechanisms to consider in addressing the hurdles of intracellular RNAi delivery to gain cellular entry is macropinocytosis. The significance of macropinocytosis as a delivery pathway is that it is broadly applicable across multiple cell types, cargos, and delivery systems. Exploitation of macropinocytosis for therapeutic delivery to cancer cells is uniquely advantageous because macropinocytosis plays a key role in sustaining cancer cell growth within the nutrient-poor by functioning as a nutrient supply route [166–170]. Frequently observed oncogenic mutations, such as *KRAS*, *PI3K*, and *PTEN*, have been demonstrated to greatly enhance macropinocytosis in cancer cells [171–174]. Thus, it is not surprising to find that numerous delivery systems, such as lipids, polymers, and peptides, use macropinocytosis to enter a broad range of cancer cells [72,175–180]. Another important aspect of employing macropinocytosis in RNAi delivery is that macropinocytic entry delivers cargo to late endosomes and multivesicular bodies, which are particularly favorable destinations for nucleic acids [181–183]. Whether the siRNAs are delivered by lipids or nanoparticles, the intracellular trafficking of siRNAs begins in early endosomal vesicles [184]. The early endosomes subsequently fuse with sorting endosomes, which in turn transfer the content to the late endosomes. The acidic late endosomes later then relocated to the lysosomes where abundant nucleases exist. Thus, late endosomes and multivesicular bodies are where most endosomal escape is reported to occur before lysosomal degradation, a major barrier to the intracellular delivery of RNAi therapeutics [185]. These reports are especially interesting considering that cancer cells use macropinocytosis for nutrient scavenging, which requires lysosomal degradation of proteins into amino acids. The studies to elucidate the exact mechanism of macropinocytic delivery have encountered numerous difficulties due to the lack of specific inhibitors, varying rates among different cell types, and high sensitivity to external stimuli, and others [186,187]. However, our rapidly growing understanding of macropinocytosis presents unique opportunities to design better strategies to target macropinocytosis for intracellular delivery of RNAi.

One major hurdle specific to pancreatic cancer is the stromal barrier that hinders the therapeutic action of current and RNAi-based drugs within tumor cells or TME. Repeated reports and studies demonstrate that the tumor's genetic aberrations cause the development of dense stromal environment and suppressive immune response, emphasizing the need for RNAi for TME reprogramming in addition to direct attack on tumor cells [188]. Targeting the infiltrated immune cells present in stromal structure is an alternative approach to cancer therapy using immunotherapeutic options. The use of nanoparticles has led to successful delivery of RNAi drugs into tumor cells; however, the uptake rate remains low for clinical significance. Similar to the immunotherapeutic drug agents, RNAi has the potential to take action on immune cells to re-sensitize the suppressive immune response as an alternative cancer therapeutic strategy [189]. Genetic mutations of signaling proteins and transcription factors triggering the immunosuppression are potential RNAi drug targets with benefits over chemical inhibitors and conventional therapies. Furthermore, the application of RNAi drugs with current checkpoint inhibitors has great potential in clinical translation to patients who developed resistance to immunotherapy.

Although milestones of RNAi-based drugs have proved clinical translation of oligo therapeutics and has opened numerous application possibilities, the systemic delivery challenge remains for specific disease targets. When designing a delivery strategy for oligonucleotide, the cargo and moiety of drug often jeopardizes the safety profile and cause adverse toxicities upon accumulation. Though RNAi-based drugs are naturally occurring biological materials, the chemical modifications and fabrication is inevitable and accompanies safety concerns. For this reason, strategies to use siRNA conjugation with more transparent and simpler structures and designs is under investigation across academia and industry for its advantages in clinical translation.

5. Conclusions

The potential to regulate selective gene expression using RNAi-based moieties holds promise for pancreatic cancer treatment. Numerous results demonstrating antitumor effects, chemosensitization, anti-metastatic activity and activation of immunogenicity against cancer cells have been shown in preclinical and even in early clinical stages. Pancreatic cancer, one of the cancer types with the lowest survival rate, is uniquely identified with dense tumor microenvironment serving as a barrier to conventional drugs. Dysregulated expression of cancer-associated genes and key signaling pathways are often the driver for acquired resistance and the discouraging outcomes of current therapies. For this reason, the RNAi-based approaches have gained attention to elicit antitumor effect or altering the nature of TME to remove barriers for chemotherapeutic action. From the studies mentioned above, it appears that RNAi-based drugs hold great hope in pancreatic cancer therapy [190]. In addition to the RNAi-based drug discovery and development, it is crucial to develop a model or system that accurately reflects the biology of human pancreatic cancer with clinically similar features. Orthotopic pancreatic cancer mouse models with an intact and functional fibrotic stroma, and genetically engineered mouse models with functional immune system and 3D tissue models have been developed and are continually being improved to advance pancreatic cancer therapy [191]. In particular, the use of 3D cultures is an innovative approach that narrows the gap between traditional 2D cell culture and animal models allowing the manipulation of individual cell population involved in TME specific to the disease type [192]. Several groups have demonstrated spheroids, hydrogel structures, and microfluidic designs to mimic the TME nature of pancreatic cancer [193–197]. These models, used in parallel with clinically relevant in vivo models, can inform the design of novel RNAi-based drugs. Since personalized medicine and targeted therapies are the next generation of cancer treatments, the unique pathophysiology of pancreatic cancer needs to be addressed. In particular, the dense fibrotic stroma presenting a large barrier is a strong consideration in introducing a novel RNAi-based approach for pancreatic cancer therapy. Despite the heterogeneity and complexity of pancreatic cancer, we believe that versatility of RNAi-based drugs with numerous targets hold promising potential for pancreatic cancer therapy in combined with chemotherapy or immunotherapy.

Author Contributions: Conceptualization, M.J.K., T.M.R. and J.H.R.; writing—original draft preparation, M.J.K., H.C., G.N., Y.K., writing—review and editing, G.N., T.M.R. and J.H.R.; supervision, T.M.R., S.H.K. and J.H.R. All authors discussed, edited, and approved the final version of the manuscript. All authors have read and agreed to the published version of the manuscript.

Funding: This study was supported by a grant of the Korea Health Technology R&D Project through the Korea Health Industry Development Institute (KHIDI), funded by the Ministry of Health & Welfare, Korea (grant number: HI19C0753) and the Intramural Research Program of KIST.

Conflicts of Interest: The authors declare no conflict of interest.

References

1. Siegel, R.L.; Miller, K.D.; Jemal, A. Cancer statistics, 2018. *CA Cancer J. Clin.* **2018**, *68*, 7–30. [CrossRef] [PubMed]
2. Rahib, L.; Smith, B.D.; Aizenberg, R.; Rosenzweig, A.B.; Fleshman, J.M.; Matrisian, L.M. Projecting cancer incidence and deaths to 2030: The unexpected burden of thyroid, liver, and pancreas cancers in the United States. *Cancer Res.* **2014**, *74*, 2913–2921. [CrossRef] [PubMed]
3. Kamisawa, T.; Wood, L.D.; Itoi, T.; Takaori, K. Pancreatic cancer. *Lancet* **2016**, *388*, 73–85. [CrossRef]
4. Roth, M.T.; Cardin, D.B.; Berlin, J.D. Recent advances in the treatment of pancreatic cancer. *F1000 Res.* **2020**, *9*. [CrossRef] [PubMed]
5. Kleeff, J.; Korc, M.; Apte, M.; La Vecchia, C.; Johnson, C.D.; Biankin, A.V.; Neale, R.E.; Tempero, M.; Tuveson, D.A.; Hruban, R.H.; et al. Pancreatic cancer. *Nat. Rev. Dis. Primers* **2016**, *2*, 16022. [CrossRef]
6. Ho, W.J.; Jaffee, E.M.; Zheng, L. The tumour microenvironment in pancreatic cancer—Clinical challenges and opportunities. *Nat. Rev. Clin. Oncol.* **2020**, *17*, 527–540. [CrossRef]
7. Clark, C.E.; Hingorani, S.R.; Mick, R.; Combs, C.; Tuveson, D.A.; Vonderheide, R.H. Dynamics of the immune reaction to pancreatic cancer from inception to invasion. *Cancer Res.* **2007**, *67*, 9518–9527. [CrossRef]
8. Provenzano, P.P.; Cuevas, C.; Chang, A.E.; Goel, V.K.; Von Hoff, D.D.; Hingorani, S.R. Enzymatic targeting of the stroma ablates physical barriers to treatment of pancreatic ductal adenocarcinoma. *Cancer Cell* **2012**, *21*, 418–429. [CrossRef]
9. Stromnes, I.M.; Brockenbrough, J.S.; Izeradjene, K.; Carlson, M.A.; Cuevas, C.; Simmons, R.M.; Greenberg, P.D.; Hingorani, S.R. Targeted depletion of an MDSC subset unmasks pancreatic ductal adenocarcinoma to adaptive immunity. *Gut* **2014**, *63*, 1769–1781. [CrossRef]
10. Provenzano, P.P.; Hingorani, S.R. Hyaluronan, fluid pressure, and stromal resistance in pancreas cancer. *Br. J. Cancer* **2013**, *108*, 1–8. [CrossRef]
11. Daniel, S.K.; Sullivan, K.M.; Labadie, K.P.; Pillarisetty, V.G. Hypoxia as a barrier to immunotherapy in pancreatic adenocarcinoma. *Clin. Transl. Med.* **2019**, *8*, 10. [CrossRef]
12. Doktorova, H.; Hrabeta, J.; Khalil, M.A.; Eckschlager, T. Hypoxia-induced chemoresistance in cancer cells: The role of not only HIF-1. *Biomed. Pap. Med. Fac. Univ. Palacky Olomouc. Czech Repub.* **2015**, *159*, 166–177. [CrossRef]
13. Bachem, M.G.; Schneider, E.; Gross, H.; Weidenbach, H.; Schmid, R.M.; Menke, A.; Siech, M.; Beger, H.; Grunert, A.; Adler, G. Identification, culture, and characterization of pancreatic stellate cells in rats and humans. *Gastroenterology* **1998**, *115*, 421–432. [CrossRef]
14. Xu, Z.; Vonlaufen, A.; Phillips, P.A.; Fiala-Beer, E.; Zhang, X.; Yang, L.; Biankin, A.V.; Goldstein, D.; Pirola, R.C.; Wilson, J.S.; et al. Role of pancreatic stellate cells in pancreatic cancer metastasis. *Am. J. Pathol.* **2010**, *177*, 2585–2596. [CrossRef]
15. Fu, Y.; Liu, S.; Zeng, S.; Shen, H. The critical roles of activated stellate cells-mediated paracrine signaling, metabolism and onco-immunology in pancreatic ductal adenocarcinoma. *Mol. Cancer* **2018**, *17*, 62. [CrossRef] [PubMed]
16. Kusiak, A.A.; Szopa, M.D.; Jakubowska, M.A.; Ferdek, P.E. Signaling in the Physiology and Pathophysiology of Pancreatic Stellate Cells—A Brief Review of Recent Advances. *Front. Physiol.* **2020**, *11*, 78. [CrossRef]
17. Waters, A.M.; Der, C.J. KRAS: The Critical Driver and Therapeutic Target for Pancreatic Cancer. *Cold Spring Harb. Perspect. Med.* **2018**, *8*, a031435. [CrossRef]

18. Hu, H.F.; Ye, Z.; Qin, Y.; Xu, X.W.; Yu, X.J.; Zhuo, Q.F.; Ji, S.R. Mutations in key driver genes of pancreatic cancer: Molecularly targeted therapies and other clinical implications. *Acta Pharmacol. Sin.* **2021**. [CrossRef] [PubMed]
19. Kristen, A.V.; Ajroud-Driss, S.; Conceicao, I.; Gorevic, P.; Kyriakides, T.; Obici, L. Patisiran, an RNAi therapeutic for the treatment of hereditary transthyretin-mediated amyloidosis. *Neurodegener. Dis. Manag.* **2019**, *9*, 5–23. [CrossRef] [PubMed]
20. Setten, R.L.; Rossi, J.J.; Han, S.P. The current state and future directions of RNAi-based therapeutics. *Nat. Rev. Drug Discov.* **2019**, *18*, 421–446. [CrossRef]
21. Hu, B.; Weng, Y.; Xia, X.H.; Liang, X.J.; Huang, Y. Clinical advances of siRNA therapeutics. *J. Gene Med.* **2019**, *21*, e3097. [CrossRef] [PubMed]
22. Chakraborty, C.; Sharma, A.R.; Sharma, G.; Doss, C.G.P.; Lee, S.S. Therapeutic miRNA and siRNA: Moving from Bench to Clinic as Next Generation Medicine. *Mol. Ther. Nucleic Acids* **2017**, *8*, 132–143. [CrossRef] [PubMed]
23. Saw, P.E.; Song, E.W. siRNA therapeutics: A clinical reality. *Sci. China Life Sci.* **2020**, *63*, 485–500. [CrossRef]
24. Singh, A.; Trivedi, P.; Jain, N.K. Advances in siRNA delivery in cancer therapy. *Artif. Cells Nanomed. Biotechnol.* **2018**, *46*, 274–283. [CrossRef]
25. Tatiparti, K.; Sau, S.; Kashaw, S.K.; Iyer, A.K. siRNA Delivery Strategies: A Comprehensive Review of Recent Developments. *Nanomaterials* **2017**, *7*, 77. [CrossRef] [PubMed]
26. Rana, T.M. Illuminating the silence: Understanding the structure and function of small RNAs. *Nat. Rev. Mol. Cell Biol.* **2007**, *8*, 23–36. [CrossRef] [PubMed]
27. Sajid, M.I.; Moazzam, M.; Kato, S.; Yeseom Cho, K.; Tiwari, R.K. Overcoming Barriers for siRNA Therapeutics: From Bench to Bedside. *Pharmaceuticals* **2020**, *13*, 294. [CrossRef]
28. Gebert, L.F.R.; MacRae, I.J. Regulation of microRNA function in animals. *Nat. Rev. Mol. Cell Biol.* **2019**, *20*, 21–37. [CrossRef]
29. O'Brien, J.; Hayder, H.; Zayed, Y.; Peng, C. Overview of MicroRNA Biogenesis, Mechanisms of Actions, and Circulation. *Front. Endocrinol.* **2018**, *9*, 402. [CrossRef]
30. Peng, Y.; Croce, C.M. The role of MicroRNAs in human cancer. *Signal Transduct. Target Ther.* **2016**, *1*, 15004. [CrossRef]
31. Gurbuz, N.; Ozpolat, B. MicroRNA-based Targeted Therapeutics in Pancreatic Cancer. *Anticancer Res.* **2019**, *39*, 529–532. [CrossRef]
32. Bjornmalm, M.; Thurecht, K.J.; Michael, M.; Scott, A.M.; Caruso, F. Bridging Bio-Nano Science and Cancer Nanomedicine. *ACS Nano* **2017**, *11*, 9594–9613. [CrossRef]
33. Hruban, R.H.; Adsay, N.V.; Albores-Saavedra, J.; Compton, C.; Garrett, E.S.; Goodman, S.N.; Kern, S.E.; Klimstra, D.S.; Kloppel, G.; Longnecker, D.S.; et al. Pancreatic intraepithelial neoplasia: A new nomenclature and classification system for pancreatic duct lesions. *Am. J. Surg. Pathol.* **2001**, *25*, 579–586. [CrossRef]
34. Hou, P.; Kapoor, A.; Zhang, Q.; Li, J.; Wu, C.J.; Li, J.; Lan, Z.; Tang, M.; Ma, X.; Ackroyd, J.J.; et al. Tumor Microenvironment Remodeling Enables Bypass of Oncogenic KRAS Dependency in Pancreatic Cancer. *Cancer Discov.* **2020**, *10*, 1058–1077. [CrossRef]
35. Bailey, P.; Chang, D.K.; Nones, K.; Johns, A.L.; Patch, A.M.; Gingras, M.C.; Miller, D.K.; Christ, A.N.; Bruxner, T.J.; Quinn, M.C.; et al. Genomic analyses identify molecular subtypes of pancreatic cancer. *Nature* **2016**, *531*, 47–52. [CrossRef]
36. Kokkinos, J.; Ignacio, R.M.C.; Sharbeen, G.; Boyer, C.; Gonzales-Aloy, E.; Goldstein, D.; Australian Pancreatic Cancer Genome Initiative, A.; McCarroll, J.A.; Phillips, P.A. Targeting the undruggable in pancreatic cancer using nano-based gene silencing drugs. *Biomaterials* **2020**, *240*, 119742. [CrossRef]
37. Dang, C.V.; Reddy, E.P.; Shokat, K.M.; Soucek, L. Drugging the 'undruggable' cancer targets. *Nat. Rev. Cancer* **2017**, *17*, 502–508. [CrossRef]
38. Morris, J.P.t.; Wang, S.C.; Hebrok, M. KRAS, Hedgehog, Wnt and the twisted developmental biology of pancreatic ductal adenocarcinoma. *Nat. Rev. Cancer* **2010**, *10*, 683–695. [CrossRef] [PubMed]
39. Kanda, M.; Matthaei, H.; Wu, J.; Hong, S.M.; Yu, J.; Borges, M.; Hruban, R.H.; Maitra, A.; Kinzler, K.; Vogelstein, B.; et al. Presence of somatic mutations in most early-stage pancreatic intraepithelial neoplasia. *Gastroenterology* **2012**, *142*, 730–733 e739. [CrossRef] [PubMed]
40. Distler, M.; Aust, D.; Weitz, J.; Pilarsky, C.; Grutzmann, R. Precursor lesions for sporadic pancreatic cancer: PanIN, IPMN, and MCN. *Biomed. Res. Int.* **2014**, *2014*, 474905. [CrossRef] [PubMed]
41. Lee, J.W.; Komar, C.A.; Bengsch, F.; Graham, K.; Beatty, G.L. Genetically Engineered Mouse Models of Pancreatic Cancer: The KPC Model (LSL-Kras(G12D/+);LSL-Trp53(R172H/+);Pdx-1-Cre), Its Variants, and Their Application in Immuno-oncology Drug Discovery. *Curr. Protoc. Pharmacol.* **2016**, *73*, 143911–143920. [CrossRef]
42. Gopinathan, A.; Morton, J.P.; Jodrell, D.I.; Sansom, O.J. GEMMs as preclinical models for testing pancreatic cancer therapies. *Dis. Model Mech.* **2015**, *8*, 1185–1200. [CrossRef] [PubMed]
43. Bardeesy, N.; Aguirre, A.J.; Chu, G.C.; Cheng, K.H.; Lopez, L.V.; Hezel, A.F.; Feng, B.; Brennan, C.; Weissleder, R.; Mahmood, U.; et al. Both p16(Ink4a) and the p19(Arf)-p53 pathway constrain progression of pancreatic adenocarcinoma in the mouse. *Proc. Natl. Acad. Sci. USA* **2006**, *103*, 5947–5952. [CrossRef] [PubMed]
44. Aguirre, A.J.; Bardeesy, N.; Sinha, M.; Lopez, L.; Tuveson, D.A.; Horner, J.; Redston, M.S.; DePinho, R.A. Activated Kras and Ink4a/Arf deficiency cooperate to produce metastatic pancreatic ductal adenocarcinoma. *Genes Dev.* **2003**, *17*, 3112–3126. [CrossRef] [PubMed]

45. Hong, D.S.; Fakih, M.G.; Strickler, J.H.; Desai, J.; Durm, G.A.; Shapiro, G.I.; Falchook, G.S.; Price, T.J.; Sacher, A.; Denlinger, C.S.; et al. KRAS(G12C) Inhibition with Sotorasib in Advanced Solid Tumors. *N. Engl. J. Med.* **2020**, *383*, 1207–1217. [CrossRef] [PubMed]
46. Zhang, Y.; Li, M.; Wang, H.; Fisher, W.E.; Lin, P.H.; Yao, Q.; Chen, C. Profiling of 95 microRNAs in pancreatic cancer cell lines and surgical specimens by real-time PCR analysis. *World J. Surg.* **2009**, *33*, 698–709. [CrossRef] [PubMed]
47. Ma, M.Z.; Kong, X.; Weng, M.Z.; Cheng, K.; Gong, W.; Quan, Z.W.; Peng, C.H. Candidate microRNA biomarkers of pancreatic ductal adenocarcinoma: Meta-analysis, experimental validation and clinical significance. *J. Exp. Clin. Cancer Res.* **2013**, *32*, 71. [CrossRef]
48. Passadouro, M.; Faneca, H. Managing Pancreatic Adenocarcinoma: A Special Focus in MicroRNA Gene Therapy. *Int. J. Mol. Sci.* **2016**, *17*, 718. [CrossRef]
49. Guo, S.; Fesler, A.; Wang, H.; Ju, J. microRNA based prognostic biomarkers in pancreatic Cancer. *Biomark. Res.* **2018**, *6*, 18. [CrossRef]
50. Sasayama, Y.; Hasegawa, M.; Taguchi, E.; Kubota, K.; Kuboyama, T.; Naoi, T.; Yabuuchi, H.; Shimai, N.; Asano, M.; Tokunaga, A.; et al. In vivo activation of PEGylated long circulating lipid nanoparticle to achieve efficient siRNA delivery and target gene knock down in solid tumors. *J. Control. Release* **2019**, *311-312*, 245–256. [CrossRef]
51. Rao, D.D.; Luo, X.; Wang, Z.; Jay, C.M.; Brunicardi, F.C.; Maltese, W.; Manning, L.; Senzer, N.; Nemunaitis, J. KRAS mutant allele-specific expression knockdown in pancreatic cancer model with systemically delivered bi-shRNA KRAS lipoplex. *PLoS ONE* **2018**, *13*, e0193644. [CrossRef]
52. Aleku, M.; Schulz, P.; Keil, O.; Santel, A.; Schaeper, U.; Dieckhoff, B.; Janke, O.; Endruschat, J.; Durieux, B.; Roder, N.; et al. Atu027, a liposomal small interfering RNA formulation targeting protein kinase N3, inhibits cancer progression. *Cancer Res.* **2008**, *68*, 9788–9798. [CrossRef]
53. Wang, F.; Zhang, Z. Nanoformulation of Apolipoprotein E3-Tagged Liposomal Nanoparticles for the co-Delivery of KRAS-siRNA and Gemcitabine for Pancreatic Cancer Treatment. *Pharm. Res.* **2020**, *37*, 247. [CrossRef]
54. Lei, Y.; Tang, L.; Xie, Y.; Xianyu, Y.; Zhang, L.; Wang, P.; Hamada, Y.; Jiang, K.; Zheng, W.; Jiang, X. Gold nanoclusters-assisted delivery of NGF siRNA for effective treatment of pancreatic cancer. *Nat. Commun.* **2017**, *8*, 15130. [CrossRef]
55. Partecke, I.L.; Kaeding, A.; Sendler, M.; Albers, N.; Kuhn, J.P.; Speerforck, S.; Roese, S.; Seubert, F.; Diedrich, S.; Kuehn, S.; et al. In vivo imaging of pancreatic tumours and liver metastases using 7 Tesla MRI in a murine orthotopic pancreatic cancer model and a liver metastases model. *BMC Cancer* **2011**, *11*, 40. [CrossRef] [PubMed]
56. Mahajan, U.M.; Teller, S.; Sendler, M.; Palankar, R.; van den Brandt, C.; Schwaiger, T.; Kuhn, J.P.; Ribback, S.; Glockl, G.; Evert, M.; et al. Tumour-specific delivery of siRNA-coupled superparamagnetic iron oxide nanoparticles, targeted against PLK1, stops progression of pancreatic cancer. *Gut* **2016**, *65*, 1838–1849. [CrossRef] [PubMed]
57. Teo, J.; McCarroll, J.A.; Boyer, C.; Youkhana, J.; Sagnella, S.M.; Duong, H.T.; Liu, J.; Sharbeen, G.; Goldstein, D.; Davis, T.P.; et al. A Rationally Optimized Nanoparticle System for the Delivery of RNA Interference Therapeutics into Pancreatic Tumors in Vivo. *Biomacromolecules* **2016**, *17*, 2337–2351. [CrossRef] [PubMed]
58. McCarroll, J.A.; Sharbeen, G.; Kavallaris, M.; Phillips, P.A. The Use of Star Polymer Nanoparticles for the Delivery of siRNA to Mouse Orthotopic Pancreatic Tumor Models. *Methods Mol. Biol.* **2019**, *1974*, 329–353. [CrossRef]
59. Zhang, N.; Lyons, S.; Lim, E.; Lassota, P. A spontaneous acinar cell carcinoma model for monitoring progression of pancreatic lesions and response to treatment through noninvasive bioluminescence imaging. *Clin. Cancer Res.* **2009**, *15*, 4915–4924. [CrossRef]
60. Pittella, F.; Cabral, H.; Maeda, Y.; Mi, P.; Watanabe, S.; Takemoto, H.; Kim, H.J.; Nishiyama, N.; Miyata, K.; Kataoka, K. Systemic siRNA delivery to a spontaneous pancreatic tumor model in transgenic mice by PEGylated calcium phosphate hybrid micelles. *J. Control. Release* **2014**, *178*, 18–24. [CrossRef]
61. Pittella, F.; Miyata, K.; Maeda, Y.; Suma, T.; Watanabe, S.; Chen, Q.; Christie, R.J.; Osada, K.; Nishiyama, N.; Kataoka, K. Pancreatic cancer therapy by systemic administration of VEGF siRNA contained in calcium phosphate/charge-conversional polymer hybrid nanoparticles. *J. Control. Release* **2012**, *161*, 868–874. [CrossRef] [PubMed]
62. Zorde Khvalevsky, E.; Gabai, R.; Rachmut, I.H.; Horwitz, E.; Brunschwig, Z.; Orbach, A.; Shemi, A.; Golan, T.; Domb, A.J.; Yavin, E.; et al. Mutant KRAS is a druggable target for pancreatic cancer. *Proc. Natl. Acad. Sci. USA* **2013**, *110*, 20723–20728. [CrossRef] [PubMed]
63. Bangoura, G.; Yang, L.Y.; Huang, G.W.; Wang, W. Expression of HIF-2alpha/EPAS1 in hepatocellular carcinoma. *World J. Gastroenterol.* **2004**, *10*, 525–530. [CrossRef] [PubMed]
64. Wang, Y.; Kumar, S.; Rachagani, S.; Sajja, B.R.; Xie, Y.; Hang, Y.; Jain, M.; Li, J.; Boska, M.D.; Batra, S.K.; et al. Polyplex-mediated inhibition of chemokine receptor CXCR4 and chromatin-remodeling enzyme NCOA3 impedes pancreatic cancer progression and metastasis. *Biomaterials* **2016**, *101*, 108–120. [CrossRef]
65. Xie, Y.; Hang, Y.; Wang, Y.; Sleightholm, R.; Prajapati, D.R.; Bader, J.; Yu, A.; Tang, W.; Jaramillo, L.; Li, J.; et al. Stromal Modulation and Treatment of Metastatic Pancreatic Cancer with Local Intraperitoneal Triple miRNA/siRNA Nanotherapy. *ACS Nano* **2020**, *14*, 255–271. [CrossRef]
66. Lin, G.; Chen, C.K.; Yin, F.; Yang, C.; Tian, J.; Chen, T.; Xu, G.; He, C.; Lin, M.C.; Wang, J.; et al. Biodegradable nanoparticles as siRNA carriers for in vivo gene silencing and pancreatic cancer therapy. *J. Mater. Chem. B* **2017**, *5*, 3327–3337. [CrossRef]
67. Taniuchi, K.; Yawata, T.; Tsuboi, M.; Ueba, T.; Saibara, T. Efficient delivery of small interfering RNAs targeting particular mRNAs into pancreatic cancer cells inhibits invasiveness and metastasis of pancreatic tumors. *Oncotarget* **2019**, *10*, 2869–2886. [CrossRef]

68. Lo, J.H.; Hao, L.; Muzumdar, M.D.; Raghavan, S.; Kwon, E.J.; Pulver, E.M.; Hsu, F.; Aguirre, A.J.; Wolpin, B.M.; Fuchs, C.S.; et al. iRGD-guided Tumor-penetrating Nanocomplexes for Therapeutic siRNA Delivery to Pancreatic Cancer. *Mol. Cancer. Ther.* **2018**, *17*, 2377–2388. [CrossRef]
69. Zeng, L.; Li, J.; Wang, Y.; Qian, C.; Chen, Y.; Zhang, Q.; Wu, W.; Lin, Z.; Liang, J.; Shuai, X.; et al. Combination of siRNA-directed Kras oncogene silencing and arsenic-induced apoptosis using a nanomedicine strategy for the effective treatment of pancreatic cancer. *Nanomedicine* **2014**, *10*, 463–472. [CrossRef] [PubMed]
70. Li, Y.; Wang, H.; Wang, K.; Hu, Q.; Yao, Q.; Shen, Y.; Yu, G.; Tang, G. Targeted Co-delivery of PTX and TR3 siRNA by PTP Peptide Modified Dendrimer for the Treatment of Pancreatic Cancer. *Small* **2017**, *13*, 10–1002. [CrossRef] [PubMed]
71. Yoo, B.; Jordan, V.C.; Sheedy, P.; Billig, A.M.; Ross, A.; Pantazopoulos, P.; Medarova, Z. RNAi-Mediated PD-L1 Inhibition for Pancreatic Cancer Immunotherapy. *Sci. Rep.* **2019**, *9*, 4712. [CrossRef]
72. Kamerkar, S.; LeBleu, V.S.; Sugimoto, H.; Yang, S.; Ruivo, C.F.; Melo, S.A.; Lee, J.J.; Kalluri, R. Exosomes facilitate therapeutic targeting of oncogenic KRAS in pancreatic cancer. *Nature* **2017**, *546*, 498–503. [CrossRef]
73. Zhou, W.; Zhou, Y.; Chen, X.; Ning, T.; Chen, H.; Guo, Q.; Zhang, Y.; Liu, P.; Zhang, Y.; Li, C.; et al. Pancreatic cancer-targeting exosomes for enhancing immunotherapy and reprogramming tumor microenvironment. *Biomaterials* **2021**, *268*, 120546. [CrossRef]
74. Feig, C.; Gopinathan, A.; Neesse, A.; Chan, D.S.; Cook, N.; Tuveson, D.A. The pancreas cancer microenvironment. *Clin. Cancer Res.* **2012**, *18*, 4266–4276. [CrossRef] [PubMed]
75. Zheng, L.; Xue, J.; Jaffee, E.M.; Habtezion, A. Role of immune cells and immune-based therapies in pancreatitis and pancreatic ductal adenocarcinoma. *Gastroenterology* **2013**, *144*, 1230–1240. [CrossRef] [PubMed]
76. Lafaro, K.J.; Melstrom, L.G. The Paradoxical Web of Pancreatic Cancer Tumor Microenvironment. *Am. J. Pathol.* **2019**, *189*, 44–57. [CrossRef] [PubMed]
77. Schnittert, J.; Bansal, R.; Prakash, J. Targeting Pancreatic Stellate Cells in Cancer. *Trends Cancer* **2019**, *5*, 128–142. [CrossRef] [PubMed]
78. Najafi, M.; Mortezaee, K.; Majidpoor, J. Stromal reprogramming: A target for tumor therapy. *Life Sci.* **2019**, *239*, 117049. [CrossRef] [PubMed]
79. Kim, N.; Yoo, W.; Lee, J.; Kim, H.; Lee, H.; Kim, Y.S.; Kim, D.U.; Oh, J. Formation of vitamin A lipid droplets in pancreatic stellate cells requires albumin. *Gut* **2009**, *58*, 1382–1390. [CrossRef] [PubMed]
80. Neesse, A.; Algul, H.; Tuveson, D.A.; Gress, T.M. Stromal biology and therapy in pancreatic cancer: A changing paradigm. *Gut* **2015**, *64*, 1476–1484. [CrossRef] [PubMed]
81. Masamune, A.; Kikuta, K.; Watanabe, T.; Satoh, K.; Hirota, M.; Shimosegawa, T. Hypoxia stimulates pancreatic stellate cells to induce fibrosis and angiogenesis in pancreatic cancer. *Am. J. Physiol. Gastrointest. Liver Physiol.* **2008**, *295*, G709–G717. [CrossRef]
82. Kollmann, T.R.; Levy, O.; Montgomery, R.R.; Goriely, S. Innate immune function by Toll-like receptors: Distinct responses in newborns and the elderly. *Immunity* **2012**, *37*, 771–783. [CrossRef]
83. Masamune, A.; Kikuta, K.; Watanabe, T.; Satoh, K.; Satoh, A.; Shimosegawa, T. Pancreatic stellate cells express Toll-like receptors. *J. Gastroenterol.* **2008**, *43*, 352–362. [CrossRef] [PubMed]
84. Zhang, Y.; Yue, D.; Cheng, L.; Huang, A.; Tong, N.; Cheng, P. Vitamin A-coupled liposomes carrying TLR4-silencing shRNA induce apoptosis of pancreatic stellate cells and resolution of pancreatic fibrosis. *J. Mol. Med.* **2018**, *96*, 445–458. [CrossRef]
85. Ishiwatari, H.; Sato, Y.; Murase, K.; Yoneda, A.; Fujita, R.; Nishita, H.; Birukawa, N.K.; Hayashi, T.; Sato, T.; Miyanishi, K.; et al. Treatment of pancreatic fibrosis with siRNA against a collagen-specific chaperone in vitamin A-coupled liposomes. *Gut* **2013**, *62*, 1328–1339. [CrossRef] [PubMed]
86. Han, X.; Li, Y.; Xu, Y.; Zhao, X.; Zhang, Y.; Yang, X.; Wang, Y.; Zhao, R.; Anderson, G.J.; Zhao, Y.; et al. Reversal of pancreatic desmoplasia by re-educating stellate cells with a tumour microenvironment-activated nanosystem. *Nat. Commun.* **2018**, *9*, 3390. [CrossRef] [PubMed]
87. DeNardo, D.G.; Ruffell, B. Macrophages as regulators of tumour immunity and immunotherapy. *Nat. Rev. Immunol.* **2019**, *19*, 369–382. [CrossRef]
88. Argyle, D.; Kitamura, T. Targeting Macrophage-Recruiting Chemokines as a Novel Therapeutic Strategy to Prevent the Progression of Solid Tumors. *Front. Immunol.* **2018**, *9*, 2629. [CrossRef]
89. Qian, B.Z.; Li, J.; Zhang, H.; Kitamura, T.; Zhang, J.; Campion, L.R.; Kaiser, E.A.; Snyder, L.A.; Pollard, J.W. CCL2 recruits inflammatory monocytes to facilitate breast-tumour metastasis. *Nature* **2011**, *475*, 222–225. [CrossRef]
90. Noy, R.; Pollard, J.W. Tumor-associated macrophages: From mechanisms to therapy. *Immunity* **2014**, *41*, 49–61. [CrossRef]
91. Qian, Y.; Qiao, S.; Dai, Y.; Xu, G.; Dai, B.; Lu, L.; Yu, X.; Luo, Q.; Zhang, Z. Molecular-Targeted Immunotherapeutic Strategy for Melanoma via Dual-Targeting Nanoparticles Delivering Small Interfering RNA to Tumor-Associated Macrophages. *ACS Nano* **2017**, *11*, 9536–9549. [CrossRef]
92. Li, M.; Li, M.; Yang, Y.; Liu, Y.; Xie, H.; Yu, Q.; Tian, L.; Tang, X.; Ren, K.; Li, J.; et al. Remodeling tumor immune microenvironment via targeted blockade of PI3K-gamma and CSF-1/CSF-1R pathways in tumor associated macrophages for pancreatic cancer therapy. *J. Control. Release* **2020**, *321*, 23–35. [CrossRef]
93. Arora, S.; Swaminathan, S.K.; Kirtane, A.; Srivastava, S.K.; Bhardwaj, A.; Singh, S.; Panyam, J.; Singh, A.P. Synthesis, characterization, and evaluation of poly (D,L-lactide-co-glycolide)-based nanoformulation of miRNA-150: Potential implications for pancreatic cancer therapy. *Int. J. Nanomed.* **2014**, *9*, 2933–2942. [CrossRef] [PubMed]

94. Patel, G.K.; Khan, M.A.; Bhardwaj, A.; Srivastava, S.K.; Zubair, H.; Patton, M.C.; Singh, S.; Khushman, M.; Singh, A.P. Exosomes confer chemoresistance to pancreatic cancer cells by promoting ROS detoxification and miR-155-mediated suppression of key gemcitabine-metabolising enzyme, DCK. *Br. J. Cancer* **2017**, *116*, 609–619. [CrossRef]
95. Schnittert, J.; Kuninty, P.R.; Bystry, T.F.; Brock, R.; Storm, G.; Prakash, J. Anti-microRNA targeting using peptide-based nanocomplexes to inhibit differentiation of human pancreatic stellate cells. *Nanomedicine* **2017**, *12*, 1369–1384. [CrossRef] [PubMed]
96. Zhao, Y.; Zhao, L.; Ischenko, I.; Bao, Q.; Schwarz, B.; Niess, H.; Wang, Y.; Renner, A.; Mysliwietz, J.; Jauch, K.W.; et al. Antisense inhibition of microRNA-21 and microRNA-221 in tumor-initiating stem-like cells modulates tumorigenesis, metastasis, and chemotherapy resistance in pancreatic cancer. *Target Oncol.* **2015**, *10*, 535–548. [CrossRef] [PubMed]
97. Li, Y.; Chen, Y.; Li, J.; Zhang, Z.; Huang, C.; Lian, G.; Yang, K.; Chen, S.; Lin, Y.; Wang, L.; et al. Co-delivery of microRNA-21 antisense oligonucleotides and gemcitabine using nanomedicine for pancreatic cancer therapy. *Cancer Sci.* **2017**, *108*, 1493–1503. [CrossRef] [PubMed]
98. Chen, W.; Zhou, Y.; Zhi, X.; Ma, T.; Liu, H.; Chen, B.W.; Zheng, X.; Xie, S.; Zhao, B.; Feng, X.; et al. Delivery of miR-212 by chimeric peptide-condensed supramolecular nanoparticles enhances the sensitivity of pancreatic ductal adenocarcinoma to doxorubicin. *Biomaterials* **2019**, *192*, 590–600. [CrossRef]
99. Uz, M.; Kalaga, M.; Pothuraju, R.; Ju, J.; Junker, W.M.; Batra, S.K.; Mallapragada, S.; Rachagani, S. Dual delivery nanoscale device for miR-345 and gemcitabine co-delivery to treat pancreatic cancer. *J. Control. Release* **2019**, *294*, 237–246. [CrossRef]
100. Hu, Q.L.; Jiang, Q.Y.; Jin, X.; Shen, J.; Wang, K.; Li, Y.B.; Xu, F.J.; Tang, G.P.; Li, Z.H. Cationic microRNA-delivering nanovectors with bifunctional peptides for efficient treatment of PANC-1 xenograft model. *Biomaterials* **2013**, *34*, 2265–2276. [CrossRef]
101. Pramanik, D.; Campbell, N.R.; Karikari, C.; Chivukula, R.; Kent, O.A.; Mendell, J.T.; Maitra, A. Restitution of tumor suppressor microRNAs using a systemic nanovector inhibits pancreatic cancer growth in mice. *Mol. Cancer Ther.* **2011**, *10*, 1470–1480. [CrossRef]
102. Contado, C. Nanomaterials in consumer products: A challenging analytical problem. *Front. Chem.* **2015**, *3*, 48. [CrossRef]
103. Tinkle, S.; McNeil, S.E.; Muhlebach, S.; Bawa, R.; Borchard, G.; Barenholz, Y.C.; Tamarkin, L.; Desai, N. Nanomedicines: Addressing the scientific and regulatory gap. *Ann. N. Y. Acad. Sci.* **2014**, *1313*, 35–56. [CrossRef]
104. Behlke, M.A. Chemical modification of siRNAs for in vivo use. *Oligonucleotides* **2008**, *18*, 305–319. [CrossRef] [PubMed]
105. Rosenblum, D.; Joshi, N.; Tao, W.; Karp, J.M.; Peer, D. Progress and challenges towards targeted delivery of cancer therapeutics. *Nat. Commun.* **2018**, *9*, 1410. [CrossRef] [PubMed]
106. Greish, K. Enhanced permeability and retention (EPR) effect for anticancer nanomedicine drug targeting. *Methods Mol. Biol.* **2010**, *624*, 25–37. [CrossRef]
107. Azzopardi, E.A.; Ferguson, E.L.; Thomas, D.W. The enhanced permeability retention effect: A new paradigm for drug targeting in infection. *J. Antimicrob. Chemother.* **2013**, *68*, 257–274. [CrossRef] [PubMed]
108. Shi, J.; Kantoff, P.W.; Wooster, R.; Farokhzad, O.C. Cancer nanomedicine: Progress, challenges and opportunities. *Nat. Rev. Cancer* **2017**, *17*, 20–37. [CrossRef]
109. Attia, M.F.; Anton, N.; Wallyn, J.; Omran, Z.; Vandamme, T.F. An overview of active and passive targeting strategies to improve the nanocarriers efficiency to tumour sites. *J. Pharm. Pharmacol.* **2019**, *71*, 1185–1198. [CrossRef]
110. Kulkarni, J.A.; Witzigmann, D.; Chen, S.; Cullis, P.R.; van der Meel, R. Lipid Nanoparticle Technology for Clinical Translation of siRNA Therapeutics. *Acc. Chem. Res.* **2019**, *52*, 2435–2444. [CrossRef]
111. Yonezawa, S.; Koide, H.; Asai, T. Recent advances in siRNA delivery mediated by lipid-based nanoparticles. *Adv. Drug Deliv. Rev.* **2020**, *154–155*, 64–78. [CrossRef]
112. Kim, Y.K. RNA Therapy: Current Status and Future Potential. *Chonnam Med. J.* **2020**, *56*, 87–93. [CrossRef] [PubMed]
113. Rao, D.D.; Maples, P.B.; Senzer, N.; Kumar, P.; Wang, Z.; Pappen, B.O.; Yu, Y.; Haddock, C.; Jay, C.; Phadke, A.P.; et al. Enhanced target gene knockdown by a bifunctional shRNA: A novel approach of RNA interference. *Cancer Gene Ther.* **2010**, *17*, 780–791. [CrossRef] [PubMed]
114. Santel, A.; Aleku, M.; Keil, O.; Endruschat, J.; Esche, V.; Durieux, B.; Loffler, K.; Fechtner, M.; Rohl, T.; Fisch, G.; et al. RNA interference in the mouse vascular endothelium by systemic administration of siRNA-lipoplexes for cancer therapy. *Gene Ther.* **2006**, *13*, 1360–1370. [CrossRef] [PubMed]
115. Khan, K.H.; Yap, T.A.; Yan, L.; Cunningham, D. Targeting the PI3K-AKT-mTOR signaling network in cancer. *Chin. J. Cancer* **2013**, *32*, 253–265. [CrossRef]
116. Jopling, C.L.; Yi, M.; Lancaster, A.M.; Lemon, S.M.; Sarnow, P. Modulation of hepatitis C virus RNA abundance by a liver-specific MicroRNA. *Science* **2005**, *309*, 1577–1581. [CrossRef]
117. Ohno, S.; Takanashi, M.; Sudo, K.; Ueda, S.; Ishikawa, A.; Matsuyama, N.; Fujita, K.; Mizutani, T.; Ohgi, T.; Ochiya, T.; et al. Systemically injected exosomes targeted to EGFR deliver antitumor microRNA to breast cancer cells. *Mol. Ther.* **2013**, *21*, 185–191. [CrossRef]
118. Wu, Y.; Crawford, M.; Mao, Y.; Lee, R.J.; Davis, I.C.; Elton, T.S.; Lee, L.J.; Nana-Sinkam, S.P. Therapeutic Delivery of MicroRNA-29b by Cationic Lipoplexes for Lung Cancer. *Mol. Ther. Nucleic Acids* **2013**, *2*, e84. [CrossRef]
119. Drury, R.E.; O'Connor, D.; Pollard, A.J. The Clinical Application of MicroRNAs in Infectious Disease. *Front. Immunol.* **2017**, *8*, 1182. [CrossRef]

120. Beg, M.S.; Brenner, A.J.; Sachdev, J.; Borad, M.; Kang, Y.K.; Stoudemire, J.; Smith, S.; Bader, A.G.; Kim, S.; Hong, D.S. Phase I study of MRX34, a liposomal miR-34a mimic, administered twice weekly in patients with advanced solid tumors. *Investig. New Drugs* **2017**, *35*, 180–188. [CrossRef]
121. Hanna, J.; Hossain, G.S.; Kocerha, J. The Potential for microRNA Therapeutics and Clinical Research. *Front. Genet.* **2019**, *10*, 478. [CrossRef] [PubMed]
122. Zhang, S.; Cheng, Z.; Wang, Y.; Han, T. The Risks of miRNA Therapeutics: In a Drug Target Perspective. *Drug Des. Devel. Ther.* **2021**, *15*, 721–733. [CrossRef]
123. Sardar, R.; Funston, A.M.; Mulvaney, P.; Murray, R.W. Gold nanoparticles: Past, present, and future. *Langmuir* **2009**, *25*, 13840–13851. [CrossRef] [PubMed]
124. Carabineiro, S.A.C. Applications of Gold Nanoparticles in Nanomedicine: Recent Advances in Vaccines. *Molecules* **2017**, *22*, 857. [CrossRef] [PubMed]
125. Jobling, P.; Pundavela, J.; Oliveira, S.M.; Roselli, S.; Walker, M.M.; Hondermarck, H. Nerve-Cancer Cell Cross-talk: A Novel Promoter of Tumor Progression. *Cancer Res.* **2015**, *75*, 1777–1781. [CrossRef]
126. Wang, Z.; Wu, H.; Shi, H.; Wang, M.; Huang, C.; Jia, N. A novel multifunctional biomimetic Au@BSA nanocarrier as a potential siRNA theranostic nanoplatform. *J. Mater. Chem. B* **2016**, *4*, 2519–2526. [CrossRef]
127. Soetaert, F.; Korangath, P.; Serantes, D.; Fiering, S.; Ivkov, R. Cancer therapy with iron oxide nanoparticles: Agents of thermal and immune therapies. *Adv. Drug Deliv. Rev.* **2020**, *163–164*, 65–83. [CrossRef]
128. Yin, F.; Hu, K.; Chen, Y.; Yu, M.; Wang, D.; Wang, Q.; Yong, K.T.; Lu, F.; Liang, Y.; Li, Z. SiRNA Delivery with PEGylated Graphene Oxide Nanosheets for Combined Photothermal and Genetherapy for Pancreatic Cancer. *Theranostics* **2017**, *7*, 1133–1148. [CrossRef]
129. Anderson, T.; Hu, R.; Yang, C.; Yoon, H.S.; Yong, K.T. Pancreatic cancer gene therapy using an siRNA-functionalized single walled carbon nanotubes (SWNTs) nanoplex. *Biomater. Sci.* **2014**, *2*, 1244–1253. [CrossRef]
130. Suk, J.S.; Xu, Q.; Kim, N.; Hanes, J.; Ensign, L.M. PEGylation as a strategy for improving nanoparticle-based drug and gene delivery. *Adv. Drug Deliv. Rev.* **2016**, *99*, 28–51. [CrossRef]
131. Hofmann-Amtenbrink, M.; Grainger, D.W.; Hofmann, H. Nanoparticles in medicine: Current challenges facing inorganic nanoparticle toxicity assessments and standardizations. *Nanomedicine* **2015**, *11*, 1689–1694. [CrossRef]
132. Sadat Tabatabaei Mirakabad, F.; Nejati-Koshki, K.; Akbarzadeh, A.; Yamchi, M.R.; Milani, M.; Zarghami, N.; Zeighamian, V.; Rahimzadeh, A.; Alimohammadi, S.; Hanifehpour, Y.; et al. PLGA-based nanoparticles as cancer drug delivery systems. *Asian Pac. J. Cancer Prev.* **2014**, *15*, 517–535. [CrossRef] [PubMed]
133. Pan, X.; Zhu, Q.; Sun, Y.; Li, L.; Zhu, Y.; Zhao, Z.; Zuo, J.; Fang, W.; Li, K. PLGA/poloxamer nanoparticles loaded with EPAS1 siRNA for the treatment of pancreatic cancer in vitro and in vivo. *Int. J. Mol. Med.* **2015**, *35*, 995–1002. [CrossRef] [PubMed]
134. Chen, C.K.; Law, W.C.; Aalinkeel, R.; Nair, B.; Kopwitthaya, A.; Mahajan, S.D.; Reynolds, J.L.; Zou, J.; Schwartz, S.A.; Prasad, P.N.; et al. Well-defined degradable cationic polylactide as nanocarrier for the delivery of siRNA to silence angiogenesis in prostate cancer. *Adv. Healthc. Mater.* **2012**, *1*, 751–761. [CrossRef] [PubMed]
135. Garg, U.; Chauhan, S.; Nagaich, U.; Jain, N. Current Advances in Chitosan Nanoparticles Based Drug Delivery and Targeting. *Adv. Pharm. Bull.* **2019**, *9*, 195–204. [CrossRef] [PubMed]
136. Patel, A.; Patel, M.; Yang, X.; Mitra, A.K. Recent advances in protein and Peptide drug delivery: A special emphasis on polymeric nanoparticles. *Protein Pept. Lett.* **2014**, *21*, 1102–1120. [CrossRef] [PubMed]
137. Alipour, M.; Baneshi, M.; Hosseinkhani, S.; Mahmoudi, R.; Jabari Arabzadeh, A.; Akrami, M.; Mehrzad, J.; Bardania, H. Recent progress in biomedical applications of RGD-based ligand: From precise cancer theranostics to biomaterial engineering: A systematic review. *J. Biomed. Mater. Res. A* **2020**, *108*, 839–850. [CrossRef] [PubMed]
138. Pan, B.T.; Johnstone, R.M. Fate of the transferrin receptor during maturation of sheep reticulocytes in vitro: Selective externalization of the receptor. *Cell* **1983**, *33*, 967–978. [CrossRef]
139. Thery, C.; Zitvogel, L.; Amigorena, S. Exosomes: Composition, biogenesis and function. *Nat. Rev. Immunol.* **2002**, *2*, 569–579. [CrossRef] [PubMed]
140. Kim, Y.K.; Choi, Y.; Nam, G.H.; Kim, I.S. Functionalized exosome harboring bioactive molecules for cancer therapy. *Cancer Lett.* **2020**, *489*, 155–162. [CrossRef]
141. Nam, G.H.; Choi, Y.; Kim, G.B.; Kim, S.; Kim, S.A.; Kim, I.S. Emerging Prospects of Exosomes for Cancer Treatment: From Conventional Therapy to Immunotherapy. *Adv. Mater.* **2020**, *32*, e2002440. [CrossRef] [PubMed]
142. Kim, G.B.; Nam, G.H.; Hong, Y.; Woo, J.; Cho, Y.; Kwon, I.C.; Yang, Y.; Kim, I.S. Xenogenization of tumor cells by fusogenic exosomes in tumor microenvironment ignites and propagates antitumor immunity. *Sci. Adv.* **2020**, *6*, eaaz2083. [CrossRef] [PubMed]
143. Liu, H.; Huang, L.; Mao, M.; Ding, J.; Wu, G.; Fan, W.; Yang, T.; Zhang, M.; Huang, Y.; Xie, H.Y. Viral Protein-Pseudotyped and siRNA-Electroporated Extracellular Vesicles for Cancer Immunotherapy. *Adv. Funct. Mater.* **2020**, *30*, 2006515. [CrossRef]
144. Liu, C.; Su, C. Design strategies and application progress of therapeutic exosomes. *Theranostics* **2019**, *9*, 1015–1028. [CrossRef]
145. O'Brien, K.; Lowry, M.C.; Corcoran, C.; Martinez, V.G.; Daly, M.; Rani, S.; Gallagher, W.M.; Radomski, M.W.; MacLeod, R.A.; O'Driscoll, L. miR-134 in extracellular vesicles reduces triple-negative breast cancer aggression and increases drug sensitivity. *Oncotarget* **2015**, *6*, 32774–32789. [CrossRef]

146. Li, H.; Yang, C.; Shi, Y.; Zhao, L. Exosomes derived from siRNA against GRP78 modified bone-marrow-derived mesenchymal stem cells suppress Sorafenib resistance in hepatocellular carcinoma. *J. Nanobiotechnol.* **2018**, *16*, 103. [CrossRef]
147. Wang, Y.; Chen, X.; Tian, B.; Liu, J.; Yang, L.; Zeng, L.; Chen, T.; Hong, A.; Wang, X. Nucleolin-targeted Extracellular Vesicles as a Versatile Platform for Biologics Delivery to Breast Cancer. *Theranostics* **2017**, *7*, 1360–1372. [CrossRef]
148. Morad, G.; Carman, C.V.; Hagedorn, E.J.; Perlin, J.R.; Zon, L.I.; Mustafaoglu, N.; Park, T.E.; Ingber, D.E.; Daisy, C.C.; Moses, M.A. Tumor-Derived Extracellular Vesicles Breach the Intact Blood-Brain Barrier via Transcytosis. *ACS Nano* **2019**, *13*, 13853–13865. [CrossRef]
149. Yang, T.; Fogarty, B.; LaForge, B.; Aziz, S.; Pham, T.; Lai, L.; Bai, S. Delivery of Small Interfering RNA to Inhibit Vascular Endothelial Growth Factor in Zebrafish Using Natural Brain Endothelia Cell-Secreted Exosome Nanovesicles for the Treatment of Brain Cancer. *AAPS J.* **2017**, *19*, 475–486. [CrossRef] [PubMed]
150. Katakowski, M.; Buller, B.; Zheng, X.; Lu, Y.; Rogers, T.; Osobamiro, O.; Shu, W.; Jiang, F.; Chopp, M. Exosomes from marrow stromal cells expressing miR-146b inhibit glioma growth. *Cancer Lett.* **2013**, *335*, 201–204. [CrossRef]
151. Gourlay, J.; Morokoff, A.P.; Luwor, R.B.; Zhu, H.J.; Kaye, A.H.; Stylli, S.S. The emergent role of exosomes in glioma. *J. Clin. Neurosci.* **2017**, *35*, 13–23. [CrossRef] [PubMed]
152. Mendt, M.; Kamerkar, S.; Sugimoto, H.; McAndrews, K.M.; Wu, C.C.; Gagea, M.; Yang, S.; Blanko, E.V.R.; Peng, Q.; Ma, X.; et al. Generation and testing of clinical-grade exosomes for pancreatic cancer. *JCI Insight* **2018**, *3*, e99263. [CrossRef] [PubMed]
153. Babu, A.; Munshi, A.; Ramesh, R. Combinatorial therapeutic approaches with RNAi and anticancer drugs using nanodrug delivery systems. *Drug Dev. Ind. Pharm.* **2017**, *43*, 1391–1401. [CrossRef]
154. Creixell, M.; Peppas, N.A. Co-delivery of siRNA and therapeutic agents using nanocarriers to overcome cancer resistance. *Nano Today* **2012**, *7*, 367–379. [CrossRef] [PubMed]
155. Jain, S.; Pathak, K.; Vaidya, A. Molecular therapy using siRNA: Recent trends and advances of multi target inhibition of cancer growth. *Int. J. Biol. Macromol.* **2018**, *116*, 880–892. [CrossRef]
156. Yu, Q.; Zhang, B.; Zhou, Y.; Ge, Q.; Chang, J.; Chen, Y.; Zhang, K.; Peng, D.; Chen, W. Co-delivery of gambogenic acid and VEGF-siRNA with anionic liposome and polyethylenimine complexes to HepG2 cells. *J. Liposome Res.* **2019**, *29*, 322–331. [CrossRef]
157. Golan, T.; Khvalevsky, E.Z.; Hubert, A.; Gabai, R.M.; Hen, N.; Segal, A.; Domb, A.; Harari, G.; David, E.B.; Raskin, S.; et al. RNAi therapy targeting KRAS in combination with chemotherapy for locally advanced pancreatic cancer patients. *Oncotarget* **2015**, *6*, 24560–24570. [CrossRef]
158. McCarroll, J.A.; Sharbeen, G.; Liu, J.; Youkhana, J.; Goldstein, D.; McCarthy, N.; Limbri, L.F.; Dischl, D.; Ceyhan, G.O.; Erkan, M.; et al. betaIII-tubulin: A novel mediator of chemoresistance and metastases in pancreatic cancer. *Oncotarget* **2015**, *6*, 2235–2249. [CrossRef]
159. Lin, Y.X.; Wang, Y.; Blake, S.; Yu, M.; Mei, L.; Wang, H.; Shi, J. RNA Nanotechnology-Mediated Cancer Immunotherapy. *Theranostics* **2020**, *10*, 281–299. [CrossRef]
160. Ma, J.; Dong, C.; Ji, C. MicroRNA and drug resistance. *Cancer Gene Ther.* **2010**, *17*, 523–531. [CrossRef] [PubMed]
161. Amrutkar, M.; Gladhaug, I.P. Pancreatic Cancer Chemoresistance to Gemcitabine. *Cancers* **2017**, *9*, 157. [CrossRef] [PubMed]
162. Frixa, T.; Donzelli, S.; Blandino, G. Oncogenic MicroRNAs: Key Players in Malignant Transformation. *Cancers* **2015**, *7*, 2466–2485. [CrossRef] [PubMed]
163. Feng, Y.H.; Tsao, C.J. Emerging role of microRNA-21 in cancer. *Biomed. Rep.* **2016**, *5*, 395–402. [CrossRef]
164. Kumar, V.; Mondal, G.; Slavik, P.; Rachagani, S.; Batra, S.K.; Mahato, R.I. Codelivery of small molecule hedgehog inhibitor and miRNA for treating pancreatic cancer. *Mol. Pharm.* **2015**, *12*, 1289–1298. [CrossRef]
165. Hidalgo, M.; Cascinu, S.; Kleeff, J.; Labianca, R.; Lohr, J.M.; Neoptolemos, J.; Real, F.X.; Van Laethem, J.L.; Heinemann, V. Addressing the challenges of pancreatic cancer: Future directions for improving outcomes. *Pancreatology* **2015**, *15*, 8–18. [CrossRef]
166. Lewis, W.H. Pinocytosis by Malignant Cells. *Am. J. Cancer* **1937**, *29*, 666. [CrossRef]
167. DeBerardinis, R.J.; Sayed, N.; Ditsworth, D.; Thompson, C.B. Brick by brick: Metabolism and tumor cell growth. *Curr. Opin. Genet. Dev.* **2008**, *18*, 54–61. [CrossRef]
168. Commisso, C.; Davidson, S.M.; Soydaner-Azeloglu, R.G.; Parker, S.J.; Kamphorst, J.J.; Hackett, S.; Grabocka, E.; Nofal, M.; Drebin, J.A.; Thompson, C.B.; et al. Macropinocytosis of protein is an amino acid supply route in Ras-transformed cells. *Nature* **2013**, *497*, 633–637. [CrossRef] [PubMed]
169. Maltese, W.A.; Overmeyer, J.H. Non-apoptotic cell death associated with perturbations of macropinocytosis. *Front. Physiol.* **2015**, *6*, 38. [CrossRef] [PubMed]
170. Recouvreux, M.V.; Commisso, C. Macropinocytosis: A Metabolic Adaptation to Nutrient Stress in Cancer. *Front. Endocrinol.* **2017**, *8*, 261. [CrossRef]
171. Bar-Sagi, D.; Feramisco, J.R. Induction of membrane ruffling and fluid-phase pinocytosis in quiescent fibroblasts by ras proteins. *Science* **1986**, *233*, 1061. [CrossRef]
172. Amyere, M.; Payrastre, B.; Krause, U.; Van Der Smissen, P.; Veithen, A.; Courtoy, P.J. Constitutive macropinocytosis in oncogene-transformed fibroblasts depends on sequential permanent activation of phosphoinositide 3-kinase and phospholipase C. *Mol. Biol. Cell* **2000**, *11*, 3453–3467. [CrossRef]

173. Kim, S.M.; Nguyen, T.T.; Ravi, A.; Kubiniok, P.; Finicle, B.T.; Jayashankar, V.; Malacrida, L.; Hou, J.; Robertson, J.; Gao, D.; et al. PTEN Deficiency and AMPK Activation Promote Nutrient Scavenging and Anabolism in Prostate Cancer Cells. *Cancer Discov.* **2018**, *8*, 866–883. [CrossRef] [PubMed]
174. Hobbs, G.A.; Baker, N.M.; Miermont, A.M.; Thurman, R.D.; Pierobon, M.; Tran, T.H.; Anderson, A.O.; Waters, A.M.; Diehl, J.N.; Papke, B.; et al. Atypical KRAS^{G12R} Mutant Is Impaired in PI3K Signaling and Macropinocytosis in Pancreatic Cancer. *Cancer Discov.* **2020**, *10*, 104. [CrossRef] [PubMed]
175. Love, K.T.; Mahon, K.P.; Levins, C.G.; Whitehead, K.A.; Querbes, W.; Dorkin, J.R.; Qin, J.; Cantley, W.; Qin, L.L.; Racie, T.; et al. Lipid-like materials for low-dose, in vivo gene silencing. *Proc. Natl. Acad. Sci. USA* **2010**, *107*, 1864–1869. [CrossRef] [PubMed]
176. Gilleron, J.; Querbes, W.; Zeigerer, A.; Borodovsky, A.; Marsico, G.; Schubert, U.; Manygoats, K.; Seifert, S.; Andree, C.; Stöter, M.; et al. Image-based analysis of lipid nanoparticle-mediated siRNA delivery, intracellular trafficking and endosomal escape. *Nat. Biotechnol.* **2013**, *31*, 638–646. [CrossRef] [PubMed]
177. Sahay, G.; Querbes, W.; Alabi, C.; Eltoukhy, A.; Sarkar, S.; Zurenko, C.; Karagiannis, E.; Love, K.; Chen, D.; Zoncu, R.; et al. Efficiency of siRNA delivery by lipid nanoparticles is limited by endocytic recycling. *Nat. Biotechnol.* **2013**, *31*, 653–658. [CrossRef]
178. Asai, T.; Tsuzuku, T.; Takahashi, S.; Okamoto, A.; Dewa, T.; Nango, M.; Hyodo, K.; Ishihara, H.; Kikuchi, H.; Oku, N. Cell-penetrating peptide-conjugated lipid nanoparticles for siRNA delivery. *Biochem. Biophys. Res. Commun.* **2014**, *444*, 599–604. [CrossRef]
179. Dong, Y.; Love, K.T.; Dorkin, J.R.; Sirirungruang, S.; Zhang, Y.; Chen, D.; Bogorad, R.L.; Yin, H.; Chen, Y.; Vegas, A.J.; et al. Lipopeptide nanoparticles for potent and selective siRNA delivery in rodents and nonhuman primates. *Proc. Natl. Acad. Sci. USA* **2014**, *111*, 3955–3960. [CrossRef]
180. Huang, J.L.; Jiang, G.; Song, Q.X.; Gu, X.; Hu, M.; Wang, X.L.; Song, H.H.; Chen, L.P.; Lin, Y.Y.; Jiang, D.; et al. Lipoprotein-biomimetic nanostructure enables efficient targeting delivery of siRNA to Ras-activated glioblastoma cells via macropinocytosis. *Nat. Commun.* **2017**, *8*, 15144. [CrossRef]
181. Hamasaki, M.; Araki, N.; Hatae, T. Association of early endosomal autoantigen 1 with macropinocytosis in EGF-stimulated A431 cells. *Anat. Rec. Part A Discov. Mol. Cell. Evol. Biol.* **2004**, *277*, 298–306. [CrossRef]
182. Hewlett, L.J.; Prescott, A.R.; Watts, C. The coated pit and macropinocytic pathways serve distinct endosome populations. *J. Cell Biol.* **1994**, *124*, 689–703. [CrossRef] [PubMed]
183. Liberali, P.; Kakkonen, E.; Turacchio, G.; Valente, C.; Spaar, A.; Perinetti, G.; Böckmann, R.A.; Corda, D.; Colanzi, A.; Marjomaki, V.; et al. The closure of Pak1-dependent macropinosomes requires the phosphorylation of CtBP1/BARS. *EMBO J.* **2008**, *27*, 970–981. [CrossRef] [PubMed]
184. Dominska, M.; Dykxhoorn, D.M. Breaking down the barriers: siRNA delivery and endosome escape. *J. Cell Sci.* **2010**, *123*, 1183–1189. [CrossRef] [PubMed]
185. Juliano, R.L. Intracellular Trafficking and Endosomal Release of Oligonucleotides: What We Know and What We Don't. *Nucleic Acid Ther.* **2018**, *28*, 166–177. [CrossRef] [PubMed]
186. Ivanov, A.I. Pharmacological inhibition of endocytic pathways: Is it specific enough to be useful? *Methods Mol. Biol.* **2008**, *440*, 15–33. [CrossRef] [PubMed]
187. Lin, X.P.; Mintern, J.D.; Gleeson, P.A. Macropinocytosis in Different Cell Types: Similarities and Differences. *Membranes* **2020**, *10*, 177. [CrossRef] [PubMed]
188. Dias Carvalho, P.; Machado, A.L.; Martins, F.; Seruca, R.; Velho, S. Targeting the Tumor Microenvironment: An Unexplored Strategy for Mutant KRAS Tumors. *Cancers* **2019**, *11*, 2010. [CrossRef]
189. Monty, M.A.; Islam, M.A.; Nan, X.; Tan, J.; Tuhin, I.J.; Tang, X.; Miao, M.; Wu, D.; Yu, L. Emerging role of RNA interference in immune cells engineering and its therapeutic synergism in immunotherapy. *Br. J. Pharmacol.* **2021**, *178*, 1741–1755. [CrossRef]
190. Hu, B.; Zhong, L.; Weng, Y.; Peng, L.; Huang, Y.; Zhao, Y.; Liang, X.J. Therapeutic siRNA: State of the art. *Signal Transduct. Target Ther.* **2020**, *5*, 101. [CrossRef]
191. Herreros-Villanueva, M.; Hijona, E.; Cosme, A.; Bujanda, L. Mouse models of pancreatic cancer. *World J. Gastroenterol.* **2012**, *18*, 1286–1294. [CrossRef] [PubMed]
192. Tomas-Bort, E.; Kieler, M.; Sharma, S.; Candido, J.B.; Loessner, D. 3D approaches to model the tumor microenvironment of pancreatic cancer. *Theranostics* **2020**, *10*, 5074–5089. [CrossRef] [PubMed]
193. Zeeberg, K.; Cardone, R.A.; Greco, M.R.; Saccomano, M.; Nohr-Nielsen, A.; Alves, F.; Pedersen, S.F.; Reshkin, S.J. Assessment of different 3D culture systems to study tumor phenotype and chemosensitivity in pancreatic ductal adenocarcinoma. *Int. J. Oncol.* **2016**, *49*, 243–252. [CrossRef] [PubMed]
194. Lin, C.C.; Korc, M. Designer hydrogels: Shedding light on the physical chemistry of the pancreatic cancer microenvironment. *Cancer Lett.* **2018**, *436*, 22–27. [CrossRef] [PubMed]
195. Shichi, Y.; Sasaki, N.; Michishita, M.; Hasegawa, F.; Matsuda, Y.; Arai, T.; Gomi, F.; Aida, J.; Takubo, K.; Toyoda, M.; et al. Enhanced morphological and functional differences of pancreatic cancer with epithelial or mesenchymal characteristics in 3D culture. *Sci. Rep.* **2019**, *9*, 10871. [CrossRef] [PubMed]
196. Beer, M.; Kuppalu, N.; Stefanini, M.; Becker, H.; Schulz, I.; Manoli, S.; Schuette, J.; Schmees, C.; Casazza, A.; Stelzle, M.; et al. A novel microfluidic 3D platform for culturing pancreatic ductal adenocarcinoma cells: Comparison with in vitro cultures and in vivo xenografts. *Sci. Rep.* **2017**, *7*, 1325. [CrossRef]
197. Norberg, K.J.; Liu, X.; Fernandez Moro, C.; Strell, C.; Nania, S.; Blumel, M.; Balboni, A.; Bozoky, B.; Heuchel, R.L.; Lohr, J.M. A novel pancreatic tumour and stellate cell 3D co-culture spheroid model. *BMC Cancer* **2020**, *20*, 475. [CrossRef]

pharmaceutics

MDPI

Review

Biomimetic Bacterial Membrane Vesicles for Drug Delivery Applications

Sajid Fazal and Ruda Lee *

International Research Organization for Advanced Science and Technology, Kumamoto University, 2-39-1 Kurokami, Chuo-ku, Kumamoto 860-8555, Japan; f.r.sajid@gmail.com
* Correspondence: aeju-lee@kumamoto-u.ac.jp; Tel.: +81-96-342-3403

Abstract: Numerous factors need to be considered to develop a nanodrug delivery system that is biocompatible, non-toxic, easy to synthesize, cost-effective, and feasible for scale up over and above their therapeutic efficacy. With regards to this, worldwide, exosomes, which are nano-sized vesicles obtained from mammalian cells, are being explored as a biomimetic drug delivery system that has superior biocompatibility and high translational capability. However, the economics of undertaking large-scale mammalian culture to derive exosomal vesicles for translation seems to be challenging and unfeasible. Recently, Bacterial Membrane Vesicles (BMVs) derived from bacteria are being explored as a viable alternative as biomimetic drug delivery systems that can be manufactured relatively easily at much lower costs at a large scale. Until now, BMVs have been investigated extensively as successful immunomodulating agents, but their capability as drug delivery systems remains to be explored in detail. In this review, the use of BMVs as suitable cargo delivery vehicles is discussed with focus on their use for in vivo treatment of cancer and bacterial infections reported thus far. Additionally, the different types of BMVs, factors affecting their synthesis and different cargo loading techniques used in BMVs are also discussed.

Keywords: biomimetics; bacterial membrane vesicles; nanoparticles; drug delivery; antibiotic therapy

Citation: Fazal, S.; Lee, R. Biomimetic Bacterial Membrane Vesicles for Drug Delivery Applications. *Pharmaceutics* **2021**, *13*, 1430. https://doi.org/10.3390/pharmaceutics13091430

Academic Editor: Lucio Barile

Received: 19 July 2021
Accepted: 5 September 2021
Published: 9 September 2021

1. Introduction

For developing translatable engineered nanomedical systems for therapeutic and diagnostic applications, it is essential to consider the various different engineering and biological roadblocks these would encounter on the path to translation. From the engineering standpoint, large scale uniform production of nanoparticle (NP) systems is difficult to achieve primarily because of the complexity in their design [1,2], which leads to manufacturing difficulties in scale-up, quality control issues, downstream purification complexities, and increased cost of production. While from the biological standpoint, the immunogenicity of the NP system as a whole as well as its individual components, nanotoxicity, and overall therapeutic efficacy can further hinder its translatability [3,4] (Figure 1). In the past decade, there has been a collective effort towards addressing these issues, especially with concerns regarding the immunogenicity and biocompatibility of the designed NPs. Specifically, there has been a rise in research related to the development of NP systems that partly resemble or mimic non-immunogenic biological entities called as biomimetic/bioinspired NPs [5–7]. Such biomimetic/bioinspired NPs are considered to not only be non-immunogenic with reduced toxicity, but also possess superior pharmacokinetic properties [8–10] (due to lower macrophage clearance), paving the way for extensive research regarding their use for various therapeutic and imaging applications.

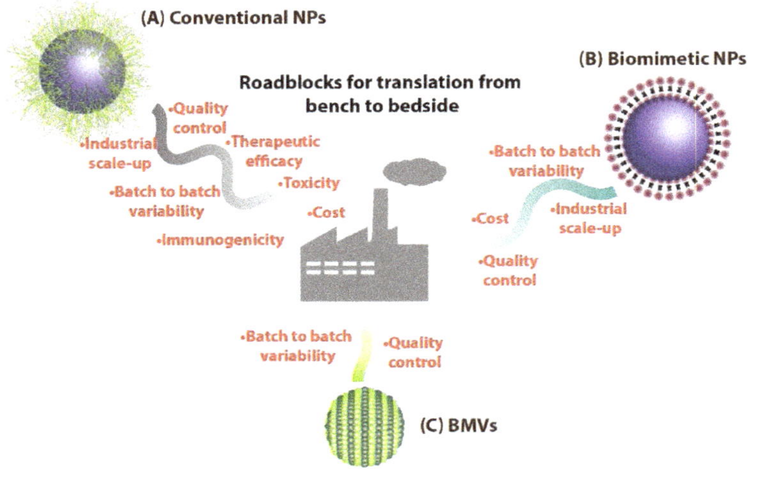

Figure 1. (**A**) Conventional NPs synthesized routinely in laboratories could face many more roadblocks on the path to clinical translation as compared to (**B**) biomimetic NPs. On the other hand, (**C**) BMV-based nanomedical systems could benefit over other biomimetic NPs as they have the potential to be easily mass produced.

These biomimetic/bioinspired nanoparticles are considered to possess several advantages as compared to conventional NP systems composed of polymeric and inorganic materials. The most important advantage that they provide is their high biodegradability and non-toxic degradation products that can be easily cleared from the body without eliciting any long-term toxicity and bioaccumulation effects. Additionally, as the building blocks for the synthesis of such NPs are biomolecules, they can be harnessed directly from biological sources, which in some cases (e.g., albumin) can be obtained in large scale at low cost. Biomimetic/bioinspired NPs have been reported to be synthesized through a number of different materials and approaches which include (a) nanoparticles derived directly from biomolecules, (b) biomaterial coated nanoparticles, and (c) cell membrane-coated nanoparticles (Table 1, Figure 2). Nanoparticles that are derived directly from biomolecules are designed using a bottom to top synthesis strategy wherein biologically derived components such as albumin [11], casein [12,13], starch [14,15], gelatin [16,17], etc. are used directly to engineer and assemble the NP system that is loaded with a cargo of interest. Of note, the FDA-approved nanomedicine Abraxane®, which consists of the chemotherapeutic drug paclitaxel bound to albumin [18], is a prime example of this type of NP. Additionally, these biomolecules have also been used for coating NP surfaces in order to combine their biocompatibility with the desired functional ability of engineered NPs [19,20]. This strategy is particularly important since the pharmacokinetic property of a drug delivery system is primarily dependent on its surface physicochemical properties. By appropriately coating a functional polymeric/inorganic NP system with suitable biomolecules (Table 1), its interaction with different blood components in the body (including proteins and macrophages) is favorably altered so as to impart improved circulation time, higher bioavailability, and reduced clearance. Another widely used approach that has gained widespread attention involves the surface modification of synthetic NPs with cell membrane components [21] obtained from cells such as erythrocytes [22–24], platelets [25–27], macrophages [28–30], etc. This leads to the creation of a phospholipid bilayer surface on the NPs, thus affording them advantageous properties similar to that of liposomal structures. This strategy can be considered to be superior as compared to biomolecule coating, since the transfer of cell membrane components and the corresponding functional proteins (including cell membrane receptors, signaling proteins, etc.) provide added functionality to the NP system as a whole such as cancer cell targeting, immune evasion, biobarrier penetration, etc.

Table 1. Different types of biomimetic/bioinspired NPs reported in literature.

Type	Biological Source	Cargo Loaded/NP	Application	Reference
Biomolecule assembly	Human Serum albumin	Indocyanine green	Active targeting and photothermal therapy of NIH-3T6.7 tumor (in vivo)	[31]
	Casein	10-hydroxycamptothecin	Drug delivery to C6 glioma tumor (in vivo)	[32]
	Human transferrin	Near-infrared dye IR-780	Photodynamic and photothermal therapy of CT26 colon carcinoma (in vivo)	[33]
	Human H-ferritin	Doxorubicin	Drug delivery to U87MG human glioma	[34]
Surface modification	Bovine serum albumin	Silver NPs	Photothermal ablation of B16F10 murine melanoma (in vitro)	[35]
	Casein	Iron-Oxide NPs	Active EGFR targeting (in vitro) and MRI contrast (in vivo)	[36]
	High density lipoprotein	gold NPs	Nucleic acid delivery to PC3 prostate cancer cells (in vitro)	[37]
Mammalian cell membrane-coated NPs	Erythrocytes	poly(lactic-coglycolic) acid NPs	Toxin removal-demonstrated in mouse sepsis model	[22]
	Neural stem cells	poly(lactic-coglycolic) acid NPs	Glyburide delivery for stroke treatment (in vivo)	[38]
	Platelets	poly(lactic-coglycolic) acid NPs	Rapamycin delivery for atherosclerosis treatment (in vivo)	[26]
	Mouse leukemia cell C1498	poly(lactic-coglycolic) acid NPs	Active targeting and delivery of dexamethasone del for treatment of lung infection (in vivo)	[39]

In contrast to the above methods of synthesis of biomimetic/bioinspired NPs, there is another approach that has recently gained widespread attention in the area of nanomedicine research. This involves the direct use of naturally synthesized extracellular vesicles [40] that are ubiquitously found to be produced by all cells (Figure 2D). Among the different types of extracellular vesicles such as microvesicles, exosomes and apoptotic bodies, the nano-sized extracellular vesicles called 'exosomes' obtained either from mammalian cells [41–43] or 'Bacterial Membrane Vesicles (BMV)' obtained from bacteria [44,45], have been demonstrated as excellent drug delivery agents particularly owing to their nanoscale size. Unlike other conventional nanoparticles that could have a solid core structure, the nanovesicle structures have a hollow hydrophilic interior (similar to liposomes) and can be used to transport drugs and other cargo. These naturally derived vesicles will exhibit a surface chemical composition identical to the parent cell from which they are obtained, and therefore would demonstrate high biocompatibility and low immunogenicity. Such favorable characteristics make them ideally suited for easy translation from a biological standpoint. Moreover, as they are biologically synthesized by cells directly, no chemical synthesis step would be required for their production and they can be produced by optimizing the cell culture and growth condition.

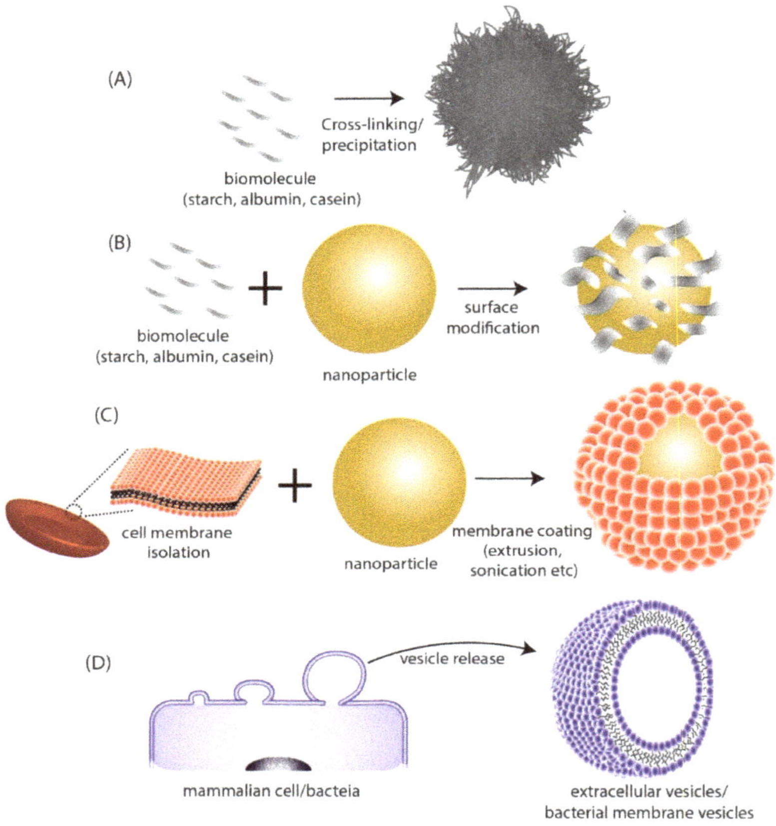

Figure 2. Depiction of different types of biomimetic/bioinspired NPs that have been reported in literature. (**A**) NPs synthesized directly from a biomolecule. (**B**) Surface modification of synthesized NPs with biomolecular structures. (**C**) Coating of synthesized NPs with cell membrane surfaces derived from mammalian cells. (**D**) Direct utilization of nano-sized extracellular vesicles/bacterial membrane vesicles isolated from mammalian or bacteria cells respectively.

Even though such biomimetic nanovesicles could make it easier to scale the biological roadblock of clinical translation, it is important to overcome the large-scale design and engineering roadblocks with regards to their synthesis. Taking this into consideration, exosomes derived from mammalian cells would be particularly difficult to translate due to the challenge involved in undertaking large-scale mammalian cell culture to obtain exosomes in large quantities, in addition to the high cost for maintaining mammalian cell culture conditions at an industrial scale [46]. In this regard, exosomes obtained from bacterial sources, i.e., BMVs, could have high translation potential. The mass production of bacteria in bacterial growth tanks would be relatively easier to accomplish and the subsequent associated costs would also be relatively low [47], as compared to mammalian cell cultures. Additionally, a unique advantage associated with utilizing bacteria for BMV synthesis is their ease of genetic engineering which can help specifically design and produce BMVs with functional moieties [48]. In this review, the biomedical applications of

BMVs are discussed with regards to their use as drug delivery vehicles for cancer therapy and antibacterial therapy. Particularly, attention will be focused on the bacterial sources for BMV production, their separation and purification, characterization techniques, drug loading strategies, and their in vivo biomedical applications reported thus far.

2. Types of BMVs and Factors Affecting Their Synthesis

Broadly, two different types of BMVs can be considered to exist based on the Gram staining of the bacterial source from which they are produced. BMVs that are secreted from Gram-negative bacteria are generally termed Outer Membrane Vesicles (OMVs), while those that are secreted from Gram-positive bacteria are simply called membrane vesicles (MVs) or Cytoplasmic Membrane Vesicles (CMVs). The reason the BMVs secreted from Gram-negative bacteria are called OMVs is because they originate from the outer membrane of the complex cell envelope [49] that encompasses the Gram-negative bacteria (Figure 3). On the other hand, MVs synthesized from Gram-positive bacteria originate directly from the cytoplasmic membrane of the simple Gram-positive bacterial cell wall [50]. Apart from these typical BMVs, several other structures such as Outer–Inner Membrane vesicles [51], Explosive Outer Membrane Vesicles [52], and Tube-shaped Membranous structures [53] have also been identified to be secreted by bacterial cells.

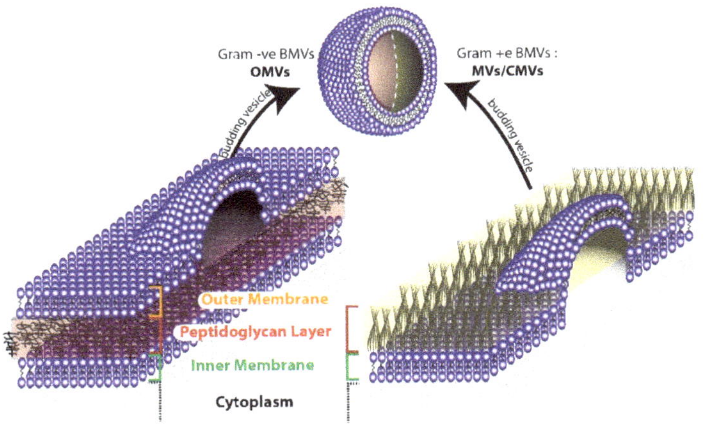

Figure 3. Origin of BMVs differ in gram negative and gram positive bacteria due to the inherent differences in the cell membrane structure.

Even though the route of synthesis of BMVs is not clearly understood, several hypotheses for the same can be made by evaluating their composition and relative concentrations of protein, lipids, and nucleic acid contents [54]. As OMVs contain lipids and proteins that are typically present in the outer membrane and periplasmic space of their parent source bacteria, they are considered to originate from the gram negative bacterial outer membrane through cell membrane blebbing mechanisms [55]. On the other hand, Outer–Inner Membrane vesicles, which consist of two membrane layers (from both the outer membrane and inner membrane), additionally contain cytoplasmic protein and DNA and are considered to originate from the inner membrane of Gram-negative bacteria, by pinching off cytoplasmic cell components [51,56,57]. Outer–Inner Membrane vesicles are sometimes also found to contain chromosomal DNA, in which case they are considered to originate due to explosive cell lysis that result in cell death [52,58]. The synthesis of OMVs in Gram-negative bacteria are thought to occur due to defects in the peptidoglycan layer of the bacterial cell wall which can lead to the dissociation of the outer membrane. These defects can arise due to a number of factors such as disrupted crosslinking between the peptidoglycan and the outer membrane [59,60], accumulation of misfolded proteins in the periplasmic space [61,62], and through 'bilayer coupling' effects that are brought about by molecules that induce membrane curvatures [63,64]. Another possible mechanism for vesicle blebbing involves the action of the endolysin/autolysin enzyme, that can degrade the peptidoglycan layer. In Gram-negative bacteria, the action of endolysin leads to membrane instability that causes explosive cell lysis and eventual vesiculation [52,65]. For Gram-positive bacteria, the cytoplasmic membrane vesicles are considered to arise either from dying cells [65] or from other conservative blebbing mechanisms [56]. Here, the action of endolysin/autolysin is also considered to play a key role in the formation of CMVs. However, as Gram-positive bacteria have a thicker peptidoglycan layer, membrane instability effects are relatively less pronounced which led to the protrusion of the cytoplasmic membrane through pores in the peptidoglycan layer eventually leading to the release of CMVs.

A number of different genetic and environmental factors can affect vesicle formation in bacteria. Genetic factors can predispose a bacterium to produce more vesicles due to the accumulation of misfolded proteins in the peptidoglycan layer or due to lipid and protein composition in the outer membrane that can affect membrane curvature or cell envelope cross-linking. An example of a hypervesiculating bacteria is the *Escherichia coli* JC8031 produced by the genetic knockout of tolRA gene, that leads to membrane instability in the *E. coli* cell envelope [66]. Due to this hypervesiculating nature, *E. coli* JC8031 has been explored for various biomedical applications including the development of vaccines [67,68] and for drug delivery (discussed below). Environmental factors including bacterial growth conditions, medium composition, and other stress factors (including thermal stress [69,70] and antibiotic stress [71]) can also increase the release of BMVs.

BMVs play important roles including intracellular communication such as horizontal gene transfer between different bacterial species [72,73], immunomodulation in a potential host [74], aiding the formation of bacterial biofilms [75], and many others. In the human body, BMVs not only play important role in bacterial infection, but also play a protective role in preventing inflammation such as from commensal bacteria that reside in the gut [76]. Information regarding their structural composition, functions and mechanism of action are not yet fully unraveled, however readers are referred to exhaustive resources [77,78] that provide up-to date knowledge regarding these.

3. BMV Source for Cargo Delivery

BMVs for cargo delivery have been reported to be procured from a wide variety of different Gram-negative (*Escherichia coli* [79], *Acinetobacter baumannii* [71], *Cystobacter velatus* [80], *Klebsiella pneumoniae* [81], *Salmonella typhimurium* [82], and *Salmonella enterica* [83]) and Gram-positive bacteria (*Staphylococcus Aureus* [84] and *Lactobacillus acidophilus* [83]) sources which have been demonstrated for use in various biomedical applications (Table 2). Unlike other biomimetic NPs such as exosomes [41] or erythrocyte membrane mimicking NPs [23], BMVs generated from bacteria have the potential to produce an immunogenic response in vivo due to the presence of LPS (Gram-negative) or LTA (Gram-positive) and other non-human bacterial proteins on the BMV surface [85]. To counter this effect or to increase their tolerability, attenuated bacterial strains such as the *E. coli* K-12 W3110 strain [86] (carrying an *msbB* mutation which produces under-acylated LPS) or the attenuated *S. Typhimurium* strains [82] have been reported, which can exhibit reduced endotoxicity when used in vivo. However, the immunogenic potential of BMVs from non-attenuated bacterial strains have also been reported successfully for different applications including immunotherapy [87] and antibiotic delivery [71]. Additionally, BMVs have also been obtained from genetically engineered bacteria such that the synthesized BMVs have surface modifications that impart specific functions (IgG [88] and anti-HER2 affibody [86]) or are pre-loaded with cargo such as enzymes (luciferase [89] and phosphotriesterase [90]) or a therapeutic agent (melanin [91]). For successful translation, however, it is imperative that BMVs be either generated in large quantities using bacterial bioreactors or through the use of hypersecreting BMV strains such as the *E. coli* JC8031 [89]. Additionally, to circumvent issues related to toxicity and safety during clinical translation, utilizing non-pathogenic commensal bacteria such as *Bacteroides thetaiotaomicron* (that are part of the intestinal microbiota) could also be a viable alternative [92].

Table 2. BMVs isolated from different sources utilized for various applications.

Bacterial Species	BMV Size (nm)	Cargo Loaded	Loading Method	Application
E. coli K-12 W3110 strain [86]	30–250	siRNA	Electroporation	Anti-tumor therapy
E. coli [88]	55 ± 1	NanoLuc Luciferase enzyme	Genetic engineering of parent bacteria	Bioluminescence Imaging
E. coli strain BL21 [90]	136 ± 67	Phosphotriesterase enzyme	Genetic engineering of parent bacteria	Environmental remediation
A. baumannii [71]	200–300	Antibiotics (ceftriaxone, amikacin, azithromycin, ampicillin, levofloxacin, ciprofloxacin, norfloxacin)	Antibiotic treatment of parent bacteria	Antibacterial Therapy
E. coli K-12 W3110 strain [91]	20–200	Melanin	Genetic engineering of parent bacteria	Cancer theranostics
E. coli JC8031 [89]	40	NanoLuc Luciferase enzyme	Genetic engineering of parent bacteria	Ability to decorate multiple functional protein moieties demonstrated
K. pneumoniae (attenuated) [81]	~70	Doxorubicin	Simple incubation of drug with BMVs	Anti-tumor therapy (drug + immunotherapy)
P. aeruginosa [93]	30–200	Gold NPs	Electroporation	Showed ability to load gold NPs in BMV lumen

4. Separation, Purification, and Storage of BMVs

The separation and purification of BMVs from bacterial culture typically involves the following steps (Figure 4): (1) Centrifugation—employed at low speeds for the separation of bacteria from the culture suspension; (2) filtration and concentration—using 0.45 µm filters followed by concentration of crude BMVs using a 100 KDa membrane; and finally, (3) ultracentrifugation at reduced temperatures. Additional steps involving multiple filtration and centrifugation steps are routinely employed to separate the BMVs. Caution on multiple centrifugation steps must be taken as excessive centrifugal forces may disrupt the vesicular structure of the BMVs or lead to vesicle clumping.

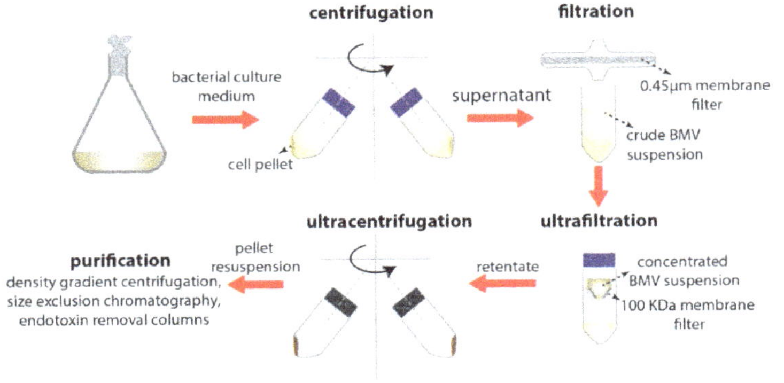

Figure 4. Separation and purification steps involved in BMV isolation.

For the purification of BMVs from extraneous proteins purification steps such as sucrose density gradient centrifugation and size exclusion chromatography techniques are utilized. However, these purification steps can affect total BMV yield, as BMVs have a wide size distribution and composition that can affect their density leading to reduction in BMV content in the final yield. Additionally, to remove the presence of free endotoxins, BMV concentrates could be purified using endotoxin removing columns.

For long-term storage of BMVs, most studies have employed low temperature storage condition of −80 °C wherein the BMVs themselves are resuspended either in PBS or water with or without anti-freeze compounds such as glycerol. However, one report which assessed the storage stability of BMVs at 4 °C, −20 °C, −80 °C, and lyophilized powder conditions (for storage between 7 and 75 days) has found that compared to the low temperature storage conditions, lyophilization of BMVs produced the lowest reduction in BMV concentration and size [80]. It is therefore imperative that a universal storage protocol be developed for long-term BMV use without affecting their size, physico-chemical stability and surface protein activity.

5. BMV Characterization Techniques

The physicochemical characterization techniques that are routinely applied for nanoparticles are also used for BMV characterization and consists of DLS and Zeta potential measurements, NTA analysis and TEM imaging. Typically, BMVs are 30–300 nm in size and are negatively charged. Figure 5A shows the TEM images of OMVs obtained from both wild type *E. coli* and its Δ*msbB* mutant [83], and were found to have a similar hydrodynamic diameter of ~38 nm. However, upon comparing the yield of OMVs obtained from both strains through protein quantification, the Δ*msbB* mutant was found to produce significantly higher yield as compared to the wild type strain.

By undertaking such protein concentration measurements to quantify the yield of BMVs obtained after separation and purification, bacterial growth culture conditions can be optimized in order to improve BMV yield, as the external growth conditions and medium can significantly alter bacterial growth rate and the corresponding vesicle release. This was interestingly depicted for OMVs obtained from *A. baumannii* (Figure 5B) when grown in the presence of different antibiotics at sub-lethal concentrations [71]. Through protein content and particle number analysis (NTA), it was found that only in the presence of levofloxacin, ~2.47-fold increase in the number of OMVs were observed. Correspondingly under these conditions, a significant increase in protein yield and OMV hydrodynamic diameter was also observed as compared to treatments with other tested antibiotics and untreated control (discussed in detail below).

Apart from the above physicochemical characterizations, BMVs are also bio-chemically characterized to determine the specific proteins that are expressed on its surface which are derived from the bacterial parent source. This is particularly important to assess the biological activity of the synthesized BMVs. These surface expressed proteins can be utilized as an anchor point to selectively express other functional proteins of interest or to load a suitable cargo. In one report, the bacterial membrane protein α-pore-forming toxin Cytolysin A was used to anchor anti-HER2 affibody [86] on its surface, so as to impart active targeting capability towards cancer cells that overexpress HER2. Similarly, in another report, the bacterial membrane protein OmPA was used to selectively load an enzyme cargo in vitro within the lumen of synthesized BMVs [90] (discussed later).

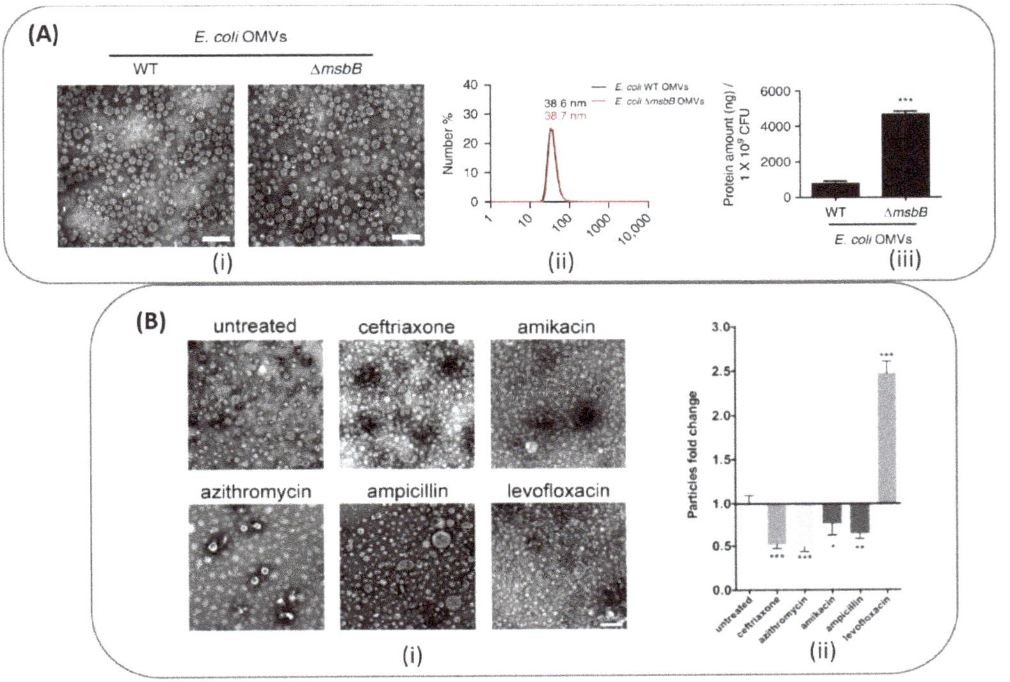

Figure 5. Top panel (**A**): (i) TEM and (ii) DLS data depicting the actual site and hydrodynamic diameter of the OMVs obtained from *E. coli* and its Δ*msbB* mutant (***: $p < 0.001$). The comparative yield of OMVs from each strain was quantified through protein concentration measurements. Reproduced from [83], Springer Nature, 2017. Bottom panel (**B**): (i) TEM image of OMVs obtained from *A. baumannii* cultured in different antibiotics and their (ii) comparative yield measured using NTA analysis (*** $p < 0.001$; ** $p < 0.01$; * $p < 0.05$). Reproduced from [71], Elsevier, 2020.

6. Cargo Loading and Surface Modification Using BMVs

The loading of cargo into BMVs have been undertaken through many different active (energy dependent) and passive (energy independent) loading mechanisms as reported in literature (Figure 6, Table 1). Among the different active loading techniques discussed below, co-extrusion and sonication have also been employed for the direct surface modification of NPs. By Np surface modification, the bio-(physicochemical) property of the parent source bacteria can be transferred onto the NPs for harnessing their function as discussed in specific examples below.

6.1. Active Cargo Loading

6.1.1. Electroporation

Electroporation is usually employed as a non-viral gene delivery technique in vitro and in vivo. The technique involves the application of short high-voltage pulses to cells to form pores within its cell membrane to create a transient state of permeability [94,95]. This state of permeability allows the entry of drugs and fluorochrome compounds and even large-sized cargo such as nucleotides, which is optimized by varying the electric pulse and its duration. If optimized correctly, the phospholipid membrane then recovers its structure once this process is completed, without incurring any irreversible damage. Being applicable to cells, particularly to cell surface membranes, this process has been naturally extended for cargo loading of other lipid bilayer structures which can act as delivery vehicles such as biomimetic exosomes [96–98].

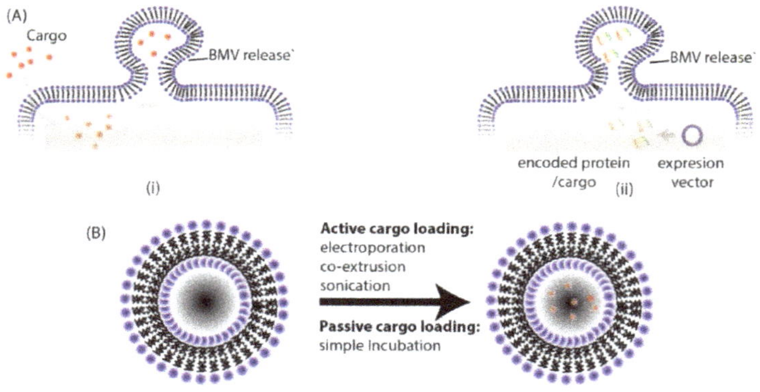

Figure 6. BMVs can be loaded either through (**A**) (i) direct exposure of cargo to parent bacteria, (ii) by transformation of parent bacteria with desired expression vector or (**B**) through different active and passive cargo loading techniques post BMV isolation.

The first report on the use of electroporation to load cargo into BMVs was published in 2014 wherein BMVs from Gram-negative *E. coli* was used for the delivery of therapeutic siRNA [86]. The loading of siRNA in this case was achieved through the electroporation technique (at 700 V and 50 μF) leading to a high loading efficiency of 15 wt% siRNA in BMVs (Figure 7A). Here, fluorescent dye-labeled siRNA was loaded into BMVs, which extended its usage as a theranostic agent for cancer. Apart from nucleotide loading, metallic gold (Au) NP has also been loaded successfully into BMVs [93] (Figure 7B). In this study, small-sized Au NPs (<10 nm) could be loaded into *P. aeruginosa* BMVs by applying an optimal voltage of 470 V and 1 pulse yielding an encapsulation efficiency of ~35%. Note here that when *P. aeruginosa* BMVs alone were subjected to a higher electroporation voltage of 1500 V, a reduction in their structural stability was observed, indicated by a reduction in its protein concentration and an overall increase in the standard deviation of its hydrodynamic diameter. Thus, it is important to optimize the parameters for electroporation in order to successfully load BMVs with a desired cargo. Nevertheless, such Au NP-loaded BMVs could have wide biomedical applications, and this process of electroporation could be used for the loading of other types of cargo such as iron-oxides for MR imaging applications or quantum dots for fluorescence imaging.

6.1.2. Co-Extrusion/Surface Modification

Co-extrusion has been reported to be widely used as a technique for surface modification of NPs, wherein a material of interest is mixed with the NP and extruded together so as to force an interaction between them [99]. Specifically, a number of different biomimetic NPs have been synthesized through this route by coating NP surface with cell membrane structures such as those obtained from red blood cells [23], leukocytes [100], cancer cells [101], platelets [102], etc. The technique mostly involves the isolation of the cell surface membranes followed by repeated mechanical extrusion of these membranes with the NP of interest through polycarbonate membranes of varying pore sizes. More recently, this has been utilized for the synthesis of exosome membrane coated NPs [103–105] and has now been extended to bacterial membranes and BMVs as discussed below.

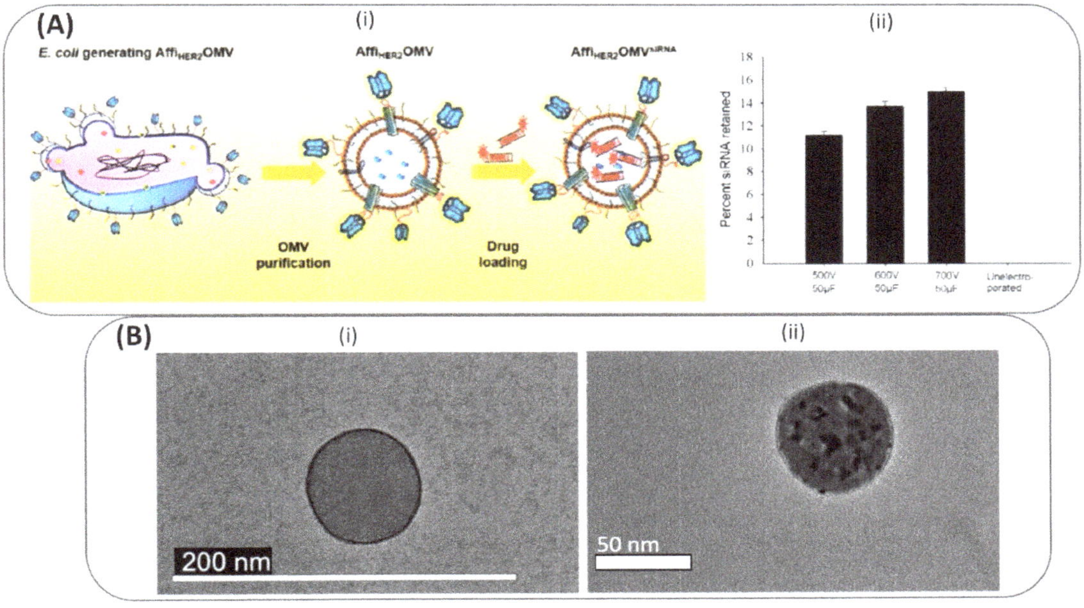

Figure 7. Top panel (**A**): (i) Schematic of electroporation mediated siRNA loading of siRNA into *E. coli* OMVs. (ii) siRNA loading was optimized at different high-voltage pulses. Reproduced from [86], American Chemical Society, 2014. Bottom panel (**B**): (i) TEM image of *P. Aeruginosa* BMV before (i) and after (ii) loading of Au NPs through electroporation. Reproduced from [93], Springer Nature, 2019.

In one such study, 30 nm citrate stabilized Au NPs were surface modified (Figure 8(Ai) using *E. coli* derived BMVs (~30–30 nm diameter) by extruding their mixture 7 times through a 50 nm polycarbonate porous membrane resulting in the production of ~42 nm surface modified Au nanoparticles [87]. This extrusion and surface coating process resulted in a natural increase in the hydrodynamic diameter of the Au NPs which was caused by the presence of BMV proteins on its surface. To further confirm the presence of surface-bound proteins on Au NPs, a fluorescence quenching analysis was done using FITC-thiol fluorochrome, as Au NP surfaces are known to be ultra-efficient quenchers (Figure 8(Aii) [106]. Due to successful surface modification, fluorescence quenching was observed only when unmodified Au NPs were mixed with FITC-thiol. It was found that the surface-modified BMVs contained ~8 wt% surface proteins as measured through the BCA protein assay, and this led to an increase in the stability of Au NPs in a physiological environment when compared to unmodified bare Au NPs. Similar to the above study, the coating of BMVs on drug loaded micelles have also been demonstrated [82]. Here, two different steps of extrusion steps were carried out to create a unique BMV modified micelle structure. BMVs isolated from *S. typhimurium* were first modified to incorporate a polymer, PEG-RGD by extrusion through a 220 nm polycarbonate membrane. This was done initially to reduce the immunogenicity of BMVs and to impart active targeting capability to them. These modified BMVs were then further extruded with tegafur-loaded F127 micelles, to obtain BMV coated micelle structures (Figure 8B).

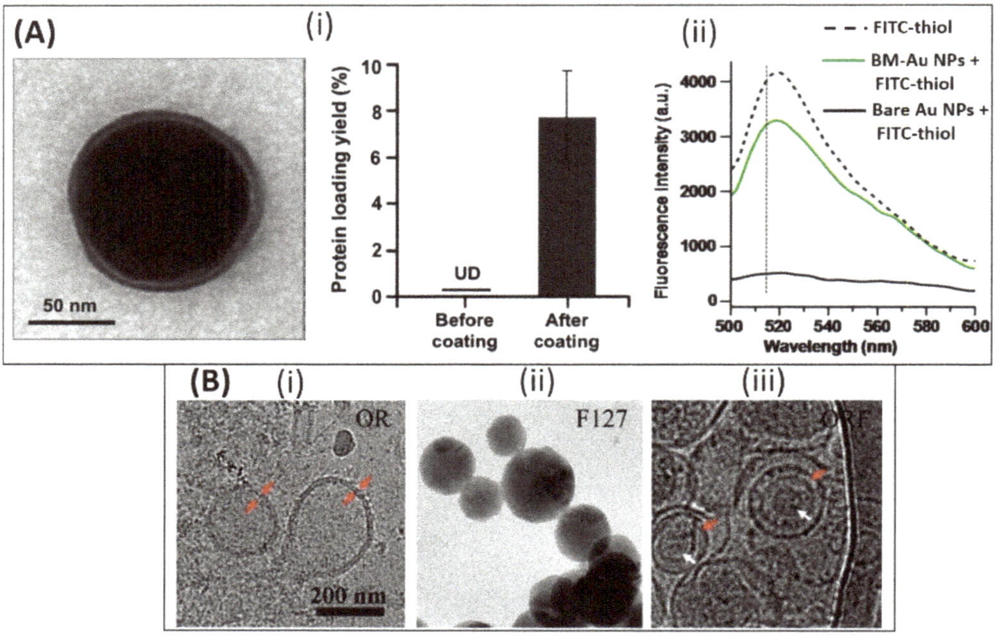

Figure 8. Top panel: (**A**) (i) TEM image of bacterial membrane coated Au NPs and measurement of protein content on Au NPs before and after coating. (ii) Fluorescence quenching assay showing the reduction in fluorescence intensity of FITC-thiol only when incubated with bare Au NPs as compared to bacterial membrane coated Au NPs (BM-Au NPs). Reproduced from [87], American Chemical Society, 2015. Bottom panel: (**B**) TEM image of (i) RGD functionalized OMV [OR], (ii) F127 polymeric micelle and (iii) OMV coated F127 micelle. Reproduced from [82], American Chemical Society, 2020.

6.1.3. Sonication/Surface Modification

Similar to the co-extrusion technique, sonication is a simpler alternative which can be used for surface modification of NPs. The application of ultrasonic frequencies to cell membrane components can lead to their disruption and subsequent attachment on the surface of NPs [22,107]. Such NPs have altered physico-chemical properties that mimic cell surface characteristics in vivo.

In another report, antibiotic-loaded polymeric PLGA NPs were coated with BMVs for investigating active delivery of antibiotics to infection sites in vivo [84]. Here, BMVs isolated from *S. aureus* and *E. coli* bacteria were coated on vancomycin and rifampicin loaded PLGA NPs by mixing the NPs and BMVs at a 2:1 mass ratio followed by bath sonication of the resultant mixture. The protein loading on the surface modified PLGA NPs were found to be ~7 wt%. An interesting observation that was observed in this study, was the specific uptake of BMV membrane coated PLGA NPs by macrophages that were previously infected with bacteria. Importantly, it was found that this uptake was dependent on the specific bacteria that the macrophage had previously been infected to, i.e., macrophages infected with *S. aureus* or *E. coli* bacteria specifically showed a significant uptake of PLGA NPs that were either coated with *S. aureus* or *E. coli* BMVs, respectively.

6.2. Passive Cargo Loading

6.2.1. Simple Incubation

BMVs can also be loaded through a passive loading technique of simple incubation with cargo. In one such study, Gram-negative *K. pneumoniae* BMVs were loaded with chemotherapeutic drug doxorubicin-hydrochloride [81] by incubating the BMVs with the drug at 37 °C for 4 h, followed by the removal of free drug using 100 KDa membrane ultrafiltration and PBS wash repeatedly. The encapsulated drug in BMVs were quantified using mass spectrometry analysis which showed that maximum encapsulation efficiency of ~78% could be obtained when the drug and BMVs were incubated at a mass ratio of 1:45. In other reported studies, the simple incubation technique was used for the fluorescent labelling of BMVs for in-vivo imaging applications. [79,83] Here, the BMVs were incubated with an NIR dye Cy7 mono NHS ester for 2 h at 37 °C followed by their separation from excess dye through ultracentrifugation.

6.2.2. Incubation with Parent Bacteria

A simple mode of loading BMVs is achieved through incubating the cargo material with the bacteria of interest during its growth phase. The bacteria in this case would engulf the cargo present in the extra cellular environment and sort the same into BMVs which are shed from the bacteria. In one such extensive study, antibiotic-loaded BMVs were synthesized from *A. baumanii* by culturing the bacteria in antibiotic containing medium [71]. Specifically, different antibiotics such as ceftriaxone, amikacin, azithromycin, ampicillin, levofloxacin, ciprofloxacin, and norfloxacin were added at different fractions of their respective Minimum Inhibitory Concentrations (1/2, 1/4, 1/8). Characterization studies post drug loading showed that the phospholipid wall in BMVs had thickened when they were loaded with ceftriaxone, amikacin, azithromycin, and levofloxacin. Additionally, it was found that the antibiotic levofloxacin at 1/8 minimum inhibitory concentration produced the highest encapsulation efficiency in the generated BMVs, with ~120 µg of Levofloxacin/10^{12} BMV particles, while, on the other hand, ciprofloxacin, azithromycin, and ampicillin antibiotics failed to be loaded into the secreted BMVs. The authors report that the reason for this wide difference and preferential loading of drugs into BMVs could be complex and further studies are therefore needed to understand the same. Interestingly, this study also reports that mere incubation of drug cargo with empty BMVs did not lead to any cargo loading. This phenomenon could suggest that the loading of antibiotics internally through bacteria in this case could have occurred through a drug efflux mechanism.

In another report with a similar objective of developing BMVs for anti-bacterial applications, BMVs were synthesized and isolated from non-pathogenic myxobacteria (soil bacteria) *C. velatus*, Sorangiineae species strain SBSr073 [80]. Here, the BMVs were studied directly for their antibacterial property since the myxobacterial species are known to be predatory towards other competing Gram-positive and Gram-negative bacteria while using them as a nutrient source.

6.2.3. Transformation of Parent Bacteria

Cargo loading of BMVs can also be employed intrinsically by transformation of the parent bacteria (genetic engineering) wherein a plasmid expressing the desired protein cargo is engineered. In one such report, multifunctional BMVs were synthesized using genetically engineered *E. coli* [88,89]. To load cargo within BMV lumen, native proteins anchored to the periplasmic side of the outer membrane were utilized as an anchor point. Specifically, a bioluminescent agent, NanoLuciferase enzyme, was loaded within BMV lumen by anchoring it to SlyB protein by co-expressing them in a bacterial expression system. Further, a protein scaffold was used to anchor IgG antibody to BMV surface by binding to an BMV surface membrane expressed Ice Nucleation protein. The loading of NanoLuciferase within BMVs was confirmed by Western blotting analysis which showed that the cargo protein degraded upon the use proteinase K only when the BMVs were lysed using SDS. Such synthesized BMVs could be used for biosensing applications for the detection of any antigen. In another report, BMVs were utilized for the packaging of a bioremediating enzyme phosphotriesterase within BMV lumen. [90] For this, a SpyCatcher/SpyTag bioconjugation system was employed in *E. coli* bacteria, wherein the native surface membrane protein, OmpA was bound to the SpyCatcher domain while the phosphotriesterase enzyme was bound to the SpyTag domain. The packaging of cargo was attempted by co-transformation using expression vectors containing the appropriately modified SpyCatcher and SpyTag genes, which could lead to the formation of the hybrid protein within the bacteria, and its release in BMVs. It was found that the cargo packaging into BMVs increased its stability and robustness, possibly allowing its usage in harsh environmental conditions for bioremediation.

An interesting study on the use of BMVs for theranostic application was recently reported, wherein genetic engineering techniques were employed for loading a theranostic agent, melanin into *E. coli* generated BMVs [91]. Here, the enzyme tyrosinase, which is responsible for the production of melanin, was encoded into an expression vector. Upon the introduction of the tyrosin substrate, the tyrosinase enzyme catalyzed its conversion to melanin within bacterial cytosol and periplasmic space which could then be packaged into released BMVs (Figure 9). Thus, the versatility of genetic engineering techniques enables it to be employed for the loading of many different types of cargo.

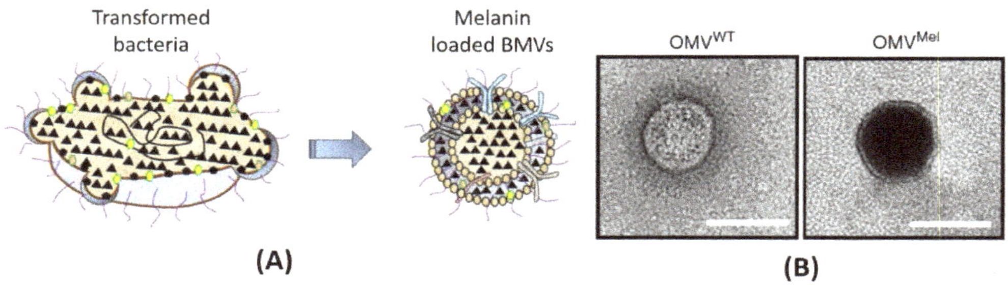

Figure 9. (**A**) Simple schematic showing the loading of melanin from transformed bacteria. (**B**) Comparative of TEM images of unloaded wild type OMV (OMV^{WT}) and melanin-loaded OMVs (OMV^{Mel}) from *E. coli*. Reproduced with permission from [91], Springer Nature, 2019.

7. Drug Delivery Applications of BMVs

The therapeutic application of BMVs has largely been explored pre-clinically for its use as an immune-modulating agent [108–110]. This is primarily because of the presence of antigenic protein molecules in BMVs which when used, may trigger a favorable immune response in the body. Until now this phenomenon has been successfully translated to clinics for the development of a vaccine against *Neisseria meningitidis* serogroup B (Bexsero® developed by Novartis) [111]. In some cases, BMVs have been demonstrated to amplify the immunogenicity of a low immunogenic protein antigen by acting as a vaccine delivery system [112,113]. Additionally, the ease of genetic modification of the bacterial source has also contributed to it being utilized as an efficient and promising immunomodulator.

The use of BMVs for drug/cargo delivery applications has only been explored recently with only a handful of published literature reports. Most of the applications for BMVs has primarily been focused either on cancer therapy or antibacterial therapy (Figure 10).

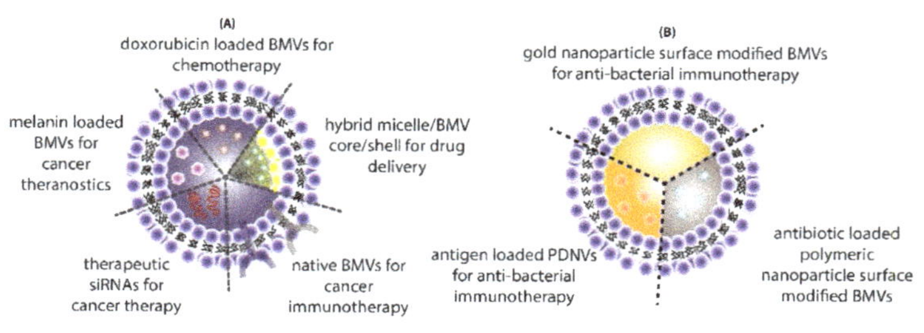

Figure 10. Different applications of BMVs reported in literature for (**A**) cancer therapy and (**B**) antibacterial therapy.

7.1. Cancer Therapy

In one such report, BMVs isolated from non-pathogenic attenuated *K. pneumoniae* were utilized for the delivery of doxorubicin [81]. Anti-tumor studies carried out using such BMVs at a dose of 2 mg/Kg of DOX (injected intraperitoneally every day for 11 days) in A549 tumor bearing BALB/c nude mice showed a significant reduction in tumor volume as compared to the use of free drug, empty BMVs, and even doxorubicin-loaded liposomes (Figure 11A). In fact, it was observed that the rate of reduction in tumor volumes was found to be greater for DOX-loaded BMVs as compared to DOX-loaded liposomes, signifying that BMVs showed a better therapeutic response. This added therapeutic response observed in DOX-loaded BMVs could be attributed to the favorable immune response that BMVs can induce in vivo which in conjunction with chemotherapeutic drugs leads to generation of a higher therapeutic efficacy. This was supported by tumor volume reduction studies in the same report that showed that the use of bare BMVs alone in vivo lead to a significant reduction in tumor volume as compared to untreated controls. Further, it was also observed that there occurred a significant accumulation of murine macrophages in tumor tissues that were treated with both doxorubicin loaded BMVs and empty BMVs. Pharmacokinetic analysis showed that the use of drug loaded BMVs lead to a greater drug retention in tumors that lasted for longer periods of time as compared to DOX-loaded liposomes with a concurrently lower retention found in the heart (Figure 11B). As a result, the cardiac toxicity (which is notable in the use of doxorubicin) was found to be significantly reduced when DOX-loaded BMVs were used (as measured through analysis of lactate dehydrogenase, aspartate aminotransferase, and creatine kinase isoenzyme in blood), which were further supported through histopathological analysis of cardiac tissues. Overall, the pharmacokinetic profile of the loaded drug was improved (characterized by an increase in the drug half-life, reduction in clearance rate, and improved bioavailability) when BMVs were utilized as a drug delivery vehicle. Immunotoxicity analysis in C57BL/6 normal mice at the therapeutic dosage (over a period of 11 days) showed that the administration of both DOX-loaded BMVs and bare BMVs lead to a significant increase in serum cytokine levels which returned back to basal levels over a period of time. These results therefore showed that BMVs could be well tolerated in vivo and could be used as an effective drug delivery vehicle.

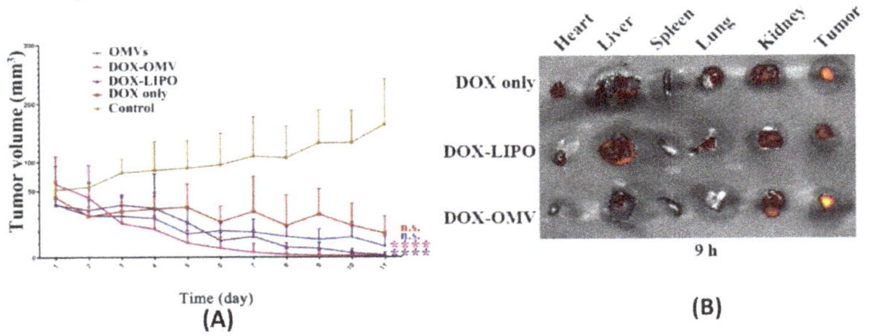

Figure 11. (**A**) Tumor volume reduction measurements and (**B**) in vivo drug distribution of doxorubicin-loaded *K. pneumoniae* OMVs as compared to controls. Reproduced from [81], Elsevier, 2020. **** $p < 0.001$ vs. control, n.s. no statistic difference vs. control.

Similar to the above, a study depicting the use of *S. typhimurium* BMVs for combined drug delivery and immunotherapy against cancer was reported [82]. In vivo tumor therapy studies carried out using hybrid BMVs (BMV/micelle/drug) in B16F10 melanoma and 4T1 mammary tumor in C57BL/6 mice at a dose of 30 μg of BMVs (once/3 days for a total of 3 injections) lead to a significant reduction in tumor volume and increase in survival as compared to controls. Furthermore, this treatment also limited the spread of cancer metastatic nodules in lung tissues, which otherwise are prevalent in the B16F10 tumor model, which could explain the increase survival of mice observed on treatment.

Interestingly, the synthesized BMVs also showed an immunoprotective effect against tumor. Mice that were pretreated with BMVs, when challenged with tumor cells later on, showed a delayed tumor growth response with significantly small tumor volumes (Figure 12). Even though the exact reasons for the same have not been elucidated in this report, these results show overall the promising effect that BMVs have towards developing a strategy for tumor prevention and treatment. Upon BMV administration, the in vivo cytokine analysis of blood and tumor samples showed that even though there occurred an increase in the cytokine levels of TNF-α, IFN-γ, IL-12, IL-4, and IL-17, the levels of these cytokine reduced to basal levels after 24 h. Overall, no blood toxicity and organ toxicity (including liver and renal functions) were found upon BMV administration.

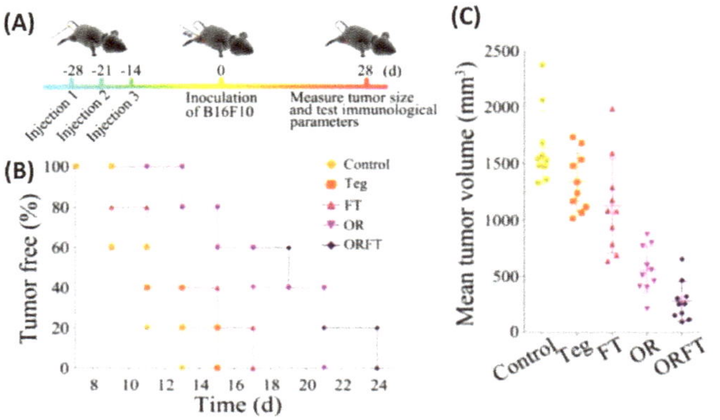

Figure 12. (**A**) Treatment timeline for evaluating the protective role of hybrid BMVs (drug-loaded micelles surface modified with BMVs)—hybrid BMVs were administered to mice before tumor challenge. (**B**) Percentage of tumor-free mice after tumor challenge (**C**) Mean tumor volume measurements of hybrid OMVs against controls (Teg-Tegafur, FT-Tegafur loaded F127 micelle, OR-RGD functionalized BMV, ORFT—Tegafur-loaded F127 micelle surface modified with RGD functionalized BMV). Reproduced with permission from [82], American Chemical Society, 2020.

Even though the above studies showed the ability of BMVs as a promising drug delivery agent, an interesting study has demonstrated its ability to be used as in its native form as a potential anti-tumor immunotherapeutic agent [83]. To demonstrate its applicability, unmodified BMVs isolated from both Gram-negative and -positive bacterial species were assessed for their ability to actuate an anti-tumor immune response in different tumor models in mice. Specifically, BMVs isolated from Gram-negative *E. coli* and *S. Enterica* and Gram-positive *S. aureus* and *L. acidophilus* were injected intravenously in BALB/c mice bearing CT26 colon adenocarcinoma at a 5 μg BMV dose (4 times at 3 days interval), significant tumor volume reductions were observed as compared to PBS injected controls (Figure 13A,B). Additionally, to demonstrate its diverse potential, BMVs from *E. coli* were assessed for their therapeutic response in CT26 colon adenocarcinoma and 4T1 mouse mammary tumor of BALB/c mice, and MC38 mouse colon adenocarcinoma and B16BL6 mouse melanoma cancer of C57BL/6 mice. At a 5 μg *E. coli* BMV dose injected intravenously (4 times at 3 days interval), significant tumor volume reductions were observed for all the treated tumor types. However, the reduction in tumor volumes in 4T1 and B16BL6 tumors were found to be less effective as compared to those observed in CT26 and MC38 tumors, which shows the important role tumor biology and characteristics have on the net therapeutic outcome. Interestingly, for the treatment of CT26 tumors, a long-term memory effect was observed for treatment using *E. coli* BMVs wherein secondary and tertiary challenges of CT-26 tumor cells were rejected in mice that recovered from the primary tumor challenge post BMV administration. These results show how BMVs have the ability to favorably modulate the immune system and possibly provide a protective environment to prevent tumor relapse as observed for other immunotherapeutic modalities [114]. Similar to other reports on the use of BMVs in vivo, an increase in the levels of cytokines and chemokines such as IL-12p40, IFN-γ, CXCL10, TNF-α, IL-6, and IL-12p70 were also observed in this study. However, it was observed that the cytokines CXCL10 and IFN-γ specifically showed elevated levels in the tumor tissues over 24 and 48 h which could imply that these cytokines play an important role in eliciting an anti-tumor immune response (Figure 13C,D).

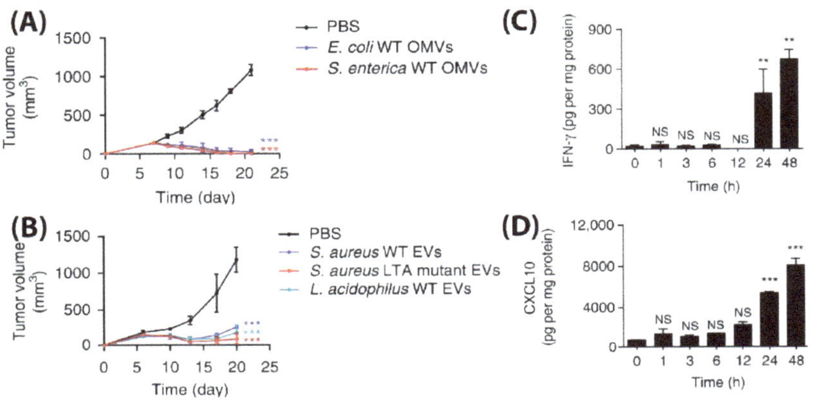

Figure 13. Tumor volume measurement study for CT26 tumors injected with (**A**) Gram-positive *S. Aureus*, lipoteichoic acid *S. Aureus* mutant, and *L. Acidophilus*. (**B**) Gram-negative *E. coli* and *S. enterica*. Measurement of (**C**) IFN-γ and (**D**) CXCL10 cytokine levels in tumor cell lysate at different time points after *E. coli* BMV administration in CT26 tumors (** $p < 0.01$, and *** $p < 0.001$). Reproduced with permission from [83], Springer Nature, 2017.

Apart from utilizing BMVs for drug loading, BMVs have also been utilized for the delivery of therapeutic nucleic acids for the treatment of cancer. By loading siRNA targeting kinetic spindle protein, a protein essential for spindle formation and continuation of cell cycle, into BMVs obtained from Δ*msbB E. coli*, in vivo treatment of liver cancer was demonstrated [86]. Significant reduction of HCC-1954 xenografts in nude mice was observed upon intravenous administration of siRNA-loaded BMVs at a 4 µg dose siRNA injected on alternate days over a 22-day treatment period as compared to controls. Serum cytokine analysis showed elevated levels of TNF alpha, IL6, and IFNγ in C57BL/6 mice upon repeated dosing (at 10–20 µg siRNA) over 4 consecutive days. However, this elevation was observed only for a brief period of 3 h post-administration and would return back to basal levels in 24 h. Note that lethal dose toxicity studies showed that the BMVs obtained from Δ*msbB E. coli* did not cause any mortality even at a single high dose of 100 µg, while, on the other hand, administration of 50 µg of BMVs obtained from wild type *E. coli* lead to mortality within 48 h post administration. This shows how the biochemical composition of the surface moieties of BMVs play a crucial role in its toxicity response. Here, the Δ*msbB* mutation in *E. coli* produces underacylated LPS which shows reduced endotoxicity.

While BMVs show great promise as a drug delivery vehicle, one report has gone further to evaluate its potential as a stimuli-responsive multifunctional theranostic agent [91]. To do so, BMVs synthesized from Δ*msbB E. coli* were loaded with melanin (which can act both as a photoacoustic and photothermal agent), by introducing the required plasmid to the bacteria. This led to the generation of melanin which is packed within the BMV lumen. Upon laser exposure, a concentration dependent thermal response was observed for melanin loaded BMVs, which when used in vitro produced significant cell death due to the photothermal effect (Figure 14A,B). When such melanin-loaded BMVs were administered in 4T1 mammary tumor-bearing FOX-N-1 nude mice intravenously at a single dose of ~150 µg protein, optoacoustic signals could be observed in tumor, liver, and kidneys, enabling the study of its biodistribution profile. The imaging results demonstrated that these BMVs accumulated in the tumor through EPR effects, underwent continuous circulation in vivo, and cleared slowly from the system over a period of 24 h. Photothermal treatment of the 4T1 tumors 3 h post-injection of a single dose of ~75 μg BMVs lead to a significant thermal response of 56 °C and 47 °C for intratumoral and intravenous administration respectively. This resulted in a significant reduction in tumor volume over an 8 days period after just a single treatment and laser therapy which its high effectiveness in cancer therapy (Figure 14C,D). As previous studies have pointed out, there occurred a significant increase in the cytokine levels of TNF-α, IL-6, and IFN-γ 2 h post administration of BMVs, which however reduced near to the baseline levels after 25 h.

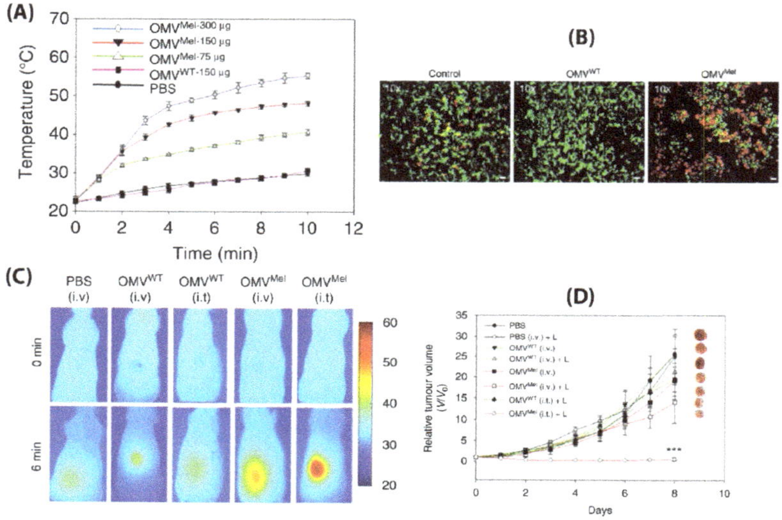

Figure 14. (**A**) Thermal response of melanin-loaded OMVs at different concentrations as compared to wild type OMVs upon laser exposure. (**B**) Live–dead (calcein AM/EthD-1) staining of 4T1 cancer cells treated with melanin-loaded OMVs and exposed to laser as compared to controls. Thermal imaging (**C**) and tumor volume reduction measurements (**D**) of 4T1 tumors after either intravenous or intratumoral administration of melanin loaded OMVs and laser exposure as compared to controls. Reproduced with permission from [91], Springer Nature, 2019. *** $p < 0.001$ vs. PBS with laser treatment.

7.2. Antibacterial Therapy

The use of BMVs for antibacterial applications has mostly been investigated by utilizing the inherent antigenic molecules present on BMV surface for eliciting a favorable immune response against the invading pathogen. In this regard, BMVs have been utilized either directly in its unmodified form [80] or modified appropriately to present the antigenic proteins with a suitable material to elicit a desired immune response. One way of presenting antigenic proteins to the immune system is through the use of NPs, which can maximize immune cell recognition owing to their large surface area and size scale that facilitates particle uptake [115–117]. Numerous literature reports have utilized this methodology to develop nanotechnology-based vaccines and similar approaches have been demonstrated with BMVs for antibacterial therapy.

To demonstrate the potential of NP-based immunomodulation by utilizing BMV antigen proteins, surface-modified Au NPs have been investigated in vivo for antibacterial therapy applications [87]. Upon intravenous administration of BMV protein-coated Au NPs (2.5 μg dose) into immunocompetent CD-1 mice, a number of highly precise immune responses were observed as compared to bare BMVs and PBS injected controls. Specifically, it was observed that BMV-coated Au NPs lead to a heighted activation and maturation of dendritic cells and T cells and an increased B cell response and a consequent increase in antibody titers. Note, in this study, that when smaller sized Au NPs (~30 nm) were used as compared to larger sized Au NPs (~90 nm) a relatively heighted dendritic cell activation and maturation was observed. This was found to be due to the better accumulation of smaller sized Au NPs in the lymph nodes of mice, thereby making them more suitable for immune activation. However, in this study the application of these BMV protein surface-modified Au NPs was not evaluated in an infection model.

In an interesting report, BMVs were utilized to surface modify polymeric NP-based antibiotic delivery systems such that target specificity could be achieved [84]. Preliminary in vitro studies undertaken by this group showed that BMVs isolated either from *E. coli* or *S. aureus* showed specific uptake in macrophages that were pre-infected with *E. coli* or *S. aureus*, respectively. As a result, antibiotic-loaded PLGA NPs surface modified with *S. aureus* BMVs were found to show significantly greater accumulation in major organs of an *S. aureus* infected BALB/c mice model as compared to *E. coli* BMV-coated PLGA NPs and liposome-coated PLGA NPs. This greater internalization was explained to be caused by the greater ability of infected macrophages to internalize the NPs first, followed by the natural biodistribution of these macrophages to the infected organs. As a result of this greater accumulation observed for the *S. aureus* BMV-coated antibiotic loaded PLGA NPs, effective reduction of bacterial CFU counts could be observed in kidneys and lungs of the infected mice. However, note here that effective therapeutic efficacy could not be achieved for all major organs, and more studies need to be carried out to exploit this specific property of BMV-coated NPs. It would be interesting to study in this case if the specific uptake of BMV-coated NPs in infected macrophages could be extended to other infected mammalian cells and cancer cells. If such a specificity could be achieved, it could open up the possibility of treating patients with tumors bearing bacterial load. Such infected tumor conditions are nowadays observed in clinical investigations and the presence of these bacteria in tumors are found to play an important role in inhibiting the efficacy of chemotherapeutic drug treatments [118,119].

Beyond the use of BMVs, one unique report demonstrated the direct use of bacterial cells for NP surface modification [120]. Here, bacterial protoplasts were first isolated by treating bacteria with lysozyme to remove the bacterial cell wall, followed by its serial extrusion through 10, 5, and 1 μm sized polycarbonate membrane filters. These protoplast-derived nanovesicles (PDNVs) show inherent advantage over BMVs as they can be directly synthesized in large amounts from the bacterial suspension. Additionally, due to removal of the bacterial cell wall, the resultant PDNVs were depleted of the outer membrane proteins OmpA and lipid A which are components of LPS, thus making them less immunogenic and more favorable for drug delivery and theranostic applications. Here, for investigating its immunotherapeutic potential, PDNVs were loaded separately either with *E. coli* antigen OmpA and *S. aureus* antigen Scoagulase by expressing the desired antigen in the parent *E. coli* bacterium. The resultant PDNVs, when administered in vivo in C57BL/6 mice, showed high specificity in developing an immune-protective response against bacterial challenge. Specifically, it was observed that PDNVs harboring either the OmpA antigen or the Scoagulase antigen showed effective immune response in vivo when challenged with lethal doses of *E. coli* or *S. aureus* respectively. This response was found to be highly specific with respect to the antigen that the PDNVs harbored such that mice that were administered with PDNVs harbouring *E. coli* antigen OmpA succumbed when challenged with *S. aureus* infection, and vice versa. The protective immunity offered by such PDNVs could be observed in mice up to 6 weeks post-administration. Interestingly, these PDNVs were found to show low in vitro and in vivo toxicity as compared to *E. coli*-derived BMVs. Specifically, it was observed that even upon administration of upto 1 mg of PDNVs in C57BL/6 mice all animals survived, while a dosage of only 25 μg of BMVs lead to the death of 80% of the animals. Overall, PDNVs show great promise towards the development of biomimetic drug delivery vehicles and show several advantages as compared to BMVs.

8. Challenges and Future Perspective of Utilizing BMVs for Drug Delivery

The use of BMVs for drug delivery applications would encounter challenges that are similar to the design and synthesis of other drug delivery systems as well as other unique challenges that pertain to it alone. One of the most important properties of drug delivery systems over and above its therapeutic efficacy is its safety and immunogenicity. Unlike in the use of BMVs for vaccine development, for drug delivery applications, it is imperative that BMVs do not elicit an immunogenic response in the body. The presence of immunogenic molecules such as LPS and LTA and other bacterial proteins can impede the utility of BMVs for such applications. In such cases, efforts need to be focused on utilizing and developing bacterial strains that can produce BMVs containing lower amounts of or attenuated immunogenic molecules (e.g., Δ*msbB E. coli* [83]). To prevent immune recognition other methods such as pegylation [121] or the incorporation of anti-phagocytic CD-47 molecules [122] on the surface of BMVs could be utilized. However, this could lead to additional synthesis/modification steps which could add to the complexity of the otherwise simple BMV system.

Another challenge that researchers face at this stage in utilizing BMVs for drug delivery applications is the lack of knowledge regarding BMV synthesis routes, mechanism of cargo packaging, and factors affecting the same. Knowledge regarding these can enable the design of suitable strategies for maximizing vesicle synthesis and cargo loading such that cargo loaded BMVs could be obtained in the first instance of BMV synthesis. This is advantageous because separation and purification procedures would be comparatively simple for preloaded BMVs as compared to post-loading strategies (after BMV separation) wherein unloaded cargo, empty BMVs, and cargo-loaded BMVs have to be separated individually. To overcome challenges related to separation, affinity-based separation could be utilized if the cargo of interest is also coupled with a desired protein tag. However, in this case the cargo and protein tag should be present on the BMV surface such that the proteins face the outer surface of the vesicles in order to interact with the affinity molecule used for separation. Magnetic separation of BMVs could also be a viable alternative if the cargo of interest is coupled to magnetic NPs which can then be loaded into BMVs. Such a separation technique has been demonstrated successfully for exosomes [123].

As the use of BMVs for drug delivery applications is still in its infancy, and BMVs have very high heterogeneity (based on the parent bacterial source and strain and growth conditions), it is absolutely essential that researchers working on this field quantify data regarding BMV synthesis, cargo loading, and culture conditions such that comparative analysis across various studies could be made effectively These standard parameters should also include details such as number of BMVs obtained/CFU, isolation protocol, storge conditions, and their effect on long-term stability, protein content, and concentration etc. For drug delivery applications particularly, it is essential that %cargo loading be evaluated and the cargo loading method be detailed. Additionally, it is also important to carry out studies on evaluating the differences on utilizing BMVs of different sizes for drug delivery obtained from the same bacterial source to understand if the size parameter plays an important role in their overall cargo delivery/pharmacokinetics. Understanding and cataloging such information would help steer and accelerate the use of BMVs for drug delivery applications in the future.

From the engineering standpoint, one major challenge that BMVs would face for their utility for biomedical application is their difficulty in separation and purification for clinical translation. At present, a number of different steps are required to separate BMVs from the parent bacterial growth culture and other extraneous proteins, which in the long-run can increase their cost of production. Other alternative low-energy separation techniques need to be developed and optimized for successful translation of BMVs for drug delivery. Ultimately, the cost-to-benefit ratio of synthesizing and purifying BMVs from bacterial culture must be assessed as compared to other promising biomimetic systems such as exosomes which could require a large-scale mammalian culture facility.

9. Conclusions

Even though the use of BMVs for drug delivery applications is still in its infancy, it has tremendous potential to become a successfully translatable drug delivery system. This is primarily because of their ability to be produced innately by bacteria in large quantities using inexpensive medium and culture conditions. Until now, BMVs have been investigated for the delivery of a number of different cargos including chemotherapeutic drugs, therapeutic nucleic acids, antibiotics, NPs, etc. and have been evaluated in vivo in a few studies with promising results. Nevertheless, more thorough and detailed investigations are required on evaluating BMVs for different biomedical applications.

Author Contributions: S.F.; writing—original draft preparation, S.F. and R.L.; writing—review and editing, R.L.; supervision, R.L.; funding acquisition. Both authors have read and agreed to the published version of the manuscript.

Funding: This research was funded by JSPS KAKENHI (Grant Number 20K20203).

Institutional Review Board Statement: Not applicable.

Informed Consent Statement: Not applicable.

Conflicts of Interest: The authors declare no conflict of interest.

References

1. Hua, S.; de Matos, M.B.C.; Metselaar, J.M.; Storm, G. Current Trends and Challenges in the Clinical Translation of Nanoparticulate Nanomedicines: Pathways for Translational Development and Commercialization. *Front. Pharmacol.* **2018**, *9*, 790. [CrossRef] [PubMed]
2. Metselaar, J.M.; Lammers, T. Challenges in nanomedicine clinical translation. *Drug Deliv. Transl. Res.* **2020**, *10*, 721–725. [CrossRef] [PubMed]
3. Dobrovolskaia, M.A.; Shurin, M.; Shvedova, A.A. Current understanding of interactions between nanoparticles and the immune system. *Toxicol. Appl. Pharmacol.* **2016**, *299*, 78–89. [CrossRef]
4. Hannon, G.; Lysaght, J.; Liptrott, N.J.; Prina-Mello, A. Immunotoxicity Considerations for Next Generation Cancer Nanomedicines. *Adv. Sci.* **2019**, *6*, 1900133. [CrossRef] [PubMed]
5. Parodi, A.; Molinaro, R.; Sushnitha, M.; Evangelopoulos, M.; Martinez, J.O.; Arrighetti, N.; Corbo, C.; Tasciotti, E. Bio-inspired engineering of cell- and virus-like nanoparticles for drug delivery. *Biomaterials* **2017**, *147*, 155–168. [CrossRef] [PubMed]
6. Sushnitha, M.; Evangelopoulos, M.; Tasciotti, E.; Taraballi, F. Cell Membrane-Based Biomimetic Nanoparticles and the Immune System: Immunomodulatory Interactions to Therapeutic Applications. *Front. Bioeng. Biotechnol.* **2020**, *8*, 627. [CrossRef] [PubMed]
7. Jin, J.; Bhujwalla, Z.M. Biomimetic Nanoparticles Camouflaged in Cancer Cell Membranes and Their Applications in Cancer Theranostics. *Front. Oncol.* **2019**, *9*, 1560. [CrossRef] [PubMed]
8. Parodi, A.; Quattrocchi, N.; van de Ven, A.L.; Chiappini, C.; Evangelopoulos, M.; Martinez, J.O.; Brown, B.S.; Khaled, S.Z.; Yazdi, I.K.; Enzo, M.V.; et al. Synthetic nanoparticles functionalized with biomimetic leukocyte membranes possess cell-like functions. *Nat. Nanotechnol.* **2013**, *8*, 61–68. [CrossRef]
9. Pasto, A.; Giordano, F.; Evangelopoulos, M.; Amadori, A.; Tasciotti, E. Cell membrane protein functionalization of nanoparticles as a new tumor-targeting strategy. *Clin. Transl. Med.* **2019**, *8*, 8. [CrossRef]
10. Hu, C.-M.J.; Fang, R.H.; Wang, K.-C.; Luk, B.T.; Thamphiwatana, S.; Dehaini, D.; Nguyen, P.; Angsantikul, P.; Wen, C.H.; Kroll, A.V.; et al. Nanoparticle biointerfacing by platelet membrane cloaking. *Nature* **2015**, *526*, 118–121. [CrossRef]
11. Elzoghby, A.O.; Samy, W.M.; Elgindy, N.A. Albumin-based nanoparticles as potential controlled release drug delivery systems. *J. Control. Release* **2012**, *157*, 168–182. [CrossRef] [PubMed]
12. Penalva, R.; Esparza, I.; Agüeros, M.; Gonzalez-Navarro, C.J.; Gonzalez-Ferrero, C.; Irache, J.M. Casein nanoparticles as carriers for the oral delivery of folic acid. *Food Hydrocoll.* **2015**, *44*, 399–406. [CrossRef]
13. Malekhosseini, P.; Alami, M.; Khomeiri, M.; Esteghlal, S.; Nekoei, A.-R.; Hosseini, S.M.H. Development of casein-based nanoencapsulation systems for delivery of epigallocatechin gallate and folic acid. *Food Sci. Nutr.* **2019**, *7*, 519–527. [CrossRef]
14. Ahmad, M.; Gani, A.; Hassan, I.; Huang, Q.; Shabbir, H. Production and characterization of starch nanoparticles by mild alkali hydrolysis and ultra-sonication process. *Sci. Rep.* **2020**, *10*, 3533. [CrossRef] [PubMed]
15. Le Corre, D.; Bras, J.; Dufresne, A. Starch nanoparticles: A review. *Biomacromolecules* **2010**, *11*, 1139–1153. [CrossRef]
16. Ethirajan, A.; Schoeller, K.; Musyanovych, A.; Ziener, U.; Landfester, K. Synthesis and optimization of gelatin nanoparticles using the miniemulsion process. *Biomacromolecules* **2008**, *9*, 2383–2389. [CrossRef] [PubMed]
17. Kuo, W.T.; Huang, J.Y.; Chen, M.H.; Chen, C.Y.; Shyong, Y.J.; Yen, K.C.; Sun, Y.J.; Ke, C.J.; Cheng, Y.H.; Lin, F.H. Development of gelatin nanoparticles conjugated with phytohemagglutinin erythroagglutinating loaded with gemcitabine for inducing apoptosis in non-small cell lung cancer cells. *J. Mater. Chem. B* **2016**, *4*, 2444–2454. [CrossRef]
18. Desai, N. Nanoparticle Albumin-Bound Paclitaxel (Abraxane®). In *Albumin in Medicine: Pathological and Clinical Applications*; Otagiri, M., Chuang, V.T.G., Eds.; Springer: Singapore, 2016; pp. 101–119. [CrossRef]

19. Mariam, J.; Sivakami, S.; Dongre, P.M. Albumin corona on nanoparticles—A strategic approach in drug delivery. *Drug Deliv.* **2016**, *23*, 2668–2676. [CrossRef] [PubMed]
20. Naha, P.C.; Liu, Y.; Hwang, G.; Huang, Y.; Gubara, S.; Jonnakuti, V.; Simon-Soro, A.; Kim, D.; Gao, L.; Koo, H.; et al. Dextran-Coated Iron Oxide Nanoparticles as Biomimetic Catalysts for Localized and pH-Activated Biofilm Disruption. *ACS Nano* **2019**, *13*, 4960–4971. [CrossRef] [PubMed]
21. Raza, F.; Zafar, H.; Zhang, S.; Kamal, Z.; Su, J.; Yuan, W.E.; Mingfeng, Q. Recent Advances in Cell Membrane-Derived Biomimetic Nanotechnology for Cancer Immunotherapy. *Adv. Healthc. Mater.* **2021**, *10*, e2002081. [CrossRef]
22. Ben-Akiva, E.; Meyer, R.A.; Yu, H.; Smith, J.T.; Pardoll, D.M.; Green, J.J. Biomimetic anisotropic polymeric nanoparticles coated with red blood cell membranes for enhanced circulation and toxin removal. *Sci. Adv.* **2020**, *6*, eaay9035. [CrossRef]
23. Hu, C.M.; Zhang, L.; Aryal, S.; Cheung, C.; Fang, R.H.; Zhang, L. Erythrocyte membrane-camouflaged polymeric nanoparticles as a biomimetic delivery platform. *Proc. Natl. Acad. Sci. USA* **2011**, *108*, 10980–10985. [CrossRef] [PubMed]
24. Xia, Q.; Zhang, Y.; Li, Z.; Hou, X.; Feng, N. Red blood cell membrane-camouflaged nanoparticles: A novel drug delivery system for antitumor application. *Acta Pharm. Sin. B* **2019**, *9*, 675–689. [CrossRef]
25. Wei, X.; Ying, M.; Dehaini, D.; Su, Y.; Kroll, A.V.; Zhou, J.; Gao, W.; Fang, R.H.; Chien, S.; Zhang, L. Nanoparticle Functionalization with Platelet Membrane Enables Multifactored Biological Targeting and Detection of Atherosclerosis. *ACS Nano* **2018**, *12*, 109–116. [CrossRef] [PubMed]
26. Song, Y.; Huang, Z.; Liu, X.; Pang, Z.; Chen, J.; Yang, H.; Zhang, N.; Cao, Z.; Liu, M.; Cao, J.; et al. Platelet membrane-coated nanoparticle-mediated targeting delivery of Rapamycin blocks atherosclerotic plaque development and stabilizes plaque in apolipoprotein E-deficient (ApoE(-/-)) mice. *Nanomedicine* **2019**, *15*, 13–24. [CrossRef]
27. Wang, S.; Duan, Y.; Zhang, Q.; Komarla, A.; Gong, H.; Gao, W.; Zhang, L. Drug Targeting via Platelet Membrane–Coated Nanoparticles. *Small Struct.* **2020**, *1*, 2000018. [CrossRef] [PubMed]
28. Gao, C.; Huang, Q.; Liu, C.; Kwong, C.H.T.; Yue, L.; Wan, J.-B.; Lee, S.M.Y.; Wang, R. Treatment of atherosclerosis by macrophage-biomimetic nanoparticles via targeted pharmacotherapy and sequestration of proinflammatory cytokines. *Nat. Commun.* **2020**, *11*, 2622. [CrossRef]
29. Zhang, Y.; Cai, K.; Li, C.; Guo, Q.; Chen, Q.; He, X.; Liu, L.; Zhang, Y.; Lu, Y.; Chen, X.; et al. Macrophage-Membrane-Coated Nanoparticles for Tumor-Targeted Chemotherapy. *Nano Lett.* **2018**, *18*, 1908–1915. [CrossRef]
30. Chen, L.J.; Zhao, X.; Liu, Y.Y.; Yan, X.P. Macrophage membrane coated persistent luminescence nanoparticle@MOF-derived mesoporous carbon core-shell nanocomposites for autofluorescence-free imaging-guided chemotherapy. *J. Mater. Chem. B* **2020**, *8*, 8071–8083. [CrossRef]
31. Lee, C.; Kang, S. Development of HER2-Targeting-Ligand-Modified Albumin Nanoparticles Based on the SpyTag/SpyCatcher System for Photothermal Therapy. *Biomacromolecules* **2021**, *22*, 2649–2658. [CrossRef]
32. Gao, C.; Liang, J.; Zhu, Y.; Ling, C.; Cheng, Z.; Li, R.; Qin, J.; Lu, W.; Wang, J. Menthol-modified casein nanoparticles loading 10-hydroxycamptothecin for glioma targeting therapy. *Acta Pharm. Sin. B* **2019**, *9*, 843–857. [CrossRef]
33. Wang, K.; Zhang, Y.; Wang, J.; Yuan, A.; Sun, M.; Wu, J.; Hu, Y. Self-assembled IR780-loaded transferrin nanoparticles as an imaging, targeting and PDT/PTT agent for cancer therapy. *Sci. Rep.* **2016**, *6*, 27421. [CrossRef] [PubMed]
34. Fan, K.; Jia, X.; Zhou, M.; Wang, K.; Conde, J.; He, J.; Tian, J.; Yan, X. Ferritin Nanocarrier Traverses the Blood Brain Barrier and Kills Glioma. *ACS Nano* **2018**, *12*, 4105–4115. [CrossRef] [PubMed]
35. Kim, D.; Amatya, R.; Hwang, S.; Lee, S.; Min, K.A.; Shin, M.C. BSA-Silver Nanoparticles: A Potential Multimodal Therapeutics for Conventional and Photothermal Treatment of Skin Cancer. *Pharmaceutics* **2021**, *13*, 575. [CrossRef] [PubMed]
36. Huang, J.; Wang, L.; Lin, R.; Wang, A.Y.; Yang, L.; Kuang, M.; Qian, W.; Mao, H. Casein-coated iron oxide nanoparticles for high MRI contrast enhancement and efficient cell targeting. *ACS Appl. Mater. Interfaces* **2013**, *5*, 4632–4639. [CrossRef] [PubMed]
37. McMahon, K.M.; Mutharasan, R.K.; Tripathy, S.; Veliceasa, D.; Bobeica, M.; Shumaker, D.K.; Luthi, A.J.; Helfand, B.T.; Ardehali, H.; Mirkin, C.A.; et al. Biomimetic high density lipoprotein nanoparticles for nucleic acid delivery. *Nano Lett.* **2011**, *11*, 1208–1214. [CrossRef]
38. Ma, J.; Zhang, S.; Liu, J.; Liu, F.; Du, F.; Li, M.; Chen, A.T.; Bao, Y.; Suh, H.W.; Avery, J.; et al. Targeted Drug Delivery to Stroke via Chemotactic Recruitment of Nanoparticles Coated with Membrane of Engineered Neural Stem Cells. *Small* **2019**, *15*, e1902011. [CrossRef]
39. Park, J.H.; Jiang, Y.; Zhou, J.; Gong, H.; Mohapatra, A.; Heo, J.; Gao, W.; Fang, R.H.; Zhang, L. Genetically engineered cell membrane-coated nanoparticles for targeted delivery of dexamethasone to inflamed lungs. *Sci. Adv.* **2021**, *7*, abf7820. [CrossRef]
40. Doyle, L.M.; Wang, M.Z. Overview of Extracellular Vesicles, Their Origin, Composition, Purpose, and Methods for Exosome Isolation and Analysis. *Cells* **2019**, *8*, 727. [CrossRef]
41. Kalluri, R.; LeBleu, V.S. The biology, function, and biomedical applications of exosomes. *Science* **2020**, *367*, aau6977. [CrossRef]
42. Wiklander, O.P.B.; Brennan, M.A.; Lotvall, J.; Breakefield, X.O.; El Andaloussi, S. Advances in therapeutic applications of extracellular vesicles. *Sci. Transl. Med.* **2019**, *11*, aav8521. [CrossRef]
43. Dai, J.; Su, Y.; Zhong, S.; Cong, L.; Liu, B.; Yang, J.; Tao, Y.; He, Z.; Chen, C.; Jiang, Y. Exosomes: Key players in cancer and potential therapeutic strategy. *Signal Transduct. Target Ther.* **2020**, *5*, 145. [CrossRef]
44. Li, M.; Zhou, H.; Yang, C.; Wu, Y.; Zhou, X.; Liu, H.; Wang, Y. Bacterial outer membrane vesicles as a platform for biomedical applications: An update. *J. Control. Release* **2020**, *323*, 253–268. [CrossRef]

45. Wang, S.; Gao, J.; Wang, Z. Outer membrane vesicles for vaccination and targeted drug delivery. *Wiley Interdiscip. Rev. Nanomed. Nanobiotechnol.* **2019**, *11*, e1523. [CrossRef] [PubMed]
46. van der Meel, R.; Fens, M.H.; Vader, P.; van Solinge, W.W.; Eniola-Adefeso, O.; Schiffelers, R.M. Extracellular vesicles as drug delivery systems: Lessons from the liposome field. *J. Control. Release* **2014**, *195*, 72–85. [CrossRef] [PubMed]
47. Seker, E. Friendly Bacteria. *Sci. Transl. Med.* **2014**, *6*, 220ec215. [CrossRef]
48. Li, R.; Liu, Q. Engineered Bacterial Outer Membrane Vesicles as Multifunctional Delivery Platforms. *Front. Mater.* **2020**, *7*, 202. [CrossRef]
49. Schwechheimer, C.; Kuehn, M.J. Outer-membrane vesicles from Gram-negative bacteria: Biogenesis and functions. *Nat. Rev. Microbiol.* **2015**, *13*, 605–619. [CrossRef] [PubMed]
50. Brown, L.; Wolf, J.M.; Prados-Rosales, R.; Casadevall, A. Through the wall: Extracellular vesicles in Gram-positive bacteria, mycobacteria and fungi. *Nat. Rev. Microbiol.* **2015**, *13*, 620–630. [CrossRef]
51. Perez-Cruz, C.; Carrion, O.; Delgado, L.; Martinez, G.; Lopez-Iglesias, C.; Mercade, E. New type of outer membrane vesicle produced by the Gram-negative bacterium Shewanella vesiculosa M7T: Implications for DNA content. *Appl. Environ. Microbiol.* **2013**, *79*, 1874–1881. [CrossRef]
52. Turnbull, L.; Toyofuku, M.; Hynen, A.L.; Kurosawa, M.; Pessi, G.; Petty, N.K.; Osvath, S.R.; Carcamo-Oyarce, G.; Gloag, E.S.; Shimoni, R.; et al. Explosive cell lysis as a mechanism for the biogenesis of bacterial membrane vesicles and biofilms. *Nat. Commun.* **2016**, *7*, 11220. [CrossRef]
53. Remis, J.P.; Wei, D.; Gorur, A.; Zemla, M.; Haraga, J.; Allen, S.; Witkowska, H.E.; Costerton, J.W.; Berleman, J.E.; Auer, M. Bacterial social networks: Structure and composition of Myxococcus xanthus outer membrane vesicle chains. *Environ. Microbiol.* **2014**, *16*, 598–610. [CrossRef]
54. Nagakubo, T.; Nomura, N.; Toyofuku, M. Cracking Open Bacterial Membrane Vesicles. *Front. Microbiol.* **2019**, *10*, 3026. [CrossRef] [PubMed]
55. Roier, S.; Zingl, F.G.; Cakar, F.; Schild, S. Bacterial outer membrane vesicle biogenesis: A new mechanism and its implications. *Microb. Cell* **2016**, *3*, 257–259. [CrossRef]
56. Toyofuku, M.; Nomura, N.; Eberl, L. Types and origins of bacterial membrane vesicles. *Nat. Rev. Microbiol.* **2019**, *17*, 13–24. [CrossRef] [PubMed]
57. Dorward, D.W.; Garon, C.F. DNA Is Packaged within Membrane-Derived Vesicles of Gram-Negative but Not Gram-Positive Bacteria. *Appl. Environ. Microbiol.* **1990**, *56*, 1960–1962. [CrossRef] [PubMed]
58. Zhou, L.; Srisatjaluk, R.; Justus, D.E.; Doyle, R.J. On the origin of membrane vesicles in gram-negative bacteria. *FEMS Microbiol. Lett.* **1998**, *163*, 223–228. [CrossRef] [PubMed]
59. Sonntag, I.; Schwarz, H.; Hirota, Y.; Henning, U. Cell envelope and shape of Escherichia coli: Multiple mutants missing the outer membrane lipoprotein and other major outer membrane proteins. *J. Bacteriol.* **1978**, *136*, 280–285. [CrossRef] [PubMed]
60. Schwechheimer, C.; Kulp, A.; Kuehn, M.J. Modulation of bacterial outer membrane vesicle production by envelope structure and content. *BMC Microbiol.* **2014**, *14*, 324. [CrossRef]
61. Schwechheimer, C.; Kuehn, M.J. Synthetic effect between envelope stress and lack of outer membrane vesicle production in Escherichia coli. *J. Bacteriol.* **2013**, *195*, 4161–4173. [CrossRef]
62. McBroom, A.J.; Kuehn, M.J. Release of outer membrane vesicles by Gram-negative bacteria is a novel envelope stress response. *Mol. Microbiol.* **2007**, *63*, 545–558. [CrossRef]
63. Florez, C.; Raab, J.E.; Cooke, A.C.; Schertzer, J.W. Membrane Distribution of the Pseudomonas Quinolone Signal Modulates Outer Membrane Vesicle Production in Pseudomonas aeruginosa. *mBio* **2017**, *8*, e1034-17. [CrossRef] [PubMed]
64. Schertzer, J.W.; Whiteley, M. A bilayer-couple model of bacterial outer membrane vesicle biogenesis. *mBio* **2012**, *3*, e297-11. [CrossRef] [PubMed]
65. Toyofuku, M.; Carcamo-Oyarce, G.; Yamamoto, T.; Eisenstein, F.; Hsiao, C.C.; Kurosawa, M.; Gademann, K.; Pilhofer, M.; Nomura, N.; Eberl, L. Prophage-triggered membrane vesicle formation through peptidoglycan damage in Bacillus subtilis. *Nat. Commun.* **2017**, *8*, 481. [CrossRef]
66. Bernadac, A.; Gavioli, M.; Lazzaroni, J.C.; Raina, S.; Lloubes, R. Escherichia coli tol-pal mutants form outer membrane vesicles. *J. Bacteriol.* **1998**, *180*, 4872–4878. [CrossRef] [PubMed]
67. Valentine, J.L.; Chen, L.; Perregaux, E.C.; Weyant, K.B.; Rosenthal, J.A.; Heiss, C.; Azadi, P.; Fisher, A.C.; Putnam, D.; Moe, G.R.; et al. Immunization with Outer Membrane Vesicles Displaying Designer Glycotopes Yields Class-Switched, Glycan-Specific Antibodies. *Cell Chem. Biol.* **2016**, *23*, 655–665. [CrossRef]
68. Stevenson, T.C.; Cywes-Bentley, C.; Moeller, T.D.; Weyant, K.B.; Putnam, D.; Chang, Y.F.; Jones, B.D.; Pier, G.B.; DeLisa, M.P. Immunization with outer membrane vesicles displaying conserved surface polysaccharide antigen elicits broadly antimicrobial antibodies. *Proc. Natl. Acad. Sci. USA* **2018**, *115*, E3106–E3115. [CrossRef]
69. de Jonge, E.F.; Balhuizen, M.D.; van Boxtel, R.; Wu, J.; Haagsman, H.P.; Tommassen, J. Heat shock enhances outer-membrane vesicle release in *Bordetella* spp. *Curr. Res. Microb. Sci.* **2021**, *2*, 100009. [CrossRef]
70. McMahon, K.J.; Castelli, M.E.; Vescovi, E.G.; Feldman, M.F. Biogenesis of Outer Membrane Vesicles in *Serratia marcescens* Is Thermoregulated and Can Be Induced by Activation of the Rcs Phosphorelay System. *J. Bacteriol.* **2012**, *194*, 3241–3249. [CrossRef]
71. Huang, W.; Zhang, Q.; Li, W.; Yuan, M.; Zhou, J.; Hua, L.; Chen, Y.; Ye, C.; Ma, Y. Development of novel nanoantibiotics using an outer membrane vesicle-based drug efflux mechanism. *J. Control. Release* **2020**, *317*, 1–22. [CrossRef]

72. Renelli, M.; Matias, V.; Lo, R.Y.; Beveridge, T.J. DNA-containing membrane vesicles of Pseudomonas aeruginosa PAO1 and their genetic transformation potential. *Microbiology* **2004**, *150*, 2161–2169. [CrossRef]
73. Domingues, S.; Nielsen, K.M. Membrane vesicles and horizontal gene transfer in prokaryotes. *Curr. Opin. Microbiol.* **2017**, *38*, 16–21. [CrossRef]
74. Kuehn, M.J.; Kesty, N.C. Bacterial outer membrane vesicles and the host-pathogen interaction. *Genes Dev.* **2005**, *19*, 2645–2655. [CrossRef]
75. Schooling, S.R.; Beveridge, T.J. Membrane vesicles: An overlooked component of the matrices of biofilms. *J. Bacteriol.* **2006**, *188*, 5945–5957. [CrossRef]
76. Shen, Y.; Giardino Torchia, M.L.; Lawson, G.W.; Karp, C.L.; Ashwell, J.D.; Mazmanian, S.K. Outer membrane vesicles of a human commensal mediate immune regulation and disease protection. *Cell Host Microbe* **2012**, *12*, 509–520. [CrossRef] [PubMed]
77. Zingl, F.G.; Leitner, D.R.; Schild, S. Biogenesis of Gram-Negative OMVs. In *Bacterial Membrane Vesicles: Biogenesis, Functions and Applications*; Kaparakis-Liaskos, M., Kufer, T.A., Eds.; Springer International Publishing: Cham, Switzerland, 2020; pp. 23–46. [CrossRef]
78. Palacios, A.; Coelho, C.; Maryam, M.; Luque-García, J.L.; Casadevall, A.; Prados-Rosales, R. Biogenesis and Function of Extracellular Vesicles in Gram-Positive Bacteria, Mycobacteria, and Fungi. In *Bacterial Membrane Vesicles: Biogenesis, Functions and Applications*; Kaparakis-Liaskos, M., Kufer, T.A., Eds.; Springer International Publishing: Cham, Switzerland, 2020; pp. 47–74. [CrossRef]
79. Jang, S.C.; Kim, S.R.; Yoon, Y.J.; Park, K.S.; Kim, J.H.; Lee, J.; Kim, O.Y.; Choi, E.J.; Kim, D.K.; Choi, D.S.; et al. In vivo kinetic biodistribution of nano-sized outer membrane vesicles derived from bacteria. *Small* **2015**, *11*, 456–461. [CrossRef] [PubMed]
80. Schulz, E.; Goes, A.; Garcia, R.; Panter, F.; Koch, M.; Muller, R.; Fuhrmann, K.; Fuhrmann, G. Biocompatible bacteria-derived vesicles show inherent antimicrobial activity. *J. Control. Release* **2018**, *290*, 46–55. [CrossRef]
81. Kuerban, K.; Gao, X.; Zhang, H.; Liu, J.; Dong, M.; Wu, L.; Ye, R.; Feng, M.; Ye, L. Doxorubicin-loaded bacterial outer-membrane vesicles exert enhanced anti-tumor efficacy in non-small-cell lung cancer. *Acta Pharm. Sin. B* **2020**, *10*, 1534–1548. [CrossRef] [PubMed]
82. Chen, Q.; Bai, H.; Wu, W.; Huang, G.; Li, Y.; Wu, M.; Tang, G.; Ping, Y. Bioengineering Bacterial Vesicle-Coated Polymeric Nanomedicine for Enhanced Cancer Immunotherapy and Metastasis Prevention. *Nano Lett.* **2020**, *20*, 11–21. [CrossRef]
83. Kim, O.Y.; Park, H.T.; Dinh, N.T.H.; Choi, S.J.; Lee, J.; Kim, J.H.; Lee, S.-W.; Gho, Y.S. Bacterial outer membrane vesicles suppress tumor by interferon-γ-mediated antitumor response. *Nat. Commun.* **2017**, *8*, 626. [CrossRef]
84. Gao, F.; Xu, L.; Yang, B.; Fan, F.; Yang, L. Kill the Real with the Fake: Eliminate Intracellular Staphylococcus aureus Using Nanoparticle Coated with Its Extracellular Vesicle Membrane as Active-Targeting Drug Carrier. *ACS Infect. Dis.* **2019**, *5*, 218–227. [CrossRef] [PubMed]
85. Qing, G.; Gong, N.; Chen, X.; Chen, J.; Zhang, H.; Wang, Y.; Wang, R.; Zhang, S.; Zhang, Z.; Zhao, X.; et al. Natural and engineered bacterial outer membrane vesicles. *Biophys. Rep.* **2019**, *5*, 184–198. [CrossRef]
86. Gujrati, V.; Kim, S.; Kim, S.H.; Min, J.J.; Choy, H.E.; Kim, S.C.; Jon, S. Bioengineered bacterial outer membrane vesicles as cell-specific drug-delivery vehicles for cancer therapy. *ACS Nano* **2014**, *8*, 1525–1537. [CrossRef] [PubMed]
87. Gao, W.; Fang, R.H.; Thamphiwatana, S.; Luk, B.T.; Li, J.; Angsantikul, P.; Zhang, Q.; Hu, C.-M.J.; Zhang, L. Modulating Antibacterial Immunity via Bacterial Membrane-Coated Nanoparticles. *Nano Lett.* **2015**, *15*, 1403–1409. [CrossRef] [PubMed]
88. Huang, Y.; Beringhs, A.O.R.; Chen, Q.; Song, D.; Chen, W.; Lu, X.; Fan, T.-H.; Nieh, M.-P.; Lei, Y. Genetically Engineered Bacterial Outer Membrane Vesicles with Expressed Nanoluciferase Reporter for in Vivo Bioluminescence Kinetic Modeling through Noninvasive Imaging. *ACS Appl. Bio Mater.* **2019**, *2*, 5608–5615. [CrossRef]
89. Chen, Q.; Rozovsky, S.; Chen, W. Engineering multi-functional bacterial outer membrane vesicles as modular nanodevices for biosensing and bioimaging. *Chem. Commun.* **2017**, *53*, 7569–7572. [CrossRef]
90. Alves, N.J.; Turner, K.B.; Daniele, M.A.; Oh, E.; Medintz, I.L.; Walper, S.A. Bacterial Nanobioreactors–Directing Enzyme Packaging into Bacterial Outer Membrane Vesicles. *ACS Appl. Mater. Interfaces* **2015**, *7*, 24963–24972. [CrossRef]
91. Gujrati, V.; Prakash, J.; Malekzadeh-Najafabadi, J.; Stiel, A.; Klemm, U.; Mettenleiter, G.; Aichler, M.; Walch, A.; Ntziachristos, V. Bioengineered bacterial vesicles as biological nano-heaters for optoacoustic imaging. *Nat. Commun.* **2019**, *10*, 1114. [CrossRef] [PubMed]
92. Carvalho, A.L.; Fonseca, S.; Miquel-Clopés, A.; Cross, K.; Kok, K.-S.; Wegmann, U.; Gil-Cardoso, K.; Bentley, E.G.; Al Katy, S.H.M.; Coombes, J.L.; et al. Bioengineering commensal bacteria-derived outer membrane vesicles for delivery of biologics to the gastrointestinal and respiratory tract. *J. Extracell. Vesicles* **2019**, *8*, 1632100. [CrossRef]
93. Ayed, Z.; Cuvillier, L.; Dobhal, G.; Goreham, R.V. Electroporation of outer membrane vesicles derived from Pseudomonas aeruginosa with gold nanoparticles. *SN Appl. Sci.* **2019**, *1*, 1600. [CrossRef]
94. Gothelf, A.; Gehl, J. What you always needed to know about electroporation based DNA vaccines. *Hum. Vaccin Immunother.* **2012**, *8*, 1694–1702. [CrossRef]
95. Young, J.L.; Dean, D.A. Electroporation-mediated gene delivery. *Adv. Genet.* **2015**, *89*, 49–88. [CrossRef]
96. Fuhrmann, G.; Serio, A.; Mazo, M.; Nair, R.; Stevens, M.M. Active loading into extracellular vesicles significantly improves the cellular uptake and photodynamic effect of porphyrins. *J. Control. Release* **2015**, *205*, 35–44. [CrossRef]
97. Pomatto, M.A.C.; Bussolati, B.; D'Antico, S.; Ghiotto, S.; Tetta, C.; Brizzi, M.F.; Camussi, G. Improved Loading of Plasma-Derived Extracellular Vesicles to Encapsulate Antitumor miRNAs. *Mol. Ther. Methods Clin. Dev.* **2019**, *13*, 133–144. [CrossRef]

98. Lamichhane, T.N.; Raiker, R.S.; Jay, S.M. Exogenous DNA Loading into Extracellular Vesicles via Electroporation is Size-Dependent and Enables Limited Gene Delivery. *Mol. Pharm.* **2015**, *12*, 3650–3657. [CrossRef] [PubMed]
99. Xu, C.H.; Ye, P.J.; Zhou, Y.C.; He, D.X.; Wei, H.; Yu, C.Y. Cell membrane-camouflaged nanoparticles as drug carriers for cancer therapy. *Acta Biomater.* **2020**, *105*, 1–14. [CrossRef] [PubMed]
100. Cao, H.; Dan, Z.; He, X.; Zhang, Z.; Yu, H.; Yin, Q.; Li, Y. Liposomes Coated with Isolated Macrophage Membrane Can Target Lung Metastasis of Breast Cancer. *ACS Nano* **2016**, *10*, 7738–7748. [CrossRef] [PubMed]
101. Xie, W.; Deng, W.W.; Zan, M.; Rao, L.; Yu, G.T.; Zhu, D.M.; Wu, W.T.; Chen, B.; Ji, L.W.; Chen, L.; et al. Cancer Cell Membrane Camouflaged Nanoparticles to Realize Starvation Therapy Together with Checkpoint Blockades for Enhancing Cancer Therapy. *ACS Nano* **2019**, *13*, 2849–2857. [CrossRef] [PubMed]
102. Jiang, Q.; Wang, K.; Zhang, X.; Ouyang, B.; Liu, H.; Pang, Z.; Yang, W. Platelet Membrane-Camouflaged Magnetic Nanoparticles for Ferroptosis-Enhanced Cancer Immunotherapy. *Small* **2020**, *16*, e2001704. [CrossRef] [PubMed]
103. Zhao, L.; Gu, C.; Gan, Y.; Shao, L.; Chen, H.; Zhu, H. Exosome-mediated siRNA delivery to suppress postoperative breast cancer metastasis. *J. Control. Release* **2020**, *318*, 1–15. [CrossRef]
104. Li, S.; Wu, Y.; Ding, F.; Yang, J.; Li, J.; Gao, X.; Zhang, C.; Feng, J. Engineering macrophage-derived exosomes for targeted chemotherapy of triple-negative breast cancer. *Nanoscale* **2020**, *12*, 10854–10862. [CrossRef]
105. Khongkow, M.; Yata, T.; Boonrungsiman, S.; Ruktanonchai, U.R.; Graham, D.; Namdee, K. Surface modification of gold nanoparticles with neuron-targeted exosome for enhanced blood-brain barrier penetration. *Sci. Rep.* **2019**, *9*, 8278. [CrossRef]
106. Lee, S.; Cha, E.J.; Park, K.; Lee, S.Y.; Hong, J.K.; Sun, I.C.; Kim, S.Y.; Choi, K.; Kwon, I.C.; Kim, K.; et al. A near-infrared-fluorescence-quenched gold-nanoparticle imaging probe for in vivo drug screening and protease activity determination. *Angew. Chem. Int. Ed Engl.* **2008**, *47*, 2804–2807. [CrossRef]
107. Zhang, Q.; Fang, R.H.; Gao, W.; Zhang, L. A Biomimetic Nanoparticle to "Lure and Kill" Phospholipase A2. *Angew. Chem. Int. Ed.* **2020**, *59*, 10461–10465. [CrossRef]
108. Kaparakis-Liaskos, M.; Ferrero, R.L. Immune modulation by bacterial outer membrane vesicles. *Nat. Rev. Immunol.* **2015**, *15*, 375–387. [CrossRef]
109. Jiang, L.; Schinkel, M.; van Essen, M.; Schiffelers, R.M. Bacterial membrane vesicles as promising vaccine candidates. *Eur. J. Pharm. Biopharm.* **2019**, *145*, 1–6. [CrossRef] [PubMed]
110. Hu, R.; Li, J.; Zhao, Y.; Lin, H.; Liang, L.; Wang, M.; Liu, H.; Min, Y.; Gao, Y.; Yang, M. Exploiting bacterial outer membrane vesicles as a cross-protective vaccine candidate against avian pathogenic Escherichia coli (APEC). *Microb. Cell Fact* **2020**, *19*, 119. [CrossRef] [PubMed]
111. Granoff, D.M. Review of meningococcal group B vaccines. *Clin. Infect. Dis.* **2010**, *50* (Suppl. 2), S54–S65. [CrossRef] [PubMed]
112. Chen, D.J.; Osterrieder, N.; Metzger, S.M.; Buckles, E.; Doody, A.M.; DeLisa, M.P.; Putnam, D. Delivery of foreign antigens by engineered outer membrane vesicle vaccines. *Proc. Natl. Acad. Sci. USA* **2010**, *107*, 3099–3104. [CrossRef] [PubMed]
113. Chen, L.; Valentine, J.L.; Huang, C.J.; Endicott, C.E.; Moeller, T.D.; Rasmussen, J.A.; Fletcher, J.R.; Boll, J.M.; Rosenthal, J.A.; Dobruchowska, J.; et al. Outer membrane vesicles displaying engineered glycotopes elicit protective antibodies. *Proc. Natl. Acad. Sci. USA* **2016**, *113*, E3609–E3618. [CrossRef] [PubMed]
114. Emens, L.A.; Ascierto, P.A.; Darcy, P.K.; Demaria, S.; Eggermont, A.M.M.; Redmond, W.L.; Seliger, B.; Marincola, F.M. Cancer immunotherapy: Opportunities and challenges in the rapidly evolving clinical landscape. *Eur. J. Cancer* **2017**, *81*, 116–129. [CrossRef]
115. Kishimoto, T.K.; Maldonado, R.A. Nanoparticles for the Induction of Antigen-Specific Immunological Tolerance. *Front. Immunol.* **2018**, *9*, 230. [CrossRef] [PubMed]
116. Ben-Akiva, E.; Est Witte, S.; Meyer, R.A.; Rhodes, K.R.; Green, J.J. Polymeric micro- and nanoparticles for immune modulation. *Biomater. Sci.* **2018**, *7*, 14–30. [CrossRef] [PubMed]
117. Dacoba, T.G.; Olivera, A.; Torres, D.; Crecente-Campo, J.; Alonso, M.J. Modulating the immune system through nanotechnology. *Semin. Immunol.* **2017**, *34*, 78–102. [CrossRef]
118. Geller, L.T.; Barzily-Rokni, M.; Danino, T.; Jonas, O.H.; Shental, N.; Nejman, D.; Gavert, N.; Zwang, Y.; Cooper, Z.A.; Shee, K.; et al. Potential role of intratumor bacteria in mediating tumor resistance to the chemotherapeutic drug gemcitabine. *Science* **2017**, *357*, 1156–1160. [CrossRef] [PubMed]
119. Lehouritis, P.; Cummins, J.; Stanton, M.; Murphy, C.T.; McCarthy, F.O.; Reid, G.; Urbaniak, C.; Byrne, W.L.; Tangney, M. Local bacteria affect the efficacy of chemotherapeutic drugs. *Sci. Rep.* **2015**, *5*, 14554. [CrossRef]
120. Kim, O.Y.; Choi, S.J.; Jang, S.C.; Park, K.S.; Kim, S.R.; Choi, J.P.; Lim, J.H.; Lee, S.W.; Park, J.; Di Vizio, D.; et al. Bacterial protoplast-derived nanovesicles as vaccine delivery system against bacterial infection. *Nano Lett.* **2015**, *15*, 266–274. [CrossRef]
121. Gulati, N.M.; Stewart, P.L.; Steinmetz, N.F. Bioinspired Shielding Strategies for Nanoparticle Drug Delivery Applications. *Mol. Pharm.* **2018**, *15*, 2900–2909. [CrossRef] [PubMed]
122. Qie, Y.; Yuan, H.; von Roemeling, C.A.; Chen, Y.; Liu, X.; Shih, K.D.; Knight, J.A.; Tun, H.W.; Wharen, R.E.; Jiang, W.; et al. Surface modification of nanoparticles enables selective evasion of phagocytic clearance by distinct macrophage phenotypes. *Sci. Rep.* **2016**, *6*, 26269. [CrossRef] [PubMed]
123. Qi, H.; Liu, C.; Long, L.; Ren, Y.; Zhang, S.; Chang, X.; Qian, X.; Jia, H.; Zhao, J.; Sun, J.; et al. Blood Exosomes Endowed with Magnetic and Targeting Properties for Cancer Therapy. *ACS Nano* **2016**, *10*, 3323–3333. [CrossRef] [PubMed]

MDPI
St. Alban-Anlage 66
4052 Basel
Switzerland
Tel. +41 61 683 77 34
Fax +41 61 302 89 18
www.mdpi.com

Pharmaceutics Editorial Office
E-mail: pharmaceutics@mdpi.com
www.mdpi.com/journal/pharmaceutics

www.ingramcontent.com/pod-product-compliance
Lightning Source LLC
Chambersburg PA
CBHW061501210926
51850CB00008B/1780
9783036574325